QUANTITATIVE PLANT ECOLOGY

STUDIES IN ECOLOGY

GENERAL EDITORS

D. J. ANDERSON BSc, PhD
Department of Botany
University of New South Wales
Sydney

P. GREIG-SMITH MA, ScD
School of Plant Biology
University College of North Wales
Bangor

and

FRANK A. PITELKA PhD
Department of Zoology
University of California, Berkeley

STUDIES IN ECOLOGY VOLUME 9

Quantitative Plant Ecology

P. GREIG-SMITH

MA, ScD
Emeritus Professor,
School of Plant Biology,
University College of
North Wales

THIRD EDITION

BLACKWELL SCIENTIFIC PUBLICATIONS

OXFORD LONDON EDINBURGH

BOSTON MELBOURNE

First published by Butterworth & Co.
1957
Second edition 1964

Third edition 1983

Set by Macmillan India Ltd,
Bangalore
Printed and bound in Great Britain
at the Alden Press, Oxford.

British Library Cataloguing in
Publication Data

Greig-Smith, P.
    Quatitative plant ecology.—
    3rd ed.—(Studies in ecology; v. 9)
    1. Botany—Ecology   2. Botani-
cal research
    I. Title   II. Series
    581.5′072   QK901

    ISBN 0–632–00142–9   cloth
    ISBN 0–632–01084–3   paper

Distributed in the United States of
America by the University of
California Press Berkeley, California

Contents

Preface to Third Edition

Methods of analysing field data on vegetation have developed so greatly since the last edition was written that some sections, especially the treatment of classification and ordination, have become completely out of date. The book has therefore been extensively revised and rewritten. Considerable additions and modifications have been made to Chapters 1–6. The previous Chapter 7, on classification and ordination, has been replaced by two new chapters. Chapter 5 has been retitled 'Correlation of Species Distribution with Habitat Factors' and a new Chapter 9, 'Vegetation and Environment' deals with the broader topic of the relation of total composition of vegetation, rather than individual species, to habitat factors. The previous final chapter on 'The Quantitative Approach to Plant Ecology', which argued the case for the use of quantitative methods, is now scarcely needed and has been replaced by a chapter on 'Practical Considerations', which attempts to provide some advice on the gathering of data and their analysis. The previous appendices on meteorological data and on area and spread of species have been omitted. A significant development of the last two decades is the use of the systems approach to investigate the functioning of ecosystems. This topic, although it does come under 'quantitative ecology', uses very different approaches and would not fit easily into the scope of this book.

There has been a proliferation of methods of classification and ordination. Bearing in mind the amount of 'background noise' always present in field data, I believe there are now methods available of sufficient sensitivity to deal with most field situations. It is no longer possible to attempt to cover every variant of methodology but I have tried to consider all the main approaches. I have discussed at some length some earlier methods, now superseded, both because they have been extensively used and because they demonstrate in a straightforward context the potential and limitations of classification and ordination. I have retained the principle of describing simple numerical operations fully but only outlining more complex procedures. The latter will, in any case, usually be carried out by computer and most ecologists will use existing programmes.

It is increasingly difficult to keep abreast of the relevant literature. I am very grateful to all those who have sent me copies of their publications, especially to those, particularly Dr M. P. Austin and Dr W. T. Williams, who have kept me informed of their work in advance of publication. I am indebted to Dr A. J. Morton for recalculating Appendix Table 1(b) and to Mr M. O. Hill, Mr D. Machin and Mr A. Vardy for dealing patiently with mathematical queries over many years. Mr Hill read a draft of Chapters 7–9 and made helpful comments. Dr R. Turkington has kindly allowed me to quote results in advance of publication. Lastly, I am again indebted to my wife for much help in the preparation of the manuscript for publication; without her support and encouragement it would not have been completed.

Bangor, May 1982   P. GREIG-SMITH

Preface to Second Edition

In the six years since the first edition was written there has been a steady interest in quantitative methods in plant ecology. An encouraging feature has been the appearance of an increasing number of publications in which the emphasis has been on the use of quantitative methods rather than on the methods themselves. At the same time there have been considerable advances in methodology, which have necessitated substantial revision for this edition. Development of methods has been particularly active in classification and ordination of communities. The account of this topic has been very largely rewritten and now appears as a separate chapter (Chapter 7). Substantial additions have been made to the remaining chapters. One of the appendix tables has been expanded and two further tables included. It has been necessary, on account of the increasing volume of literature, to be more selective in citing references than in the first edition, but I believe the reference list again includes the majority of references, up to the end of 1962, which are likely to be useful to anyone concerned with choice of methods. Goodall (1962) has recently provided a more comprehensive bibliography, including papers reporting results based on the use of quantitative methods.

I am indebted to Professor W. T. Williams and Dr J. M. Lambert for making their account of nodal analysis available before publication and for reading a draft of Chapter 7 and making various helpful suggestions. I am particularly grateful to Professor Williams for allowing me to quote from work still in progress by himself and Mr M. B. Dale. I am also grateful to Dr D. J. Anderson, Dr A. W. Ghent and Mr R. T. Gittins for allowing me to quote conclusions in advance of their publication, to Mr R. I. J. Tully and his assistants, who have again been most helpful in tracing references, and to Mrs P. M. Venesoen, who has typed successive drafts. My wife has again given me invaluable assistance in the preparation of the manuscript for the press.

Bangor, May, 1963   P. GREIG-SMITH

Preface to First Edition

Change from a qualitative to a quantitative approach is characteristic of the development of any branch of science. As some understanding is achieved of the broader aspects of phenomena, interest naturally turns to the finer detail of structure or behaviour, in which the observable differences are smaller and can only be appreciated in terms of measurement. It is not surprising that a quantitative outlook has been attained earlier in most branches of physical science than in biological science. Perhaps the greatest single difference in methodology between the physical and biological sciences is that in the former it is generally possible to isolate one variable at a time for study, whereas in the latter this is rarely possible. Thus, in the physical sciences broad outlines of phenomena are more readily seen from a relatively simple programme of qualitative investigation, and the way cleared for the more exact quantitative approach. In biology not only is it rarely possible to isolate variables for study, but the subjects of investigation are themselves commonly so complex that they are difficult to measure. In some branches of biology, therefore, attainment of the quantitative stage is perhaps little more than an ideal unlikely to be achieved in the near future. Other branches are more tractable, notably plant physiology, where many 'unwanted' variables can at least be minimized by use of controlled environments and clonal material, and investigation is now very largely in the quantitative stage. Plant ecology is at present in a transitional stage, and great advances can be expected from the quantitative techniques now being developed.

The general impossibility of controlling 'unwanted' variables in biology leads to a much greater degree of error variability in measurement than in the physical sciences. In the physical sciences differences among replicate measurements are generally attributable to deficiencies of technique, whereas in biological observations differences may be due not only to these deficiencies, but also, and commonly to a much greater extent, to fluctuation in variables not under investigation and assumed to be constant. Put another way, it is very much more difficult to obtain truly replicate samples in biological measurements than in physical measurements. If measurements are made in two or more contexts with the object of determining if there is any difference in the variable measured, the means may be different but the ranges of individual measurements may overlap. Thus the problem arises whether an observed difference is significant or not, i.e. whether it reflects any real difference between the two groups sampled, or is due to chance. In the physical sciences the immediate reaction is to improve the technique to obtain more accurate measurements, when either the means remain different but the ranges no longer overlap, indicating real difference, or the means converge with ranges still overlapping, suggesting that there is no real difference. In biology, however, there is often little scope for improvement of technique, and the biologist is therefore forced to turn for help in judging significance of difference to the techniques of statistical analysis. These are based on probability theory and permit determination of the probability of observed differences arising by chance in different samples of the same population. Thus arises the apparent paradox that while the physical sciences make much greater use of a quantiative approach than the biological sciences, they are much less dependent on the techniques of statistical analysis for the interpretation of their quantitative data.

There is at the present time a growing awareness among ecologists of the need to place their science on a more exact basis. The impetus given by the pioneers of ecology, which led to rapid advances in the first three or four decades of this century, is dying down and it is clear that the emphasis is changing from extensive work on vegetation to intensive work on selected aspects. If this is to go forward, more exact techniques of examination are necessary. The need for improved technique is made the more urgent on the one hand by the rapid depletion of natural vegetation, the only source of data on many of the more fundamental aspects of ecology, and on the other hand by the realization that advances in many branches of agriculture and forestry depend upon answers to ecological questions.

Although aware of the value of the quantitative approach and of the valuable tools available in the techniques of statistical analysis, many ecologists, faced with the rapidly expanding literature on quantitative methods in ecology, are reluctant to adopt a fully quantitative approach. This reluctance is, perhaps, reinforced by the apparent disregard of practical ecological problems in some theoretical studies. There is a need, therefore, for an assessment of the practical potentialities of various methods and techniques. I have attempted to make such an assessment in this book, and it is hoped that an ecologist faced with a particular problem in the field will find here guidance on the most profitable means of obtaining and handling his data, as well as a broad survey of the quantitative approach to plant ecology.

Chapter 1 is concerned with the different methods of describing vegetation in quantitative terms. Chapter 2 follows closely on the first, and deals with the positioning and number of samples to be used and with the comparison of the results of different sets of observations.

In Chapter 3 a hypothesis is developed of the significance, in relation to determining factors, of pattern, i.e. departure from randomness of distribution of individuals within the plant community. This leads to consideration of the techniques of detection and analysis of pattern. Chapter 4, on association between species, considers pattern from another aspect, the relationship between the patterns of different species.

Chapter 5 deals with correlation between vegetation and the level of environmental factors, the type of data on which most conclusions on the main factors determining the distribution of plants have been based.

Chapter 6 is concerned with the delineation of plant communities and assessment of difference between vegetation stands. This inevitably leads to consideration of the classification of vegetation, whether it is possible and if so, how it may be placed on an objective basis.

Chapter 7 is to some extent speculative. Believing that the quantitative approach has its own distinctive contribution to make to ecological theory, as well as putting existing practices on a sounder basis, I have devoted the greater part of this final chapter to consideration of this theme.

The appendices include brief discussions of the handling of meteorological data and of the area occupied by species, topics which do not fit conveniently into the main text, and a few tables of functions which are not readily available.

I have assumed that the reader, if he intends to make serious use of quantitative methods, will have access to Fisher and Yates' *Statistical Tables for Biological, Agricultural and Medical Research* or to some other source of the commoner statistical tables, and to one or other of the elementary textbooks of statistical methods for biologists such as Mather's *Statistical Analysis in Biology* or Snedecor's *Statistical Methods Applied to Experiments in Agriculture and Biology*. At the same time I have not hesitated to illustrate and discuss procedures at length where experience suggests that the biologist who is not very mathematically minded has difficulty in grasping them. By contrast I have given no examples of computation at all for more complicated statistical procedures, but have merely outlined their principles, believing that in such cases the biologist should take advice from a competent statistician, at least until he has had very considerable experience of statistical methods.

The reference list is strictly a list of references cited and makes no attempt at completeness. At the same time I believe that it includes the majority of references on methodology which are of current practical value rather than historical interest. The literature pertaining to the subject matter of Chapters 1 to 4 and Chapter 6 has been listed fairly completely by Goodall (1952a).

The quantitative ecologist is very dependent upon sound statistical advice. It therefore gives me great pleasure to express here my especial thanks to Professor M. S. Bartlett, of the University of Manchester, who has given freely of his time to advise me on statistical methods on various occasions since I first became interested in quantitative ecology. He has very generously read the whole of this book in manuscript, except Chapter 7, and corrected a number of mis-statements and ambiguities.

To Professor P. W. Richards I am indebted for his constant encouragement during the writing of the book and for reading parts in manuscript and making a number of suggestions for improvement. I am grateful to Mr R. I. J. Tully, of the Library of the University College of North Wales, for help in obtaining literature from other libraries, and to Dr A. D. Q. Agnew, Dr K. A. Kershaw and Dr W. S. Lacey for allowing me to quote data prior to their publication.

Lastly, I am grateful to my wife for much help in preparation of the manuscript for publication.

Bangor, April, 1957   P. GREIG-SMITH

CHAPTER I

Quantitative Description of Vegetation

The description of vegetation, with or without concurrent recording of factors of the environment, has played a major part in the development of plant ecology and continues to be important. It is essential, therefore, to place the description of vegetation on as sound a basis as possible. Two methods of community description predominated in the past and are still used to some extent. The first involves the making of a complete list of the species present in a community with the assignment of 'frequency symbols' or numerical ratings by inspection. This developed from the subjective assessment of species as rare, occasional, common, etc. in floras and represents an essentially similar process applied to a much smaller and more closely defined area. The second method derives from the work of Raunkiaer (1909, see Raunkiaer, 1934), and depends on the recording of presence or absence of species in small samples of the community under investigation. The sampling unit is a square, or less commonly a rectangle or a circle, of defined area, which may be placed either at random or in some regular manner. The results are expressed as the *percentage frequency*, i.e. the percentage of samples in which each species has been found. The species may be grouped for convenience of comparison into a number of frequency classes, with the limits of the classes forming either an arithmetic or a geometric series. The five classes 0–20%, 20–40%, 40–60%, 60–80%, and 80–100% have commonly been used. This procedure has sometimes been referred to as *valence analysis*.

These two methods have been widely used and it is important to consider their validity and limitations. The species list with frequency symbols or ratings is so well established that its value is too rarely questioned. The surprising feature is the degree of consistency of results obtained by experienced field workers. Several attempts have been made to assess the importance of the personal factor in deciding the rating assigned. Hope-Simpson (1940) has shown that one observer may give markedly different assessments on different occasions, particularly at different seasons. Smith (1944) investigated personal error affecting the simpler technique of visual assessment of percentage cover of total vegetation in plots. He showed that individual observers out of a group of eight deviated in their assessments from

1

the group mean by as much as 25 %. There was little evidence of any tendency to give consistently lower or higher values than the mean. There is little doubt that similar results would be obtained if the assignments of frequency symbols by a number of observers for the same sample area were compared. Indeed, in view of the complex of factors affecting assignment of a symbol, considered below, the discrepancy might well be greater. A further source of personal error lies in the mental state of the observer. Every ecologist with some experience of frequency estimation is aware that rare and inconspicuous species tend to be rated lower when the observer is tired than when he is fresh and fully alert. Conversely, familiarity with a vegetation type and the species involved tends to produce higher ratings.

The difficulties introduced by personal factors are perhaps less important to an independent worker than to teams of workers. Once he is sufficiently experienced to attain reasonable consistency in repeated assessments of the same vegetation his ratings are likely to give a fairly reliable comparison between different communities, e.g. Scott (1966), working with the admittedly simple structure of epiphytic bryophyte communities, established a consistent relationship between frequency symbols and percentage cover. It is still impossible to attach any absolute value to an individual's results. Moreover, results of different workers cannot be compared except in very broad terms, unless ratings for at least one and preferably several communities in common are available for the several observers. This drawback is so serious that, in general, frequency symbols should be used as the sole description of a community only when lack of time prevents the use of any more exact measure.

The difficulties of comparison of the results of several workers are more obvious in co-operative work. Unfortunately it is often in such work, e.g. in broad-scale surveys of the vegetation of large areas, that a rapid method of description is required. The errors introduced by personal factors can, however, be considerably reduced by careful standardization on the same vegetation between members of a team before the work is started and at intervals during its course.

There is a more serious objection to the use of frequency symbols. Several factors influence the observer in his assignment of a frequency symbol. Those uppermost in the minds of most observers are probably *density* or number of plant units per unit area and *cover* or percentage of the total area covered by the aerial parts of plants of a species, rather than true *frequency*, which is itself a complex character (see below). It is, however, difficult to avoid being influenced by the differing growth forms of different species and by the varying pattern of distribution of the individuals of different species on the ground, two factors greatly affecting the relative conspicuousness of different species. Even if density and cover alone are taken into consideration, an ideal probably impossible to attain, an attempt is being made to assess on one scale two largely independent variables. The problem is made clear by inspecting a weedy lawn, which includes at one extreme grasses, such as *Poa annua*, which have a high density of shoots but relatively low cover value per shoot, and at the other rosette weeds such as *Plantago* spp., each shoot of which covers a relatively large amount of ground. Cover and density each represent only one aspect among several of the contribution made by a species to the community. With practice in one vegetation type an observer can establish for himself an arbitrary scale of relative importance attached to cover and density but this relationship can scarcely be standardized and cannot readily be communicated to others. Moreover it has to be established afresh for communities of different physiognomy.

The confusion between cover and density is unavoidable. Other sources of error may with experience be reduced, though not eliminated. Species vary in conspicuousness and it is difficult to avoid overrating conspicuous species and underrating inconspicuous ones. Even the same species may vary greatly in conspicuousness between flowering and non-flowering states, e.g. *Deschampsia flexuosa* growing in comparatively small quantity amongst *Eriophorum vaginatum* gives a distinctive appearance to the community when flowering but is picked out only with difficulty when purely vegetative. Seasonal variation in assessment by the same observer on the same community, other than for annual species, results mainly from this variation in conspicuousness between different states. A further complication is introduced by the varying patterns that individuals of different species may form within the community. Individuals may be distributed more or less randomly through the community or they may be markedly aggregated into groups more or less clearly separated by areas in which the species is lacking or sparse. A high degree of aggregation may be indicated by rating the species as 'locally abundant', 'locally frequent', etc. When the aggregation is less marked overrating is liable to occur through increased conspicuousness of a species whose individuals are found in groups. The nature of these spatial patterns is of great importance to an understanding of community structure and is discussed in detail in a later chapter.

The use of frequency symbols has been treated at some length because their nature is often not fully understood, and because the method is commonly regarded as an elementary and straightforward one not needing any detailed consideration. It is evident, however, once its basis is examined, that care and experience are necessary before results of value can be obtained and that at its best the method is subject to considerable error. It is more satisfactory in description and comparison of communities of similar and uniform physiognomy than in those including diverse growth forms.

Frequency symbols raise a further point of importance: the highest category normally used is 'dominant'. This term is rather an unfortunate one, but probably too well established to be replaced. In practice it generally represents nothing more than the highest grade of density plus cover in the vegetation under examination, but there is a tendency to confuse this use of the term 'dominant' with the concept of dominance as a degree of influence exerted over other species of the community (by a variety of competitive and other effects). It is clear that some degree of such controlling influence has often been attached to species figuring as 'dominant' in species lists, without any evidence other than their having been assigned the highest frequency rating. A species may be dominant in both senses in a community but is not necessarily so. It would certainly lead to clarity if another term could be substituted for 'dominant' as a frequency symbol. The dominant species of a community might then be defined as that species which exerts the greatest influence on other species of the community and is least influenced by them. The determination of the dominant species of a community in this restricted sense would involve prolonged investigations in most, if not all, communities, including autecological studies of all the more important species. It would probably be least difficult in forest with a single species only in the canopy layer. A third sense for 'dominant' is occasionally found in ecological literature, viz. to describe any individual tree of the canopy in forest whose crown is more than half exposed to full illumination. It is widely used in this sense by foresters together with the complementary term 'predominant' (better, 'emergent') for individuals rising above the continuous canopy. Richards *et al.* (1940) suggested that 'dominant (ecol.)' should be used for the species with the highest frequency rating where there is any doubt about the meaning intended.

The percentage frequency method is more conveniently considered after discussion of the various quantitative measures of vegetation. There are many such measures and it will be necessary to consider in detail only the more important. They fall into two categories, those in which the figure obtained is independent, within the limits of observational error, of the method used to determine it, and those in which it is dependent on the mode of sampling and has meaning only when coupled with a statement of the method used. Of the former, which may be described as *absolute measures*, the more important are density, cover and the various measures of yield and biomass. Of the *non-absolute measures* the only one of importance is frequency.

Density is the measure of number per unit area. The objects counted may be either whole plants or portions of a plant, depending on the morphology of the species involved. Thus individuals of trees or annual herbs are usually clearly distinguishable but definition of an individual in many perennial herbs is difficult or impossible, e.g. many rhizomatous and rosette-forming species. Even if individuals are recognizable they may not be the most useful unit owing to their wide range of size, e.g. tussocking grasses in which tillers are more appropriate units. The term *mean area*, introduced by Kylin (1926) and defined as the reciprocal of density, is sometimes useful. Density is readily determined by direct counts in suitable areas. Alternatively, the inverse relationship between density and the distance between individuals allows density to be determined from measurements of distance from a point to the nearest plant, or from a plant to its nearest neighbour. Unfortunately, such methods are complicated by variation in the spatial distribution of individuals (see pp. 47–51).

Cover is defined as the proportion of ground occupied by perpendicular projection on to it of the aerial parts of individuals of the species under consideration. Its nature is perhaps most clearly brought out by noting that if a community on level ground composed of one species only were illuminated vertically the proportion of ground in shadow would represent the cover of the species. Cover is usually expressed as a percentage, and it should be noted that the total cover for all species in a community may exceed 100% and normally does so in all except open communities. This follows from the overlying or underlying of a part of one individual by parts of one or more others of different species, obvious in any closed community. *Top-cover*, the proportion of ground for which a species provides the uppermost cover, is a more easily determined measure and it is clear that some references in the literature to 'cover', particularly in vegetation without clearly defined layers, in fact refer to top-cover.

Cover may be either estimated or measured. Estimations are subject to the personal errors already discussed in relation to frequency symbols, though easier to make since they involve a single characteristic only. Various techniques have been devised to assist in estimation of cover. The literature has been reviewed by Brown (1954). Measurement of cover may be made by the point quadrat method, which depends on recording the presence or absence of a species vertically above a number of points in the community being described. The percentage of points above which the species is present represents the percentage cover. The theoretical basis of this method is simple. If a sample area has a finite but small size it may be completely covered by the projection of the aerial parts of individuals of a species under consideration, incompletely covered or not covered at all. As the size of the sample area is reduced it becomes more likely that it is either completely covered or not covered at all, until, when it is infinitely small, i.e. a point, it is always either completely covered or not covered. As long as the sample area is finite the whole area under examination may be considered as consisting of a large but finite number of such sample areas, each falling into one of the categories completely covered, partially covered and not covered. As the sample area decreases in size the proportion of the total number of sample areas which are either partially or completely covered approaches more nearly to the value of the species cover. At the limit, when the sample area becomes a point, the proportion which is covered of the infinitely large total number of sample areas equals the cover of the species. The points actually examined in an investigation, if properly selected, are an unbiased sample of the infinitely large number of possible points and give an estimate of the true value of cover, the accuracy of which can be increased to any desired degree of precision by increasing the sample size.

In practice the sample size used cannot, of course, be a true point. Warren Wilson (1959a,b) has used a fine needle mounted on the end of a rod, recording only contacts with the tip, and Poissonet et al. (1972) a 'bayonet', sharpened along one edge, recording contacts with the edge; Winkworth and Goodall (1962), Morrison and Yarranton (1970) and Reynolds and Edwards (1977) have used sighting devices incorporating cross-wires; these methods give a very close approximation to a point. Sampling has, however, usually been by means of long pins which are lowered through the vegetation. The use of a pin of finite diameter will give a value of cover greater than the true value because plants will be touched that would not make contact with the axis of the pin. The magnitude of this effect is frequently not realized (Warren Wilson, 1963b). It is demonstrated by the data in Table 1 and by Fig. 1, both taken from Goodall (1952b), who gives a critical discussion of the point-quadrat method. The values given in the table for pin diameter 0 were obtained by an apparatus of cross wires. Sighting devices directed vertically upwards may be used in a similar manner to determine the canopy cover in forest (Garrison, 1949; Morrison & Yarranton, 1970).

Basal area, as generally understood, is a measure somewhat similar to cover, being the proportion of ground surface occupied by a species. (The term has also been used instead of cover, e.g. West, 1937.) It is of particular value in dealing with species of tussock form. Its estimation and measurement involve similar considerations to those of cover. In measurement presence of basal parts of the plant at the sampling point is substituted for presence of aerial parts.

Table 1. Frequency (%) of contact between foliage and pins of different diameters (From Goodall, 1952b, by courtesy of *Australian Journal of Scientific Research*)

Locality	Species	No. of points	Pin diameter (mm)			Significance (P) for difference in pin diameter	
			0	1·84	4·75	0–1·84 mm	1·84–4·75 mm
Seaford	Ammophila arenaria	200	39·0	66·5	71·0	<0·001	>0·05
	Ammophila arenaria	200	60·5	74·0	82·0	0·001–0·01	>0·05
Black Rock	Ehrharta erecta	200	74·5	87·0	93·5	0·001–0·01	0·01–0·05
	Lepidosperma concavum	200	19·5	22·0	27·5	>0·05	>0·05
Sorrento	Spinifex hirsutus	200	35·0	48·5	61·0	0·001–0·01	0·01–0·05
Carlton	Fumaria officinalis	200	20·5	31·5	30·0	0·01–0·05	>0·05
	Ehrharta longiflora		24·5	25·5	37·5	>0·05	0·01–0·05
	No contact		53·0	42·5	38·5	0·01–0·05	>0·05
	Lolium perenne	200	65·0	85·5	82·5	<0·001	>0·05

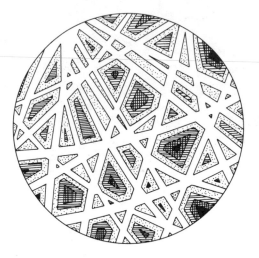

1 cm

☐ 0 mm
▦ 1 mm
▤ 2 mm
▨ 3 mm
■ 4 mm

Fig. 1. Projection of part of tussock of *Ammophila arenaria*, showing areas over which contact would be made with the foliage by pins of the diameters stated; each zone is understood as including those with less dense shading. (From Goodall, 1952b, by courtesy of *Australian Journal of Scientific Research*)

An alternative method of measuring cover is to record total length of interception made by plants of a species on line transects. The proportion of the total length of the transect intercepted by a species gives a measure of the cover of that species. This method has an evident advantage in speed of working and in eliminating the exaggeration introduced by using a point of finite size, but involves some approximation in that an individual plant must be assumed to have a definite boundary within which it has 100% cover. It is thus well adapted to measurement of basal area, and of cover of densely tussocking species, where this condition is nearly true, but of little use for vegetation where species are intermingled with one another. It is also appropriate to the measurement of the canopy of trees and shrubs, a feature sometimes of importance and not necessarily equivalent to their cover. How nearly cover and canopy correspond depends on the proportion of gaps in the leaf mosaic. If small units along the transect are recorded separately for species intercepted, the data are more comparable to point quadrat data, though more time-consuming to collect than measurements of the lengths of intercepts by individuals or patches of a species (cf. Fisser & Van Dyne, 1966).

Measures of *yield* and *biomass* call for little specific comment. They are determinations of quantity of material present per unit area and are made in a similar way to density determinations, the harvest of the relevant material being substituted for a count of plant units. An indirect estimate of bulk of shoot material can be readily obtained in some types of grassland by a modification of the point-quadrat method. If the total number of hits made on each species is recorded instead of only their presence or absence at each sampling point, it has been found empirically that the proportions of hits on the different species corresponds closely to the weight of shoot material. The validity of interpreting the proportions of hits in this way clearly depends on the species having similar growth form. Where it can be applied it is particularly valuable because it permits of an estimate of biomass without destroying the vegetation, so that observations can be repeated on subsequent occasions without allowance having to be made for interference.

The measures so far considered are straightforward in conception, though they may present difficulties in determination. *Frequency* is usually the easiest of the quantitative measures to determine, but its meaning in biological terms is not so clear-cut. Frequency of a species, determined by a particular size of sample area, is the chance of finding the species within the sample area in any one trial. It is determined by examining a series of sample areas placed at random within the vegetation being described and recording the species present in each sample area. The number of samples in which a species occurs, expressed as a proportion or percentage of the total, is an estimate of the chance of its occurring in any one sample, i.e. the frequency. The accuracy of the estimate can be increased to any desired extent by increasing the number of samples. It is evident from the definition that a frequency value has meaning only in relation to the particular size and shape of sampling area used. Increase in size of sampling area will necessarily result in an increase in the chance of a species occurring in any particular sample.

The ease of determination of frequency, in comparison with density and cover, has weighed heavily with ecologists. Frequency has one disadvantage in that a value for one particular position on a transect line or a grid cannot be obtained by the normal method of random throws. A small quadrat in which a count of density is made, or a frame of, say, ten or twenty pins for cover determination may be placed in a definite position on a line or grid. The resulting value, though subject to a relatively large sample error, can be localized in relation to the frame of reference provided by the line or grid. Areas within which frequency is determined by random throws must be relatively larger, and the value obtained, though subject to proportionately less sample error than single determinations of density or cover, cannot be so precisely localized. Correlation with habitat factors which vary over small areas is thus made more difficult to detect. The difficulty may be largely overcome by using a closely similar measure which may be termed *local frequency*. Instead of a simple quadrat in which presence or absence is recorded, the sampling unit consists of a quadrat divided into a number of smaller squares, e.g. a 25 cm square quadrat divided into twenty-five squares each 5 cm × 5 cm. Presence or absence is recorded for each sub-unit. The value obtained, though subject to a relatively large sampling error, can be localized to an area considerably smaller than any within which random placing of quadrats would be practicable.

'Presence' of species in sample areas may be interpreted in two different ways. It is commonly taken to mean that an individual or part of an individual is rooted in the sample area. Alternatively the occurrence of any aerial part of the plant may be taken as indication of presence. When it is necessary to distinguish the two different usages they may be referred to as 'rooted frequency' and 'shoot frequency' respectively. Most species will clearly show a much higher shoot frequency than rooted frequency in the same community. In the following pages frequency will be taken to mean rooted frequency unless shoot frequency is specifically stated.

Rooted frequency is clearly not independent of the absolute measure of density. In two communities, otherwise comparable, one of which has a higher density of a particular species than the other, the frequency of the species will be higher in the community with the greater density. There is a similar relationship between shoot frequency and cover. Various attempts have been made to establish an exact relationship between frequency and density, both empirically and theoretically. Difficulty arises because frequency is dependent not only on the number of individuals in the area but also on the way in which they are distributed over the area or, in other words, on the pattern. The effect of pattern is illustrated diagrammatically in Fig. 2. Each square contains the same number of dots, representing individual plants, so that density is the same in all three squares. In Fig. 2(a) the individuals are uniformly spaced. If the community were sampled by a quadrat of the size shown, 100 % frequency would be found because the maximum distance between adjacent individuals is less than the shortest dimension of the quadrat. In (b) the opposite extreme is shown with all the individuals occurring close together in one part of the area; here sampling with the same size quadrat would give a low frequency value. Fig. 2(c) represents an intermediate condition, more comparable to the type of situation commonly found in the field, with individuals forming a

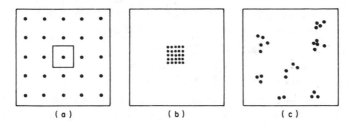

Fig. 2. Three different distributions having the same density. A quadrat in position in (a) (see text).

number of groups more or less clearly separated from one another; here sampling would give a frequency intermediate between the first two figures.

Frequency is thus dependent partly on density and partly on pattern. This has both advantages and disadvantages. There is the disadvantage found in the use of frequency symbols that two different properties are being assessed on the same scale, with the result that the same frequency may be shown by a species in different communities in which it plays quite different parts in the make-up of the vegetation. On the other hand frequency does integrate to some extent two important aspects of vegetation, without any subjective comparison between them being necessary. If both aspects were readily assessed without much additional work being required, there would be little excuse for using frequency. Density, as we have already seen, is readily determined provided the species under consideration present units that can be counted, but the determination is often time-consuming. A variety of techniques of describing pattern are available (Chapter 3) but all are relatively laborious compared with a simple frequency estimate. For many purposes the loss of possible information when frequency is used is counterbalanced by the great gain in speed of description of vegetation. Hence frequency determination is likely to remain an important ecological tool.

If the pattern of individuals is of a form that can be completely described in mathematical terms, then the corresponding frequency for a particular density and size of quadrat can be calculated, though it may be a laborious operation to do so, and, moreover, the result is of little practical value. If, however, the individuals are strictly randomly distributed in the community there is a definite and easily-calculated relationship between density and frequency. This likewise is of little practical value for determining density from frequency values, as some earlier workers had hoped, because plants are most commonly not randomly distributed. It is important, however, in view of the biological implications of the occurrence of non-random distributions, to know the corresponding frequency for any particular density value and quadrat size in a random distribution as a basis of comparison for distributions actually found.

The nature of random distribution is not always clearly understood. In such a distribution the probability of finding an individual at a point in the area is the same for all points. Fig. 3 shows a number of points randomly distributed within a square. This figure was constructed by using two sides of the square as axes and drawing random co-ordinates from a table of random numbers. Consideration of a uniform distribution (Fig. 2(a)) such as that of trees in an orchard, shows that the probability of finding an individual is not uniform over the area but rises at the corners of an imaginary grid. Put another way, in a random distribution the presence of one individual does not either raise or lower the probability of another occurring near by. In a uniform distribution the probability is lowered and in a clumped distribution, e.g. Fig. 2(c), it is raised.

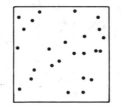

Fig. 3.   A random distribution.

We may now turn to the relationship between density and frequency for a random distribution. Suppose an area A contains n individuals and is sampled by a quadrat of area a.

Let the ratio $A/a = r$.

The density of individuals is $x = n/A$.

The chance of finding one particular individual in a throw of the quadrat is $a/A = r^{-1}$.

The chance of not finding that individual is therefore $1 - r^{-1}$.

Therefore the chance of finding none of the n individuals is $(1 - r^{-1})^n$.

But $n = Ax = a(A/a)x = axr$, and $(1 - r^{-1})^n = (1 - r^{-1})^{axr}$.

Now if r is large

$$(1 - r^{-1})^r = e^{-1}$$

where
$$e = 1 + 1 + (1/2!) + (1/3!) + \ldots,$$

the base of natural logarithms; and the chance of the quadrat containing no individuals becomes e^{-ax}.

The chance of a quadrat containing one particular individual and no others is $r^{-1}(1-r^{-1})^{n-1}$.

Any one of the n individuals may be present in the quadrat alone so that the chance of a quadrat containing one individual is

$$n \cdot r^{-1}(1-r^{-1})^{n-1}$$

$$= n \cdot r^{-1} \cdot \frac{1}{(1-r^{-1})} \cdot (1-r^{-1})^n$$

$$= axr \cdot r^{-1} \cdot \frac{1}{(1-r^{-1})} \cdot e^{-ax}.$$

If r is large r^{-1} approaches 0 and $1-r^{-1}$ approaches 1 and this becomes $ax \cdot e^{-ax}$.

Similarly the chance of finding two particular individuals and no others in a quadrat throw is $r^{-2} \cdot (1-r^{-1})^{n-2}$.

Two individuals may be selected from n individuals in $n(n-1)/2!$ ways so that the chance of obtaining two individuals in a quadrat is

$$\frac{n(n-1)}{2!} \cdot r^{-2} \cdot (1-r^{-1})^{n-2}$$

$$= \frac{n(n-1)}{2!} \cdot r^{-2} \cdot \frac{1}{(1-r^{-1})^2} \cdot (1-r^{-1})^n$$

$$= \frac{axr(axr-1)}{2!} \cdot r^{-2} \cdot \frac{1}{(1-r^{-1})^2} \cdot e^{-ax}$$

$$= \frac{ax(ax-r^{-1})}{2!} \cdot \frac{1}{(1-r^{-1})^2} \cdot e^{-ax}$$

$$= \frac{(ax)^2}{2!} \cdot e^{-ax} \quad \text{(when } r \text{ is large).}$$

Similarly the chance of a quadrat containing three individuals can be shown to be $((ax)^3/3!) \cdot e^{-ax}$ and in general the probabilities of a quadrat containing 0, 1, 2, 3, . . . , n, . . . individuals are given by the series

$$e^{-ax}, \quad axe^{-ax}, \quad \frac{(ax)^2}{2!} e^{-ax}, \quad \frac{(ax)^3}{3!} e^{-ax}, \quad \cdots \quad \frac{(ax)^n}{n!} e^{-ax}, \quad \cdots.$$

This converging series is known as the Poisson series. The total probabilities must add up to 1 and this is easily demonstrated

$$e^{-ax} + axe^{-ax} + \frac{(ax)^2}{2!}e^{-ax} + \frac{(ax)^3}{3!}e^{-ax}, + \cdots$$

$$+ \frac{(ax)^n}{n!} \cdot e^{-ax} + \cdots$$

$$= e^{-ax}\left(1 + ax + \frac{(ax)^2}{2!} + \frac{(ax)^3}{3!} + \cdots + \frac{(ax)^n}{n!} + \cdots\right)$$

$$= e^{-ax} \cdot e^{ax}$$

$$= 1.$$

In practice the Poisson series is often more usefully expressed

$$e^{-m}, me^{-m}, \frac{m^2}{2!}e^{-m}, \frac{m^3}{3!}e^{-m}, \cdots \frac{m^n}{n!}e^{-m}, \cdots$$

where $m = ax$ is the mean number of individuals per quadrat.

Frequency is the proportion of quadrats containing at least one individual, i.e. $1 - e^{-m}$, or percentage frequency $F = 100\,(1 - e^{-m})$. Thus density (per quadrat)

$$m = -\ln\left(1 - (F/100)\right).$$

The relationship between density and frequency is thus logarithmic and not, as some early workers assumed, linear. This was pointed out by Svedberg (1922), Kylin (1926) and others but has often not been realized.

Provided that distribution of individuals is random, density may thus be obtained from frequency counts. Further, frequency for any desired quadrat size may be calculated from a determination for one size. If F_1 is the frequency at quadrat size a_1 and F_2 the frequency at quadrat size a_2,

$$F_1 = 100\,(1 - e^{-a_1 x}) \text{ and } F_2 = 100\,(1 - e^{-a_2 x})$$

$$\frac{F_1}{F_2} = \frac{1 - e^{-a_1 x}}{1 - e^{-a_2 x}},$$

or, better, using percentage absence

$$100 - F_1 = 100\,e^{-a_1 x} \text{ and } 100 - F_2 = 100\,e^{-a_2 x}$$

$$\frac{100 - F_1}{100 - F_2} = \frac{e^{-a_1 x}}{e^{-a_2 x}}$$

$$\ln\left(\frac{100 - F_1}{100 - F_2}\right) = (a_2 - a_1)x$$

$$\text{or } \log_{10}\left(\frac{100 - F_1}{100 - F_2}\right) = 0{\cdot}4343\,(a_2 - a_1)x.$$

In view of the relative rarity of random distributions these relationships are not of great importance. It is essential to examine the type of distribution before using frequency to determine density and normally this will be more laborious than direct density determination. It might perhaps be useful to use frequency to estimate density in a few problems such as the study of minor changes in density in the same community over a period of time. In such cases even if distribution is not random it may be possible to establish an empirical relationship between density and frequency. In some cases the logarithms of percentage absence and density are proportional but density calculated from percentage absence must be multiplied by a constant to obtain the true value (Blackman, 1942). Lynch and Schumacher (1941) have shown, for seedlings in Western Pine Forest of North America, a linear relationship between the probit of frequency and logarithm of density. Goodall (1952a) points out that such relationships imply the existence of particular types of non-random distribution.

After this consideration of the nature of frequency the percentage frequency method of community description needs little further comment. We now realize that Raunkiaer's method, instead of giving a measure of the bulk of material contributed by each species, as was first believed, is an uncertain assessment of several different characteristics, the principal value of which lies in its speed of determination. From earlier results obtained Raunkiaer (1918, see Raunkiaer, 1934) deduced his Law of Frequencies. If the total number of species in a community is divided into the following five classes

$$
\begin{array}{lll}
A & 0\text{--}20 & \% \quad \text{frequency} \\
B & 21\text{--}40 & \% \quad \text{frequency} \\
C & 41\text{--}60 & \% \quad \text{frequency} \\
D & 61\text{--}80 & \% \quad \text{frequency} \\
E & 81\text{--}100 & \% \quad \text{frequency}
\end{array}
$$

the law of distribution of frequencies states that $A > B > C \gtrless D < E$. This is shown graphically in Fig. 4.

The general fall in the first three or four frequency classes accords with the experience of field botanists that there are more rare species than common ones, whatever the area under consideration. The rise in the fifth class was unexpected and various explanations of it have been advanced. It has played a notorious part in some attempts at definition of plant associations. The explanations advanced are largely invalidated, as Kylin (1926) pointed out, by the fact that the shape of the resultant curve is largely determined by the method of sampling (see also Ashby, 1935). These will not be discussed here. Table 2 shows the densities corresponding to the boundaries of Raunkiaer's

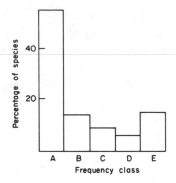

Fig. 4. 'Law of Frequencies.' Total data for a number of Scandinavian plant communities. Frequency classes: A, 1–20%; B, 21–40%; C, 41–60%; D, 61–80%; E, 1–100% (data from Raunkiaer, 1934).

frequency classes. It will be seen that the third class covers a range of density of 0·405 and the fourth class of 0·693 but the fifth class includes all densities from 1·609 to the maximum found. This maximum value is limited theoretically by the size of an individual, which is small compared with the size of quadrats commonly used. (Many workers, following Raunkiaer, have used $0.1\,m^2$ quadrats, i.e. a square of side 316 mm.) In practice, densities many times 1·609 are commonly found. The fifth class thus covers a range of density exceeding that of the remaining classes put together and a correspondingly large number of species fall into it, the increase in number due to greater range more than counterbalancing the fall off in number of species with increasing density*. If

Table 2. Density corresponding to certain frequencies in random distributions (see also Appendix Table 6)

Frequency (%)	Density (number of quadrat)	Range of density included in Raunkiaer's frequency classes
0		0·223
20	0·223	0·288
40	0·511	0·405
60	0·916	0·693
80	1·609	(1·609 to maximum
100	∞	present)

* Increase in size of quadrat used will obviously change the proportions of species falling into the different classes and thus the shape of the frequency curve (cf. Dagnelie, 1962).

distribution is more uniform than random expectation, the density limits of the frequency classes will be lowered and the rise in the fifth class will be still more pronounced. If species are aggregated, the limits will be increased but the effect will still be found unless the aggregation is very pronounced. Preston (1948), discussing the relative proportion of rare and common species, shows that the results expressed as Raunkiaer's Law of Frequencies follow from an assumption that the numbers of species of different degrees of rarity and commonness fall on a lognormal curve. (Here 'rare' and 'common' refer to number of individuals per association or other statistical universe.) The results may also follow from other assumptions of the relation between numbers of species of differing degrees of rarity (cf. Williams, 1950). It is clear that the exact form of the curve of frequencies depends on a complex of factors, including the make-up of the population in terms of relative numbers of individuals of different species, pattern assumed by individuals of different species, and size and, possibly (Williams, 1950), number of quadrats used.

The relationship between shoot frequency and other measures of vegetation has been little considered. We might expect that there would be a similar relation between shoot frequency and cover to that between rooted frequency and density. Blackman (1935) has shown a close correlation between cover and percentage absence for *Trifolium repens* in grassland communities. It is evident that cover cannot be randomly distributed, but always must show aggregation with areas of high cover at least the size of individual plants. A theoretical relationship between cover and shoot frequency for random distribution of cover would therefore be of no practical value. Blackman's observations show that an empirical relationship may be obtained in at least some cases, which is of similar utility in intensive studies to the empirical relationship obtainable between density and rooted frequency. Aberdeen (1954, 1958), following a suggestion of Archibald (1952), has drawn together the two concepts of rooted and of shoot frequency by considering frequency of plant units of radius r and density d in a circular quadrat of radius R. Assuming random distribution, shoot frequency $F = 100[1 - \exp(-\pi(R+r)^2 d)]$ and estimates of both density and plant size may be obtained from frequency data for two quadrat sizes. He has applied a similar argument to volume samples for estimations of soil fungi, when $F = 100[1 - \exp(-\frac{4}{3}\pi(R+r)^3 d)]$ (Aberdeen, 1955). As in other frequency approaches, the utility of the relationship is limited by the general departure from randomness found in the field.

Dirven *et al.* (1969) have considered 'percentage dominance', the percentage of quadrats in which a species is the one present in greatest amount. This represents a compromise between a frequency estimate and the more useful but more laborious estimate of the amount of a species in each quadrat. They show that, in grassland, as the size of individual increases relative to quadrat size, dominance percentage of a species approaches the percentage contribution of the species to the dry weight of the sward. Smartt *et al.* (1974, 1976) have examined the properties of an essentially similar measure, the percentage of quadrats in which a species forms the top-cover over at least half the area.

A distinction has been drawn above between absolute and non-absolute measures. Lambert (1972; Smartt *et al.*, 1974, 1976) has usefully contrasted bounded measures, which are constrained to an upper limit, e.g. cover, top-cover and frequency, and unbounded measures, which, at least theoretically, have no upper limit, e.g. density and biomass. When the total species complement of an area of vegetation is considered, this results in a threefold distinction between unbounded data, e.g. total biomass partitioned between species, bounded data, e.g. top-cover and partially bounded data, in which the individual species values are bounded, but that for the total species complement is not, e.g. cover and frequency.

In intensive studies of particular species it is often necessary to obtain some measure of the performance of the species under different conditions. The possible measures are numerous and the most suitable one to use must depend on the growth form of the species and the particular aspect of its behaviour under investigation. Direct measures of size, or number of parts, or estimates of yield, may all be appropriate under different conditions. Cover repetition, the mean number of hits by pins, at points where at least one hit is made, is basically a measure of performance, as it reflects the number of layers of foliage produced by an individual. It is not necessary to enlarge on the possible measures here, but it is perhaps worth emphasizing that less obvious measures may be valuable indicators of performance. Thus Phillips (1954b) found that the ratio of solid tip to grooved blade in the leaves of *Eriophorum angustifolium* reflects the rates of production of leaf primordia relative to rate of leaf elongation, and used the ratio to correlate these growth rates with environmental conditions.

CHAPTER 2

Sampling and Comparison

The value of quantitative data on the composition of vegetation depends on the sampling procedure used to obtain them. Since collection of quantitative data in the field is at best a time-consuming task, it is imperative that the samples taken should be such as give the maximum amount of information in return for the effort and time involved. With few exceptions the object in making quantitative estimates of vegetation will fall into one or other of three categories: (a) an estimate of the overall composition of the vegetation within certain boundaries, with a view to comparison with other areas or with the same area at another time, (b) the investigation of variation within the area, or (c) correlation of vegetation differences with differences in one or more habitat factors. This division is not absolute but it is worth making, because, for instance, the most useful procedure for sampling for overall composition may not be the most satisfactory for examination of variation within the area. Sometimes a compromise is possible whereby more than one objective may be attained, although perhaps at a lower level of precision, by suitable sampling procedures. This chapter is mainly concerned with the first category of sampling overall composition, though some of the general principles to be mentioned are applicable to all sampling. For a more comprehensive account of the theory of sampling, reference may be made to accounts such as those of Sampford (1962) and Cochran (1963).

There must inevitably be an element of subjectivity in sampling procedure because the boundaries within which a set of samples is taken are fixed by the ecologist on the basis of his judgement of what can suitably be described as one unit for the purpose in hand. This handicap is less serious in agricultural ecology, where man-made influences play a large part in determining vegetation and have commonly been applied uniformly over defined areas. But even when the grassland worker uses fields as his units he is depending on his own judgement that the influence of identical management over the field overrides any diversifying effect of soil or microtopography. This subjective element cannot be eliminated, but steps can be taken to minimize its effects. Firstly, parts of an area may be sampled separately if there is doubt as to its homogeneity. The data can be lumped to give a single value if comparison of the results for the different parts shows them to be sufficiently alike by whatever standards seem appropriate to the particular problem, and here again subjective judgement is involved. Secondly, the sampling may be carried out in such a way that information may be obtained on homogeneity and that some division of the data into separate portions corresponding to parts of the area may be possible after sampling has been completed. Common sense must be used in applying these precautions. Clearly any considerable fragmentation of an area for descriptive purposes cannot be carried out without so reducing the information on each part that it has little value by itself, or else so increasing the number of samples that the labour involved is out of proportion to the information to be obtained. Sampling procedure must always be related to the importance of the information sought to the problem under investigation, and to the degree of precision necessary.

Within an area selected and defined on this basis, samples might be placed in three ways: by selecting sites considered typical of the area as a whole, by placing samples randomly or by placing them systematically in some regular pattern, or some combination of these methods. Selection of typical sites for samples is clearly inappropriate to a quantitative approach, as their choice is dependent on the observer's preconceived ideas of the character of the vegetation, and data from such samples cannot be considered an unbiased estimate of the vegetation of the area. The alternatives of random or systematic sampling merit rather more consideration. If samples are taken at random there is available not only an estimate of the mean value of the density, or whatever measure may be used, but also an estimate of the precision of this mean. This, normally expressed as the standard error of the mean, allows a statement that, within any desired probability, the true value lies within a certain range. Thus, suppose the mean density from sixty quadrat samples is 11·5 individuals per unit area with a standard error of 1·5. For 59 degrees of freedom the 1 % level of t is (from tables) approximately 2·7, so that there is a 1 % probability of obtaining a deviation of $2·7 \times 1·5 = 4·05$, or greater, from the true value by chance, or 99 % probability that the true value lies within the range $11·5 \pm 4·05$. Such an estimate of the precision of the mean is desirable even if the data for one area are considered alone. If we wish to compare the densities in two different areas it becomes essential; only if the accuracy of the two estimates is known is it possible to assess the likelihood of getting by chance such divergent means as those observed in two different sets of samples from one population. Suppose a density of 10 individuals per unit area is observed in one community and 19 in another. An observer ignorant of the implications of sampling might say that since one value is nearly twice the other there must be a real difference in density. To do so is quite unjustified.

Without knowledge of the precision of the two figures no conclusion can be drawn as to whether there is any real difference in density between the two communities. If, however, we know the standard errors of the two means, then a statement of the probability of the difference arising by chance is possible. Suppose the standard error of the first mean is 3 and that of the second is 4, then the standard error of their difference is $\sqrt{(3^2 + 4^2)} = 5$. The observed difference is 9 and $t = 9/5 = 1.8$. Since the 5% point for t based on a reasonably large number of observations is about 2, the probability of getting such a large difference by chance, even if the two populations were identical, would be more than 5%, and we can scarcely consider this adequate evidence of real difference. It must be emphasized that the final judgement is subjective. A statistical test can indicate only the probability, on a given hypothesis, of a particular kind of result arising by chance.

If sampling is systematic an estimate of the mean is available which may, in some circumstances at least, deviate less from the true value than that given by random samples, but there is no indication of its precision and no possibility of assessing the significance of its difference from the mean in another area. In spite of the latter disadvantage systematic sampling has been preferred by many workers on the grounds that it is more representative of variations over the area and hence likely to give a better estimate than random samples, and that it is easier to carry out in the field. Various comparisons of random and systematic sampling have been made. Hasel (1938) and Finney (1948) for timber volume, and Bourdeau (1953) for density and basal area of forest trees, found that any gain in accuracy from systematic sampling was slight. Finney (1950) found that, if there was a strongly marked pattern of variation over the area, there was little advantage in systematic sampling, and, if the pattern was periodic, the values obtained by systematic sampling were likely to be less accurate than those from random. In these investigations the variance of the observed systematic samples was used as a measure of their accuracy, but it must be emphasized that the variance is here being used as a measure of conformity between successive samples over the same area and cannot be justified as a basis of tests of significance between samples taken over different areas. Whether systematic sampling gives a more accurate value than random depends on the pattern of variability within the area. Systematic sampling does have the advantage that, if the area being sampled is relatively large, it is likely to take less time in the field than random sampling of the same number of samples, particularly in tall vegetation.

If systematic sampling is desired for a particular investigation, it is important that the pattern of sampling adopted should be such as to give uniform representation over the area. Otherwise the advantage that may result from its use will be lost. This point is not always appreciated. For example, Brown (1954) quotes an investigation in which plots, presumably rectangular, were examined by points placed at equal distances along the two diagonals of the plot and along lines joining the midpoints of the opposite sides. This results in a much greater intensity of sampling towards the centre of the plot; half the samples are in fact taken from a quarter of the area.

We have so far considered only systematic or random sampling. The choice is, however, not necessarily between entirely systematic and entirely random samples. The essential feature of random sampling, which allows the observed variance of the data to be used as the basis of tests of significance, is that any point within the area has an equal chance of being represented in the samples. Randomization may be restricted in any way such that this is still true. Thus, if the area is divided up into a number of blocks of the same size, even if they are not of the same shape, and the same number of samples taken at random within each block the condition will be satisfied (stratified random sampling, Fig. 5(a))*. Provided the blocks are not comparable in size to the scale of variability within the area, this will result in no loss of precision compared with random sampling, even if the variability present is periodic, and will give an increase in precision comparable to that from entirely systematic sampling under conditions where systematic sampling would do so. Theoretically the greatest advantage will be obtained by subdivision as far as possible, so that only one sample is taken from each block. In practice a balance must be struck between this and the increasing labour of laying down large numbers of boundary lines. This increased precision of restricted as against unrestricted random sampling has been demonstrated by a number of workers, e.g. Pechanec and Stewart (1940) for yield, Bourdeau (1953) for density and basal area, and Goodall (1952b) for cover. Division into blocks has a further advantage. After the samples have been taken it may become apparent that the area should have been regarded as two or more distinct units. If so, it is possible to assign observations to one or other of the units and obtain a mean value for each of the latter. Indeed, especially if the blocks are all of the same shape, the arrangement facilitates judgement on this by allowing one portion of the area to be tested against another.

---

* Strata need not necessarily be of equal area, provided the number of samples taken from each stratum is proportional to its area. Thus, strata might be based on topography, with, for example, parts of the area representing flat ground forming different strata from those on slopes. Stratification based on features of the vegetation itself is undesirable because it involves assumptions about the nature of that vegetation (Smartt & Grainger, 1974).

Completely systematic sampling, involving a perfectly regular grid of samples, may interact with periodic variation in the vegetation, to give a mean value varying widely according to the exact placing of the grid, e.g. grassland showing 'ridge and furrow' topography sampled by a grid with its interval corresponding to the distance between furrows. 'Stratified systematic un-aligned sampling' (Quenouille, 1949) overcomes this difficulty, while retaining the advantage of systematic sampling in ensuring a more or less even spread of samples over the area (Fig. 5(b)). A point a is located in the first stratum by random coordinates. Remaining strata in the first row have the points located by the same x coordinate as in the first stratum, but with random y coordinates. The remaining strata have points located by coordinates corresponding to their row and column positions. (See Smartt & Grainger, 1974.)

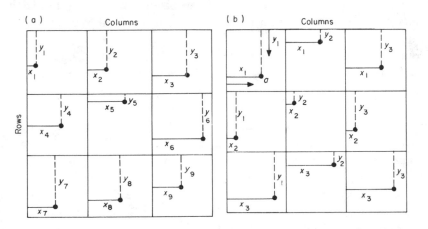

Fig. 5. (a) Stratified random sampling for an area subdivided into nine strata. (b) Stratified unaligned sampling for an area subdivided into nine strata. (From Smartt & Grainger, 1974, by courtesy of *Journal of Biogeography*.)

Random placing of a quadrat or other sample is not as straightforward as is sometimes thought. It is not uncommonly believed to be sufficient to walk over an area throwing the quadrat over one shoulder or throwing with the eyes shut or in some other way that apparently avoids any deliberate choice of the exact position of the quadrat. It is an instructive exercise to do this, at the same time plotting the position of each quadrat. If the positions are tested for randomness (e.g. by the method of grid analysis described in Chapter 3), it will almost always be found that the samples are not randomly distributed over the area. Commonly some portions of the area will have been omitted and those that have been sampled will show too regular a pattern of samples. Even if small plots a few feet square only are being sampled and care is taken to throw the quadrats over a larger area, ignoring those falling outside the boundaries, it is difficult to avoid a consistent over-representation of the centre of the plots (cf. Greig-Smith, 1952a). The extra trouble of some more objective method of randomizing is, therefore, generally well worth while. This is most readily done by laying down two lines at right angles as axes (which for rectangular plots are most conveniently two adjacent sides of the plot) and using a pair of random numbers as co-ordinates to position each sample. Measurement of the distances from the axes to sample positions need not be exact—pacing is quite sufficient in large plots. Selections of random numbers are readily available (e.g. Fisher & Yates, 1943; Snedecor, 1946). It is important to choose an appropriate scale for the coordinate axes. If too coarse a scale is used, so that only a limited number of possible positions are available, the system degenerates into a random sub-set of points on a very limited grid which has the same disadvantage of possible bias in any one set of samples as systematic sampling (Lambert, 1972).

An alternative means of achieving randomization is that of the 'random walk'. From an arbitrary starting point a random distance is travelled in random direction and a sample taken; from that point a new random distance is travelled in a new random direction, and so on until the desired number of samples have been obtained. Lambert (1972) has tested this method, but found serious difficulty both in determining a suitable distance scale and in finding a suitable method for reflection when the boundary of the area is reached that does not result in oversampling of the outer parts of the area.

Smartt & Grainger (1974) have made a careful investigation of the relative efficiency of several sampling strategies in assessing the proportions of several vegetation types in an area. They conclude that overall accuracy increases in the order random, systematic, stratified random and stratified unaligned systematic sampling. Stratified unaligned systematic was consistently the most accurate, but the ranking of the remaining strategies varied somewhat according to the area occupied by a vegetation type and its degree of fragmentation. These results are relevant rather to placing of stands* in a vegetation survey than to sampling most quantitative characteristics within one stand, and are discussed further in that context (pp. 304–6). They are applicable, however, to measures which record only the presence or absence of a species in a sample, i.e. cover or top-cover determined by independent placing of point-quadrats and frequency.

---

* Stand—an area considered as a unit for purposes of description of vegetation (see p. 149).

Detailed examination of an area may be intended as a basis for study of changes in its vegetation over a period rather than for a comparison with the vegetation of other areas. The usual practice is to take a suitable series of samples at the beginning of the period, preferably by some scheme of restricted randomization, and another series at the end of the period, using the same sampling procedure. Goodall (1952b) has pointed out that if the same samples (point quadrats in the case he is considering) are examined at the beginning and end of the period, a large proportion of the sampling error of individual points is removed from the estimate of change. He quotes data for percentage cover changes over a year obtained from permanent point quadrats and from point quadrats located afresh at the second examination. Out of data for eight species and percentage bare ground, in all cases except one the variance of the difference is less for permanent points (in most cases markedly so); for one species it is less than one-fifth the corresponding value for independent points. It is important that the marking of points is done in such a way that it does not interfere with the vegetation; a peg at the point of sampling is undesirable—fixing by measurement from two pegs a little distance away is better. Goodall notes a difficulty that may arise with fixed samples if several successive records are to be taken. This is the possibility that successive changes at one point may be correlated. It is not certain how likely this is in natural vegetation, but the possibility clearly has to be taken into account. The difficulty may be overcome by recording at the end of the first period a second set of samples which are used to estimate changes in the second period, and so on, i.e. samples A are enumerated on the first occasion, samples A and B on the second, samples B and C on the third, etc.

One further general consideration on sampling procedure remains before turning to the different measures of vegetation. In large-scale projects a number of observers may be involved. The differing bias of different observers estimating vegetation subjectively, referred to in Chapter 1, is well known. That observers enumerating objective measures of vegetation may also exhibit personal bias is more commonly overlooked. With most measures and types of sampling it is likely to be small and unlikely to be significant unless small differences are important in the investigation. Ellison (1942) found considerable differences between estimates of cover (from 17 to 22%, using 2400 points) of the grass *Buchloë dactyloides* by different observers using a frame of pins in the same position, but each dropping the pins afresh. Goodall (1952b) concluded from similar experiments, but with the pins remaining in position for all observers, that part of the error was due to movement of the pins, and that observers experienced in the method could obtain considerably better agreement. However, even under these conditions, although observers A and B agreed in their estimates for all seven species examined, observer C significantly exceeded the estimates of A and B for two species and fell below them for a third. In cover estimates the personal factor derives from judgement whether a pin is or is not touched. In other measures similar effects can be expected, e.g. in deciding whether an individual on the boundary of a quadrat does or does not lie within it.* There is likewise scope for personal judgement in deciding whether an axillary bud starting growth has developed far enough to be considered a new tiller. In a large-scale investigation employing a number of observers, particularly if it is prolonged, it is worth while for the observers to check their assessments against one another on the same samples from time to time. If persistent differences of the same sign appear, it may be feasible to calculate correcting factors to make their estimates strictly comparable. If several observers are working together on the same areas, it is certainly worth while to divide the sampling among them in such a way that personal error is included in the sampling error and not in between-plot or between-area differences, e.g. if two observers are recording a number of plots, they should each record half the samples on each plot rather than each recording all the samples on half the plots. It is scarcely necessary to emphasize the importance of establishing beforehand the exact details of sampling procedure to be adopted in any investigation involving several observers, as even slight deviations in method between observers may invalidate comparisons of their data.

Turning to considerations peculiar to different measures we may consider first density. Three decisions have to be made before sampling is begun. These concern the size and shape of quadrat and the number of samples to be used, considerations which are not necessarily independent of one another. If the individuals counted are randomly distributed over the area, i.e. in this context, if the numbers per quadrat of the size and shape used fall on a Poisson series, the accuracy of the estimate obtained can readily be shown to depend only on the number of individuals counted. Suppose x individuals are enumerated in n quadrats. Then the mean number per quadrat is x/n and, since the variance of a Poisson series is equal to its mean, the variance of a single quadrat is also x/n.

Variance of the mean of n quadrats then $= n^{-1}(x/n)$.

Standard error of the mean $= \sqrt{x}/n$.

Ratio of standard error of mean to mean $= \dfrac{\sqrt{x}}{n} \cdot \dfrac{n}{x} = \dfrac{1}{\sqrt{x}}$.

* Beshir (1968) has demonstrated, by comparing density estimates from quadrats of different size and shape, the importance of consistent inclusion or exclusion of individuals on the boundary. He has also shown that an observer may be an 'includer' of some species and an 'excluder' of others.

Thus the standard error of the estimate obtained will be the same for the same number of individuals counted, whether many small quadrats or few large ones are used, or even a single sample only counted. That this relationship applies even to a single quadrat is not perhaps immediately obvious, but it follows from the fact that one single count of x random individuals in a quadrat can be regarded as one sample from a whole series of such samples. It has therefore a variance equal to its mean (x) and a standard error $\sqrt{x}$. With a single quadrat the estimate of accuracy is in fact approximate only as the observed value x may deviate considerably from the mean of the series from which it is drawn. Indeed the ratio $1/\sqrt{x}$ can only be used for reasonably large numbers of quadrats. If the individuals of the species under consideration were randomly distributed, any convenient size of quadrat could be used and counts repeated with it until sufficient individuals had been enumerated to give the required accuracy.

The theoretical relationship for density data between the mean and its standard error is in practice of little assistance in deciding the number of quadrats necessary. In the field individuals are almost always found not to be randomly distributed but to show contagious distribution. In the present connection the important feature of this is that the variance is greater than the mean. Very rarely individuals may be regularly distributed, with variance less than the mean. Thus the calculated value of $1/\sqrt{x}$ for the ratio of standard error of mean to mean for x individuals counted is of very limited use as a guide to the number of samples that should be used. At the most it can be predicted that the accuracy attained is most unlikely to be as great as that indicated by the theoretical value and may be very much less.

When individuals are not randomly distributed, not only is the variance not equal to the mean but, not surprisingly, it is found not to be proportional to it either. Under these conditions the size of quadrat used and possibly its shape also will affect the accuracy of the estimate of density obtained. Table 3 shows data for mean and variance of density obtained by two different sizes of quadrat on the same area, and illustrates the kind of difference commonly found in the field. Here a reduced number of a larger size of quadrat gave an increase in the standard error expressed as a proportion of the mean. This is the commonest result to find, at least within the range of quadrat sizes that are practicable in the field, but it is possible to obtain the reverse effect. Fuller examination of these alternatives must be postponed to the discussion of pattern within communities in Chapter 3. We may anticipate briefly this discussion by noting that if a non-random population is sampled by quadrats of a size very much smaller than the average size of patches of individuals, then the variance of an observation will not be much, if any, greater than the mean. As quadrat size increases and approaches the size of the patches, variance

Table 3. Density estimates of shoots of *Mercurialis perennis* made in the same area with two different sizes of quadrat

	Quadrat size	
	625 cm^2	2 500 cm^2
Mean number per quadrat	0·13	0·35
Variance	0·2343	1·1432
Variance of mean for same total area sampled $\left(\begin{array}{l}200 \text{ quadrats of } 625 \text{ cm}^2 \\ 50 \text{ quadrats of } 2500 \text{ cm}^2\end{array}\right)$	0·00117	0·02286
Standard error of mean	0·0342	0·1512
Standard error of mean/mean	0·263	0·432
Estimate of density/m^2	2·08	1·40
Standard error of estimate	0·547	0·605

relative to the mean will rise sharply. If the patches are regular, it will then fall off again, ultimately reaching, or even falling below, the mean. If, however, the patches are themselves randomly or contagiously distributed the high variance will be maintained. Unfortunately it is rarely possible to determine by inspection whether the patches are regularly arranged, especially as regular arrangement most usually occurs as a mosaic of areas of slightly different density, which, being contiguous, must be regular in the sense in which the term is used here. The position may be and often is complicated further by the occurrence of heterogeneity on several different scales in the area. Any exact investigation of pattern is far more time-consuming than direct determination of density, so that there can be no question of investigating pattern before making a density determination. The safest procedure thus appears to be to use for density determinations the smallest quadrat that is practical or desirable on other grounds.

It has long been the custom to use a square sampling area and many ecologists have probably never given thought to its efficiency compared with other possible shapes. Clapham (1932) showed for one particular case that the variance between rectangular strips was markedly less than between squares, and that the variance was least, i.e. the efficiency was greatest, if the strips were orientated at right angles to the boundaries between any obviously different parts of the area sampled. This conclusion has been confirmed in various communities (see, for example, Bormann, 1953). This could not apply to perfectly random distributions where the dependence of standard error only on number counted would hold, and its general occurrence is in itself evidence of the widespread non-randomness of vegetation. It is easy to see why it should be so in non-random populations; a more elongated unit is more likely to include portions of more than one of the density phases which make up the population. Clapham also pointed out the greater ease in the field of working with strips than with squares, in avoiding trampling on part of the sample area while examining another part, and in dividing it up for counting. Too great an elongation of the strip, however, carries with it disadvantages of increased edge-effect similar to those found with very small square quadrats, and the exact shape must be determined with the growth-form of the species being counted in mind.*

The minimizing of variance is one consideration determining the size of quadrat to be used, but there are others also. The first is a practical one; the smaller the quadrat the greater the length of quadrat boundary per unit area and consequently the greater the chance of significant edge-effects due to the observer consistently including individuals that ought to be excluded or vice versa. For this reason alone, particularly if individuals of the species under consideration have a large area or ill-defined boundary at ground level, it is advisable that the quadrat should not be too small.

The second consideration is less obvious; like the last it applies equally to random and non-random populations. Consider first a random population in which the occurrence of individuals in a quadrat follows a Poisson distribution. Figs 6–9 show the frequency distribution of samples with 0, 1, 2, etc., individuals per quadrat for different means. It will be seen that for low values of the mean the distribution curve is highly asymmetric. Now the usual statistical procedures for the comparison of means are based on the assumption that the samples being compared are drawn from populations showing normal distribution and in which the variances are independent of the means. Poisson distributions satisfy neither of these conditions. As the mean increases, the form of the distribution curve approaches that of the normal (cf. the Poisson distribution for a mean of 9 with the normal distribution of mean and variance both 9 in Fig. 9) but the condition of independence of variance is still not satisfied. These difficulties can be overcome, provided the mean is sufficiently high (appreciably greater than 1), by making a suitable transformation which should always be done before comparisons between means of random populations are tested for significance. For a Poisson distribution this takes the form of substituting for each quadrat reading its square root. If the mean

* Myers and Chapman (1953) claimed to have found that rectangles showed *greater* variance than squares of the same area in *Leptospermum* scrub, but their analysis of variance in fact demonstrates the greater variance of squares.

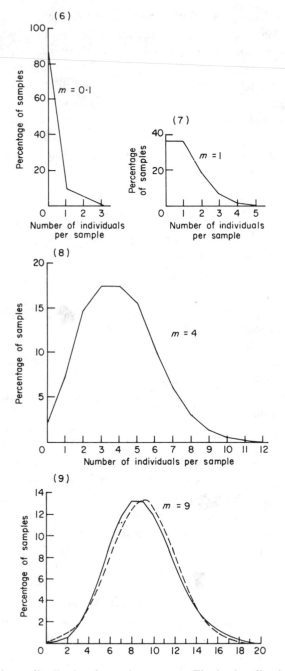

Figs 6–9. Poisson distribution for various means. The broken line in Fig. 9 shows the normal distribution with the same mean and variance.

number is less than 10 this tends to overcorrect and a more satisfactory transformation is $\sqrt{}$(observed number $+ 0.5$); this should be used if any of the values being compared are less than 10.

If the distribution of individuals is contagious the asymmetry of the distribution curve for low means will be still more pronounced, but the dependence of the variance on the mean is likely to be less pronounced, on account of the varying factors affecting pattern in different areas. If means are made reasonably large by a suitable size of quadrat, the degree of asymmetry will usually be insufficient seriously to affect tests of significance. In deciding quadrat size, therefore, reasonable symmetry of the distribution curve should be aimed at. In practice this may be interpreted as selecting a size which will not give more blank quadrats than quadrats with one individual. Before making comparisons the data should be scrutinized to see if any sets are very asymmetric or if the variances of the different sets tend to be proportional to the means. If so the square root transformation should be used.

The factors affecting the accuracy of the mean in determinations of density in the field are thus varied and complex and the theoretical relationship for random populations between number of individuals counted and accuracy is of little practical value. It remains broadly true, however, that the larger is the number counted the greater is the accuracy of the resulting mean. It is possible to calculate the variance from time to time as sampling proceeds but this course scarcely commends itself and is unlikely to be adopted unless a certain minimum level of accuracy is essential to the investigation. A rough indication may be obtained while the sampling is in progress whether further samples will greatly increase the accuracy of the mean by calculating successive means. The mean of, say, the first five, ten, fifteen, twenty, etc., observations is calculated and plotted against the number of observations. The mean will at first oscillate violently but gradually the oscillations will become less as the sample size increases. The graph can be judged in a subjective way only but does give the negative indication that if large oscillations are still occurring the sampling should be continued. Fig. 10 shows an example of such a graph together with the percentage standard error of the mean. The theoretical value of the standard error, assuming individuals to be randomly distributed, is also shown and emphasizes the lower degree of accuracy usually found in field data. The low standard error found for the first ten observations results from their chancing to be unduly uniform. The mean of the complete set of 100 observations has a standard error of 8.14%, compared with a calculated value of 5.31%.

Another useful approximate check in the field is derived from the relationship between range and standard error. If successive sets of samples are

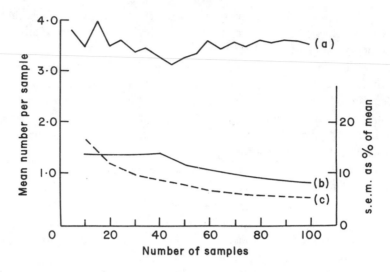

Fig. 10.   Number of individuals of *Endymion nonscriptus* in 100 random quadrats (10 cm square). (a) Mean of first 5, 10, 15, . . . 100 samples. (b) Observed and (c) calculated (for random distribution) standard error of mean, expressed as percentage of mean, for the first 10, 20, 30, . . . 100 samples.

taken from a normal population, the mean range of the set is a definite mutiple of the standard error of the population, e.g. samples of five observations have, on the average, a range 2·326 times the standard error. Thus the percentage standard error of the mean of any total number of samples, or alternatively, the number of samples necessary to give any desired degree of accuracy, may be estimated from the mean range. Appendix Table 2 shows the value of the constant for different sizes of sample. In the data shown in Fig. 10 the ranges of the successive groups of five observations were 5, 6, 2, 5, 11, 5, 13, 4, 4, 5, 7, 12, 8, 6, 6, 3, 7, 11, 6, 5. The mean value is 6·55 and the estimate of the standard error is $6·55/2·326 = 2·82$. From this the standard error of the mean of the 100 observations is estimated as 0·282, or 7·93 %, compared with the observed value of 8·14 %.

When concurrent determinations of density for more than one species are desired, the optimum size of quadrat and the number of quadrats necessary are often not the same for different species. A compromise must then be made in deciding the size and possibly even the shape of quadrat to be used. There is no reason, however, why the number of quadrats used for different species should be the same. There will clearly be a saving of labour if counting of the more abundant species is stopped after the required number of quadrats and only the rarer species recorded in the later samples.

Sampling yield in its various forms is more straightforward than sampling density, as the data obtained will usually show an approximately normal distribution with variance independent of the mean and will vary continuously (in the mathematical sense, that a reading can take any one of a large number of values. Beware, however, of very small ranges with a low degree of accuracy in weighing or measuring, e.g. if yields range from 0 to 3 g and it is possible only to weigh to within 0·5 g, only seven values are possible). Since distribution of data is normal, no prediction can be made of the number of samples necessary for a given degree of accuracy in the mean, unless there is information on the variance found for the same species in other closely similar communities. Otherwise the number to be used must be decided from the data as sampling proceeds. It is to be expected that the smaller the sampling unit the lower will be the variance of the mean for comparable areas sampled. This indicates the use of a small quadrat, but, as with density, edge-effects become more prominent as the quadrat is reduced in size. A balance must thus be struck between the lesser edge-effect (and the convenience of sampling) of larger quadrats and the greater efficiency per unit area of small quadrats. No guide but experience and trial of different sizes can be offered towards effecting this compromise. Small sample size may give pronouncedly skewed distribution curves with some yield data but this may often be corrected by logarithmic transformation (replacement of a reading x by $\log x$) as Blackman (1935) suggests.

Comparison of mean density or yield in two areas may be made by the t test, or joint comparison of several areas by the usual analysis of variance procedure, after transformation of the data, if this is necessary. It is important to realize the exact nature of the null hypothesis when these procedures are used. They are designed to test whether two or more samples can be regarded as *drawn from the same normal population*. This implies not only that they have the same mean but also the same variance. In most experimental work in botany the conditions of experiments are such that the variance remains the same for different treatments, and there is a tendency to regard the t test as testing difference of mean only. In the measurement of vegetation there is no reason to suppose that the variance will be the same in different areas, and this complicates the interpretation of the t test as usually applied, as a large value of t may result either from a difference of means or a difference of variance or both.

Full discussion of the testing of the difference of means (which is generally the main objective) when variance is heterogeneous is beyond the scope of this book but some account is given below of approximate treatments sufficient for most ecological work. An introduction to the subject, with references to more detailed accounts, is given by Snedecor (1946, § 4.6). In making a t test, if there is no evidence of heterogeneity of variance, a pooled estimate of the variance of an observation is obtained from the two sets of data. From this an estimate of the variance of a mean is obtained by dividing by the number of samples contributing to the mean. The variance of the difference of two means is the sum of the variances of the two means. Thus, for two sets of data

$$x'_1, x'_2, x'_3, x'_4, - - - x'_{n'}, \text{ and } x_1'', x_2'', x_3'', x_4'', - - - x''_{n''}.$$

The pooled estimate of variance is

$$V = \frac{\Sigma(x' - \bar{x}')^2 + \Sigma(x'' - \bar{x}'')^2}{n' + n'' - 2}.$$

Variance of difference of means $\dfrac{V}{n'} + \dfrac{V}{n''} = V\left(\dfrac{1}{n'} + \dfrac{1}{n''}\right).$

Standard error of difference of means $= \sqrt{\left\{V\left(\dfrac{1}{n'} + \dfrac{1}{n''}\right)\right\}}$ or, if $n' = n''$, the

standard error of difference is

$$\sqrt{\left[\frac{2}{n} \cdot \frac{\Sigma(x' - \bar{x}')^2 + \Sigma(x'' - \bar{x}'')^2}{2n - 2}\right]}$$

$$= \sqrt{\left[\frac{1}{n} \cdot \frac{\Sigma(x' - \bar{x}')^2 + \Sigma(x'' - \bar{x}'')^2}{n - 1}\right]}.$$

Note that if the number of samples in the sets is the same, the same value is obtained by calculating the variance of the two sets separately and adding the resulting variances of the two means, thus:

Variance of first mean $= \dfrac{1}{n} \cdot \dfrac{\Sigma(x' - \bar{x}')^2}{n - 1}.$

Variance of second mean $= \dfrac{1}{n} \cdot \dfrac{\Sigma(x'' - \bar{x}'')^2}{n - 1}.$

Variance of difference of means $= \dfrac{1}{n} \cdot \dfrac{\Sigma(x' - \bar{x}')^2 + \Sigma(x'' - \bar{x}'')^2}{n - 1}.$

The ratio of the observed difference of means to its standard error is referred to the table of t with $(n' + n'' - 2)$ degrees of freedom.

If the variances of the sets are markedly different, and it is desired to test the significance of difference of means, this procedure is modified (Snedecor, 1946).

If the number of samples is the same in the two sets, t is referred to the table with $n - 1$ instead of $2n - 2$ degrees of freedom. The probability obtained is accurate to two places of decimals.

If the number of samples is different in the two sets, an approximate method due to Cochran and Cox (1944) may be used. The variances of the two means are calculated separately and the variance of the difference obtained by addition. To obtain the value of t corresponding to any probability level the values of t at that probability for the numbers of degrees of freedom in the two sets are combined in a weighted mean, the respective weightings being the variances of the corresponding means. For example, suppose one mean has a variance of 0·14 and is derived from 10 samples and another has a variance of 0·35 and is derived from 25 samples. The 5% value of t for 9 degrees of freedom is 2·262 and for 24 degrees of freedom is 2·064. The 5% value of t for the comparison of the two means is then

$$\frac{0\cdot14 \times 2\cdot262 + 0\cdot35 \times 2\cdot064}{0\cdot14 + 0\cdot35} = 2\cdot121.$$

In practice, if both means are based on a large number of degrees of freedom, as is often the case at least for density and yield data, it will rarely be necessary to calculate the weighted mean of t as the two values will be so close that the observed t will be likely to be either above or below both values.

The way in which difference in variance may mislead in comparison of means, if the usual t test based on a pooled estimate of variance is used, may be illustrated by a hypothetical example.

In community A, 100 samples showed the following data for density:

$$\Sigma x = 960, \ n = 100, \ \bar{x} = 9\cdot6, \ \Sigma x^2 = 9711,$$
$$\Sigma(x - \bar{x})^2 = 9711 - (960^2/100) = 495,$$
$$V_x = 495/99 = 5, \ V_{\bar{x}} = 5/100 = 0\cdot05.$$

In community B, 50 samples showed the following data:

$$\Sigma x = 570, \ n = 50, \ \bar{x} = 11\cdot4, \ \Sigma x^2 = 8948,$$
$$\Sigma(x - \bar{x})^2 = 8948 - (570^2/50) = 2450,$$
$$V_x = 2450/49 = 50, \ V_{\bar{x}} = 50/50 = 1\cdot00.$$

The pooled estimate of variance for the two sets is

$$(495 + 2450)/148 = 19\cdot899.$$

Variance of difference of means $= 19\cdot899 \ (1/100 + 1/50) = 0\cdot5970$.
Standard error of difference of means $\sqrt{0\cdot5970} = 0\cdot773$.
$t = (11\cdot4 - 9\cdot6)/0\cdot773 = 2\cdot33$ with 148 degrees of freedom and probability less than 5%.

On the basis of this test a real difference between the two means might be suspected. Examination of the two means and their standard errors, however, indicates that both might be estimates of a true value of, say, 9·7, and the absence of any real difference is confirmed by the modified test

$$\text{Variance of mean for A} = 0\cdot05$$
$$\text{Variance of mean for B} = 1\cdot00$$
$$\text{Variance of differences of means } 0\cdot05 + 1\cdot00 = 1\cdot05$$
$$\text{Standard error of differences of means } \sqrt{1\cdot05} = 1\cdot025$$
$$t = 1\cdot8/1\cdot025 = 1\cdot76.$$

t for 49 degrees of freedom at 5% level of probability is about 2·01 and for 99 degrees of freedom about 1·99. There is thus no need to calculate the weighted value as the observed t is below both values and there is no indication of difference of means. The significant value of t by the normal test thus derives from a difference in variance and not one in mean.

Difference between the variances of density in different areas is unlikely to be of direct interest, beyond its effect on testing differences of means. It may sometimes be important where measures of individuals are used as indicators of performance, and the testing of significance of difference of variance is therefore worth mentioning. If two estimates only are involved, the ratio of the larger to the smaller may be referred to tables of variance ratio (F), entered with the larger number of degrees of freedom at n_1 of the table. Since the larger variance is always made the numerator of the ratio the tabular probabilities must be doubled. If several variances are to be compared, Bartlett's test is available (Bartlett, 1937). Reference may be made to Snedecor (1946, § 10.13) for an account of this test.

In sampling frequency the problems are rather different from those met with in density and yield estimates. A figure for frequency expresses the proportion of samples in which a species occurs. Hence successive sets of samples from the same population will be binomially distributed. This applies whatever the pattern of individuals within the community sampled, provided that the samples are completely random. For such a set of random samples the variance of an observed number of occupied quadrats is estimated directly from the appropriate binomial series, and is npq for a binomial series $(p + q)^n$ where n is the number of quadrats in the set, p the chance of the species being present in any one quadrat and $q = 1 - p$. Thus if 25% frequency is found from 200 samples, i.e. the observed number of quadrats containing the species is 50, the variance of the observed number is $200 \times 0\cdot25 \times 0\cdot75 = 37\cdot5$ and the standard error is $\sqrt{37\cdot5} = 6\cdot12$, so that the percentage frequency of 25% has a standard error of 3·06. The binomial distribution, unless n is large or p close to 0·5, is very asymmetric. For small values of n confidence limits cannot therefore be determined from the table of t. Mainland *et al.* (1956) give tables of the exact confidence limits at 95 % and 99 % probability levels for various values of n. A less extensive table is reproduced by Snedecor (1946, § 1.3). A small selection from the tables is given in Table 4 together with the values calculated by the use of t and the standard error. Two features appear from this table, the inadvisability of using the standard error in assessing the accuracy of frequency values at the two extremes unless they are based on a very large number of throws, and the low level of precision of values derived from a relatively small number of samples. The latter point is often overlooked. It is not uncommon in the literature to find communities described in terms of percentage frequency values of different species obtained from twenty or even less examples. It will be seen from Table 4 that, for instance, even taking the not very extreme probability level of 95 %, an observed value of 50 % from 20 samples may arise from a true value anywhere between 27 and 73 %. It is clear that a sample number of at least 100 and preferably higher should be aimed at. If circumstances

Table 4. 95 % confidence limits for binomial distribution (after Snedecor, 1946) and limits obtained by using the table of t (in brackets)

Percentage frequency observed	Number of observations				
	10	20	50	100	1 000
0	0 31	0 17	0 7	0 4	0 0
10	0 45	1 31	3 22	5 18	8 12
	(0 31)	(0 24)	(0 19)	(4 16)	(8 12)
20	3 56	6 44	10 34	13 29	18 23
	(0 49)	(1 39)	(6 31)	(12 28)	(18 22)
30	7 65	12 54	18 44	21 40	27 33
	(0 63)	(8 51)	(17 43)	(21 39)	(27 33)
40	12 74	19 64	27 55	30 50	37 43
	(5 75)	(17 63)	(20 54)	(30 50)	(37 43)
50	19 81	27 73	36 64	40 60	47 53
	(14 86)	(27 73)	(36 64)	(40 60)	(47 53)
60	26 88	36 81	45 73	50 70	57 63
	(25 95)	(37 83)	(46 74)	(50 70)	(57 63)
70	35 93	46 88	56 82	60 79	67 73
	(37 100)	(49 92)	(57 83)	(61 79)	(67 73)
80	44 97	56 94	66 90	71 87	77 82
	(51 100)	(61 99)	(69 84)	(72 88)	(78 82)
90	55 100	69 99	78 97	82 95	88 92
	(69 100)	(76 100)	(81 100)	(84 96)	(88 92)
100	69 100	83 100	93 100	96 100	100 100

prevent this, it must be accepted that only gross differences between communities will be detectable. In view of the generally asymmetric form of the binomial distribution and the correlation between its variance and mean, it is not advisable to use a t test to compare two frequencies. Instead they may be compared by a contingency table. Suppose frequencies of 51 and 62% respectively have been found from 200 samples in each of the two areas. The contingency table is as follows.

	Species present		Species absent		
Area A	a	102	b	98	200 $(a+b)$
Area B	c	124	d	76	200 $(c+d)$
	$(a+c)$	226	$(b+d)$	174	400 $(a+b+c+d=n)$

In this case the expected values, if the true frequencies are the same, are 113 in cells a and c, and 87 in cells b and d. The χ^2 testing departure from this expectation may be calculated in the normal way as the sum of (deviation^2/expected) for each of the four cells. For such a 2×2 table it is, however, simpler to use the direct formula:

$$\chi^2 = \frac{(ab-bc)^2 n}{(a+b)(c+d)(a+c)(b+d)}.$$

The resulting value is referred to the table of χ^2 with one degree of freedom. If n is small the numbers of possible tables with the same marginal totals is relatively small so that the distribution is discontinuous whereas χ^2 is a continuous distribution. The inaccuracy introduced by this may be allowed for by applying Yates's correction for continuity. For a 2×2 table this takes the form of subtracting 0·5 from each of the two values greater than expectation and adding 0·5 to those less than expectation, which is equivalent to calculating χ^2 as

$$\frac{(|ad-bc|-\frac{1}{2}n)^2 n}{(a+b)(c+d)(a+c)(b+d)}.$$

This correction should be applied if any expected value is less than 500 (Fisher & Yates, 1943). Thus, for the example quoted

$$\chi^2 = \frac{(98 \times 124 - 102 \times 76 - 200)^2\, 400}{200 \times 200 \times 226 \times 174} = 4\cdot486$$

(uncorrected $\chi^2 = 4\cdot923$)

which corresponds to a probability of between 0·05 and 0·02, indicating a significant difference at the conventional level of 5% probability.*

There is a further possible source of inaccuracy. If any expected value is small (say less than 100), there is a marked difference in probability of a given deviation in the two directions. This asymmetry is allowed for in Fisher and Yates's (1943) Table VIII where 2·5% and 0·5% points for χ_c (square root of χ^2, corrected for continuity) are tabulated for the two tails of the distribution separately. 2·5% probability for one tail, i.e. deviation in one direction is equivalent to 5% probability in the table of χ^2, where the probability given is for a particular deviation in either direction. For certain regions of the table no values can be given, owing to relatively large differences in probability for different marginal totals, and the exact solution (Fisher, 1941, § 21.02) must be used. The probability of getting one particular table with given marginal totals is, using the notation shown in the example above,

$$\frac{(a+b)!\,(c+d)!\,(a+c)!\,(b+d)!}{n!}\cdot\frac{1}{a!\,b!\,c!\,d!}.$$

Thus it is possible to calculate exactly the probability of getting a difference as extreme, or more extreme than that observed if the two areas have in fact the same frequency. Consider two estimates of 93% and 99% frequency, each derived from 100 samples. Only one more extreme table, corresponding to frequencies of 92% and 100%, is possible and the required probability is

$$\frac{100!\,100!\,192!\,8!}{200!}\left(\frac{1}{99!\,7!\,93!\,1!}+\frac{1}{100!\,8!\,92!}\right)=0\cdot032$$

or 3·2%, equivalent to 6·4% in the table of χ^2. There is no indication that the two estimates represent different frequencies.

The methods of calculating the probability of observed differences have been outlined but in practice the probability can be obtained directly in nearly all cases from tables e.g. Finney *et al.* (1963), Pearson and Hartley (1954).

If frequencies from several areas are to be compared to determine if they can reasonably be regarded as estimates of one value, a similar approach may be used. χ^2 is calculated from the contingency table, which will have degrees of freedom one less than the number of areas being compared. If comparisons of

---

* Gilbert and Wells (1966) point out that the likelihood expression

$$\ln\frac{n^3}{2\pi(a+b)(c+d)(a+c)(b+d)}+2\ln\frac{n!\,a!\,b!\,c!\,d!}{(a+b)!\,(c+d)!\,(a+c)!\,(b+d)!}$$

gives a more accurate value for χ^2 and is more quickly calculated by computer. For this table it has the value 4·912. Yates's correction tends to overcorrect.

the means of sets of estimates with one another are to be made, the data may be subjected to transformation and the ordinary analysis of variance procedure used. The appropriate transformation is to θ where $\sin \theta = \sqrt{\text{frequency}}$ expressed as a proportion. A short table of this angular transformation for steps of 1% is given by Fisher and Yates (1943) and a fuller table by Bliss is reproduced by Snedecor (1946, § 16.7). The variance of θ is largely independent of p, depending mainly on the number of samples. The approximate value is $820 \cdot 7/n$ and this may be used if n is reasonably large (50 or over). If n is small this approximation deviates widely from the true value, particularly if p is very high or low. A table of variance for a selection of lower values of n is given in Appendix Table 1. Freeman and Tukey (1950) have suggested an arc-sine transformation, which is somewhat more effective in stabilizing the variance and may usefully be employed if n is relatively small. This takes the form

$$\theta = \tfrac{1}{2}\left\{ \arcsin \sqrt{\left(\frac{x}{n+1}\right)} + \arcsin \sqrt{\left(\frac{x+1}{n+1}\right)} \right\}$$

with variance tending to $820 \cdot 7/(n+\tfrac{1}{2})$ (x = number of successes out of n trials). The transformation has been tabulated by Mosteller and Youtz (1961) for values of n up to 50.

As with density determinations, the accuracy of a frequency estimate can often be increased by suitable restriction of randomization. This can readily be seen by consideration of a hypothetical area, one half of which is completely blank and in the other half of which there are individuals so closely spaced that a quadrat of a certain size placed in it will always include at least one individual. The true value of the frequency with that size of quadrat will be 50%. Sets of random samples taken from it will show a range of value having a mean, within the limits of chance variation, of 50% and falling on a binomial series $(\tfrac{1}{2}+\tfrac{1}{2})^n$. If the randomization is restricted in such a way that half the samples are taken randomly from the occupied half of the area and half from the unoccupied half (and this still gives every part of the area an equal change of being represented) the value obtained will always be exactly 50%.

If an area over which frequency is not uniform is divided up into a number of blocks of the same size, and if an equal number of quadrats is recorded in each block, a more accurate estimate of mean frequency over the whole area is likely to result than if the same number of samples were taken entirely randomly within the area. This procedure, which is little more laborious than complete randomization, is adequate if the frequency is being determined for purely descriptive purposes. It has the serious drawback that no estimate of the variance of the resulting frequency value is available, for the distribution is no longer strictly binomial. It cannot be greater than the corresponding figure for completely random samples. If it is desired to make exact comparisons with other areas, the procedure must be modified. If several sets, each of a smaller number of samples, are taken in the same way the variance of their mean may be calculated from the data. Since the data are of similar type to binomial distribution, angular transformation is still advisable, unless the number of occupied quadrats recorded in each set is large (say 100 or more). The treatment of the data may be illustrated by an example. Suppose five sets of 100 samples were taken, equal numbers of each set from the several blocks in the manner described, and showed 18, 18, 20, 21, 23 quadrats occupied respectively. The mean frequency is 20 %. The transformed values for the five sets are 25·1, 25·1, 26·6, 27·3, and 28·7 respectively. From these figures $\Sigma x = 132·8$, $\bar{x} = 26·56$, Σx^2 $= 3536·56$, $\Sigma(x - \bar{x})^2 = 3536·56 - (132·8^2/5) = 9·392$, variance of one set $9·392/4$, variance of mean of five sets $9·392/(4 \times 5) = 0·4696$, standard error of mean $\sqrt{(0·4696)} = 0·685$. (For comparison, the variance of frequency from 500 random observations is, on the transformed scale, $820·7/500 = 1·64$, and the standard error 1·28.) Note that the calculated standard error applies only to the mean on the transformed scale, and no standard error is available for the untransformed mean. Any desired confidence limits may, however, be calculated on the transformed scale and the corresponding frequency values obtained from the table of the angular transformation. Thus, for the example quoted the value of t for 5 % probability and four degrees of freedom is 2·78 which gives as the 95 % confidence limits for the transformed mean 26·56 $\pm (0·685 \times 2·78)$ or 24·66 and 28·46, equivalent to percentage frequencies of 17·4 % and 22·7 %. For 500 random samples the corresponding limits are (from tables) 16·6 and 23·8 %.

Extensive use of completely systematic samples for frequency has been made by a group of Dutch workers concerned with artificial grassland (see De Vries & De Boer, 1959). Of particular interest is the demonstration, by Nielen and Dirven (1950), that the accuracy of such samples for grass fields is equivalent to that of random samples using the same number of quadrats.

The principles of sampling percentage cover are essentially the same as for frequency.* Indeed, in recording hits by point quadrats we are recording frequency sampled by quadrats of, theoretically, infinitesimally small size. Thus considerations of sample size, transformation of data before analysis and restriction of randomization are exactly the same as for frequency as long as points are positioned independently of one another. Three considerations only call for comment: (a) the effect of pin diameter; (b) the common practice of

* The principles of the method have been thoroughly and critically discussed by Goodall (1952b).

using frames of pins, often ten arranged in a line, instead of single pins; (c) the effect of inclining pins from the vertical. The effect of pin diameter has been mentioned in Chapter I. The sample obtained by the use of a pin is not a point but a circular area of the same radius as the pin. Thus the true percentage cover is exaggerated, sometimes very considerably (Goodall, 1952b—cf. Fig. 1 and Table 1). The error introduced is not the same for different species, as the percentage exaggeration will be greater for species having small or much elongated or dissected leaves than for species with large and more or less isodiametric leaves. The presence or absence of gaps in the leaf mosaic of an individual will also influence the degree of exaggeration, an individual with few gaps between leaves tending to behave as a large single leaf. There is little prospect of any method being devised to allow in any exact manner for this varying exaggeration of cover between different species. Fortunately, in most circumstances it remains broadly constant for any one species and so will not affect comparisons between cover values for the same species in different communities. The same considerations apply more strongly to the 'loop-frequency' method of Parker (1950, 1951) in which presence in a ring three quarters of an inch in diameter is used as a measure of cover or basal area. Hutchings and Holmgren (1959) found for crown area of the shrub *Eurotia lanata* an exaggeration of up to 99% compared with the value obtained by measurement of a sample of individuals. Johnston (1957) concluded that, for grassland, the loop-frequency method, while more rapid, gave less accurate results for comparable sampling intensity than either point-quadrats or line-intercept methods.

Following Levy (Levy, 1933; Levy & Madden, 1933) most workers have used frames of ten pins. If cover were random, or aggregated only in such small units that the reading for one pin was independent of those for pins adjacent to it, this procedure would not seriously affect the accuracy of data thus obtained, though even then open to some theoretical objection because of the lack of statistical independence of the observations. In fact plants are commonly so arranged that there is pronounced interdependence between pins of the same frame. Thus, if the number of hits per frame for a series of frames are compared with binomial expectation, there tends to be an excess of high and low values and deficiency of values around the mean. This is to be expected from the clumped nature of much vegetation and it has been demonstrated by Goodall (1952b). It follows that the variance is greater than that from the same number of random samples and that the cover estimate is less precise. Goodall (1952b) has shown for one community, not apparently with an unusual amount of non-randomness, that, in comparison with 2000 points in random frames of ten points, from a quarter to a half this number of random single points, according to species, will give the same degree of precision. This consideration should be borne in mind in deciding whether to use frames or single points. In some circumstances it may be more economical of time to examine a greater number of points in frames than fewer points separately, but, in general, in view of the uncertainty before the observations are made of how much loss of precision will result from association of points into frames, separate points are to be preferred unless there is strong evidence against their use (e.g. from previous experience with similar vegetation). If frames are used variance must be calculated from the data, no prediction based on binomial theory being possible. The use of a completely systematic arrangement of points has been considered by Tidmarsh and Havenga (1955) in connection with the use of a 'wheel-point' apparatus in which one of the spokes of a rimless wheel is used as the sampling point for basal area. They showed, both for model populations of cards and in the field, that the variance of the mean obtained from systematic points tended to that for random points provided the spacing between points exceeded the size of individuals or clusters of individuals (cf. the similar conclusion of Nielen and Dirven (1950) on systematic samples for frequency). They therefore concluded that data gathered in this way, provided the spacing does not correspond with a repetitive pattern in the vegetation, could be treated as if they were obtained from the same number of random points. This has evident advantages in view of the generally greater speed of systematic sampling. They showed, further, that if the interpoint distance was less than the individual or cluster size, the variance between repeated independent systematic samples was less than that for random points. This greater accuracy is not spurious, as they claim, but is due to an effect comparable to that of restriction of randomization, discussed on p. 40. There is, however, no means of estimating the variance of a single set of systematic samples. The contrast may be noted with the use of random frames, where the variance is increased because the frame is normally small in relation to the individuals or clusters being recorded.

The use of inclined pins was suggested by Tinney et al. (1937) on the grounds that the larger sample of leaf contacts would lead to greater accuracy, and this suggestion has been fairly widely followed. It is true that the range of confidence limits for a given probability, expressed as a proportion of the cover value, decreases as cover increases (see Table 4), so that greater accuracy of the estimate is obtained from a higher cover. At the same time it must be remembered that the proportional increase in cover value obtained for different species by inclining the pins will vary according to the morphology of their shoots, as Winkworth (1955) has demonstrated in the field. Moreover, if shoots tend to be orientated in one direction, the cover value obtained may vary according to the orientation of the inclined pins. Thus, for the clearly understood, if somewhat arbitrary, character of vertical projection of the aerial parts, is substituted a much vaguer one. On balance there seems little advantage to be gained from inclining the pins, except perhaps when interest centres on species of very low cover.

If point quadrats are used to determine contribution of different species to the vegetation (proportion of total hits) as a measure of yield, or to determine cover repetition as a measure of performance, much the same considerations apply as in simple determination of cover. Single points will give a more accurate estimate than the same number of points in groups. Increasing the diameter of the pins will increase the values for cover repetition and may alter those for relative contribution to the vegetation, owing to the differential effect of pin diameter on species of different morphology. Cover repetition is likely to be increased for most species by inclining the pins, though this is not necessarily so and it may in fact be decreased (Winkworth, 1955). Winkworth has shown that, although mean values for percentage contribution to the vegetation in the heathland he examined do not differ significantly between determinations by vertical and inclined pins, the variance is liable to be greater. This is contrary to the apparent assumption of Tinney et al. (1937) that the increased number of contacts necessarily gives greater accuracy. Warren Wilson (1959a, b, 1960; see also Warren Wilson, 1963a; Philip, 1965a, 1966; Miller, 1967) has made a critical examination of the effect of inclining point quadrats. He assumes that the objective in point quadrat techniques is always to determine the foliage area of a species per unit area of ground and is thus concerned primarily with cover repetition. He points out that vertical quadrats can in theory record any proportion of actual foliage area between 100% (for horizontal foliage) and 0% (for vertical foliage) but that an inclined quadrat never has such a wide range. He presents a formula giving the relation between apparent and true foliage area for different foliage angles and different angles of inclination of the quadrat. The angle of inclination giving an estimate of foliage area least affected by foliage angle is 21·5 degrees with the horizontal; with this inclination an estimate of foliage area per unit area of ground is obtained by multiplying the mean frequency of hits per quadrat by a factor of 1·1. He has further suggested analysing the foliage area at successive levels to bring out the relative amounts of foliage at different heights. To do this it is necessary to record the position at which a hit is made as the quadrat is moved into the vegetation. The average foliage angle (α) in any given layer can be obtained from records for vertical quadrats together with those for horizontal quadrats moving in the same layer, and is given by

$$\tan \alpha = \frac{\pi}{2} \left(\frac{\text{Contact frequency/cm of quadrat for horizontal quadrats}}{\text{Contact frequency/cm of quadrat for vertical quadrats}} \right).$$

Warren Wilson (1965) and Philip (1965b) have described the application of point quadrats to the estimation of the foliage density of single plants.

A difficulty sometimes arises when the number of contacts per pin is recorded for tussock species. In the centre of a tussock the number of contacts may be too great to count. Goodall (1953c), arguing from the fact that where counts can be made at all points the data for number of points with one, two, three, etc. contacts can be fitted to a negative binomial series, has suggested using the negative binomial series calculated from the counted points to estimate the mean number of contacts at uncounted points. As he himself admits, fitting the negative binomial series to such data is laborious, and scarcely practical for routine work. In any case the assumption that the type of distribution in tussocking species is the same as in non-tussocking species is not necessarily true.

Measurement of cover or canopy by intersects on line transects calls for little comment. Randomization may conveniently be made by laying parallel transects from random points on an arbitrary base line or lines. An estimate of standard error is available from the variance of the total length of intercept on different transects. Greater accuracy will thus be obtained from more short transects than few long lines. Each transect should, however, be long enough to include all phases of any mosaic pattern that may be present. McIntyre (1953) (see also Strong, 1966; Westman, 1971) has considered the possibility of making a concurrent estimate of density (especially of tussock or similar growth forms) from the number of separate intercepts on the transects. He concludes that, in most circumstances, there is little to be gained over making separate counts for density determination.

Comparison of estimates of cover, if they are based on random point quadrats, may be made by a 2×2 contingency table in the same way as comparison of frequency estimates. If frames of pins have been used, the assumption of binomial distribution is no longer valid and this procedure is liable to exaggerate the significance of differences. Instead the values from individual frames should be subjected to angular transformation and the variance of the means for areas calculated and used as the basis of a t test. Data for percentage contribution to the vegetation may be handled in the same way. Data for cover repetition present a different problem. Here the distribution is very skew. Goodall (1952b) has found empirically that square root transformation will put such data into a form satisfactory for the standard tests of significance.

In sampling performance by measures such as height, cover repetition, etc., which refer to the individual plant, samples should ideally be taken in such a way that all plants have an equal chance of being selected. Unless density is uniform over the area, this can only be achieved by enumerating the individuals and selecting individuals by use of random numbers. This is clearly not practicable in most circumstances. If density is reasonably uniform over the area, the best approximation is to select the individuals nearest to a number of random points. If density is obviously variable over the area, it is advisable to divide the area into portions each having roughly constant density, determine separate means for each portion, and combine the values in a weighted mean, each value being weighted by the number of individuals present in the sub-area, i.e. by the product of density and area. The effect of this may be illustrated by an example. Suppose counts are made of number of inflorescences per individual, and the number, as is likely, varies inversely with the density of individuals, results such as the following might be obtained in different parts of the area.

	A	B	C	D
Relative area	1	2	1	4
Density	5	12	18	25
Counts of inflorescences per individual	3, 4, 2, 2, 5 1, 4, 3, 2, 4	5, 3, 1, 3, 4 4, 3, 2, 2, 1	2, 4, 3, 2, 3 1, 3, 3, 1, 2	1, 3, 2, 1, 1 2, 4, 2, 1, 3
Mean numbers of inflorescences per individual	3·0	2·8	2·4	2·0
Weighting (relative area × density)	5	24	18	100

The weighted mean will then be

$$\frac{3\cdot0 \times 5 + 2\cdot8 \times 24 + 2\cdot4 \times 18 \times 2\cdot0 \times 100}{5 + 24 + 18 + 100} = \frac{325\cdot4}{147} = 2\cdot21.$$

The arithmetic mean of the 40 observations is

$$(30 + 28 + 24 + 20) \div 40 = 2\cdot55$$

which gives a quite misleading value for the average number of inflorescences, being unduly influenced by the relatively few plants with a large number of inflorescences.

There is a useful approximate test, derived from the rank of separate observations, of the significance of differences between means. This may be illustrated from an example. Ten frames of ten pins gave the following readings for cover of a species in two different areas

(A) 10, 9, 9, 8, 7, 6, 5, 4, 4, 2.   (B) 7, 6, 5, 4, 4, 3, 3, 2, 1, 1.

Is there evidence of a difference in percentage cover? The values observed are rearranged in descending order

Rank	1	2	3	4	5	6	7	8	9	10	11	12	13	14	15	16	17	18	19	20
Observed value	10	9	9	8	7	7	6	6	5	5	4	4	4	4	3	3	2	2	1	1

The total rankings for the two areas are

(A) $1 + 2 \times 2 \cdot 5 + 4 + 5 \cdot 5 + 7 \cdot 5 + 9 \cdot 5 + 2 \times 12 \cdot 5 + 17 \cdot 5 = 75.$
(B) $5 \cdot 5 + 7 \cdot 5 + 9 \cdot 5 + 2 \times 12 \cdot 5 + 2 \times 15 \cdot 5 + 17 \cdot 5 + 2 \times 19 \cdot 5 = 135.$

Where there is a tie between two or more ranks each value is allotted the mean ranking. If there is no difference in mean cover, the expected total rankings for the two areas are the same, 105. The 5% and 1% points for the lower total are approximately $(9N^2/10) - (3N/2) + 3$ and $(4N^2/5) - 9$, where N is the number of values in each set.* In this case the 5% point is 78 and the 1% point 71, so that significance at the 5% level, but not at the 1% level, is indicated. The t test, carried out after angular transformation, and entering the table at 9 degrees of freedom to allow for the markedly different variances of the two sets, indicates a probability of approximately 2%.

This ranking test is quickly performed and not only provides a rough test in the field and an indication whether more laborious tests are worth applying, but it has the additional advantage that it is independent of the distribution of the variables being compared. It can thus be used even where it is impossible to transform the data into approximately normal form and the more usual tests are therefore not available.

Considerable attention has been paid to a method of plotless sampling, particularly adapted to forest vegetation, where there are practical difficulties

* Dixon and Massey (1957) have a useful discussion of this and other ranking tests and give a table of the probability of various departures from expected total rankings. The approximate 5% and 1% points are quoted from Moroney (1951).

in delimiting the relatively large quadrats necessary for sampling trees (Cottam, 1947; Cottam & Curtis, 1949, 1955, 1956; Cottam *et al.*, 1953). From a number of randomly selected points certain measurements are made. Four different procedures have been used (Fig. 11).

(1) *Closest individual method.* The distance from the sampling point to the nearest individual is measured (Fig. 11(a)).

(2) *Nearest neighbour method.* The distance from this individual to its nearest neighbour is measured (Fig. 11(b)).

(3) *Random pairs method.* From the sampling point a line is taken to the nearest individual and a 90° exclusion angle erected on either side of it. The distance from this individual to the nearest one lying outside the exclusion angle is measured (Fig. 11(c)).

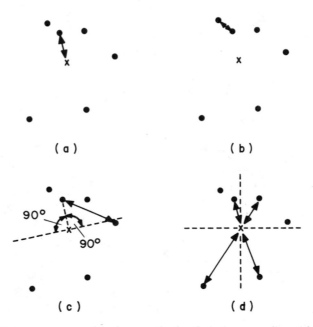

Fig. 11. Distances measured in four methods of plotless sampling. (a) Closest individual. (b) Nearest neighbour. (c) Random pairs with 180° exclusion angle. (d) Point-centred quarter. X is the sampling point in each case.

(4) *Point-centred quarter method.* The distance from the sampling point to the nearest individual in each quadrant is measured. The orientation of the quadrants is fixed in advance. Cottam and Curtis (1956), sampling at random distances along a transect line, orientate them two on either side of the line (Fig. 11(d)).

There is an extensive literature on the use of such distance measures in estimating density. Several estimates of density from the measured distance are available. Persson (1971), who provides a useful survey and introduction to the literature, distinguishes four types of estimator. If r_{ij} is the distance of an individual from the ith out of n sampling points or sampling individuals for the jth out of k sectors, and c is a constant, the four types are

$$(1) \quad c \bigg/ \bigg\{ \sum_{i=1}^{n} \sum_{j=1}^{k} r_{ij}/nk \bigg\}^2$$

$$(2) \quad c \bigg/ \bigg\{ \sum_{i=1}^{n} \sum_{j=1}^{k} r_{ij}{}^2/nk \bigg\}$$

$$(3) \quad \frac{c}{n} \sum_{i=1}^{n} \sum_{j=1}^{k} \bigg(\frac{1}{r_{ij}{}^2} \bigg)$$

$$(4) \quad \frac{c}{n} \sum_{i=1}^{n} \bigg\{ k \bigg/ \sum_{j=1}^{k} r_{ij}^2 \bigg\}.$$

He also suggests a further type, c/z^2, based on the median, z, of the observed distances. The estimators, with appropriate values of c, can also be applied to measurements of the second or subsequent nearest neighbours rather than to the nearest neighbour, but such measurements are unlikely to be used in estimating density. He provides a table of values of c for a number of enumerators in a random population, some of which had been derived by previous workers. For closest individual the type 1 and 2 estimators are $0.25/(\Sigma r/n)^2$ and $0.3183/(\Sigma r^2/n)$ respectively. For the nearest neighbours of a random individual rather than the nearest neighbours of that individual nearest to a random point (Fig. 11(b)), the same formulae apply. Earlier workers had expressed the type one estimator in terms of mean area, i.e. mean distance is half the square root of the mean area (Morisita, 1954; Clark & Evans, 1954a).

For the nearest neighbour of an individual nearest a random point the type 1 estimator is $0.3556/(\Sigma r/n)^2$. The difference arises because this method of sampling does not give a random sample of nearest neighbour distances; isolated individuals are likely to be sampled more frequently.

For the point-centred quarter method the type 1 and type 2 estimators are $1/(\Sigma r/n)^2$ and $1.273/(\Sigma r^2/n)$ respectively.

The random pairs method has apparently not been examined theoretically. Cottam and Curtis (1955) found empirically that the mean value of the distance between random pairs was 0.8 times the square root of the mean area, corresponding to a type 1 enumerator of $0.64/(\Sigma r/n)^2$.

A difficulty in using plotless methods is that if the density is low, it may be impractical to search beyond a certain distance for the nearest individual to a random point. J. H. Darwin (in Batcheler, 1971) has shown that for a maximum searching distance of R, a type 2 estimator is $0.3183\,a/\{\Sigma r^2 + (n - a)R^2\}$ where a is the number out of n points at which a distance of R or less is recorded. R should be set large enough to provide measurements at half the number of points sampled.

These methods all have one serious limitation; they assume that individuals are randomly distributed. Comparison by Cottam and Curtis (1956) of estimates for total tree density with complete enumerations of woodlands indicates that this assumption is justified, at least for the areas examined. It would not necessarily be so in all cases and almost certainly not if the methods were applied to a single species in mixed woodland. Dix (1961) has used the point-centred quarter method in grassland, with individual shoots as the unit of density. To do so may be valid in some grasslands but, in view of the evident non-randomness of many grassland species, the method should be tested against direct counts for any grassland type to which it is to be applied (see Risser & Zedler, 1968).

Morisita (1957) has suggested a type 3 estimator applicable to certain types of non-random distribution. The distance r to the nth nearest individual ($n \geqslant 3$) in each of k sectors at N points is measured. He derives two estimates of density.

$$d_1 = \frac{1}{\pi} \cdot \frac{n-1}{N} \sum \left(\frac{1}{r^2} \right)$$

$$d_2 = \frac{1}{\pi} \cdot \frac{nk-1}{N} \sum_{i=1}^{N} \frac{k}{\sum_{j=1}^{k} r_{ij}^2}.$$

If $d_1 < d_2$ the best estimate of density is $(d_1 + d_2)/2$. If $d_1 > d_2$, d_1 is the best estimate.

This approach is based on the assumption that the whole area sampled can be regarded as made up of the number of sub-areas within which individuals are randomly or uniformly distributed. It is thus likely to be applicable to a mosaic of patches of different densities but not to a distribution of clusters each composed of relatively few individuals. Morisita has tested the method with $n = 3$ and $k = 4$ on artificial populations and obtained satisfactory estimates of density. No field test has apparently been made.

Batcheler and Bell (Batcheler & Bell, 1970; Batcheler, 1971; see also Warren & Batcheler, 1979) have suggested a type 2 estimator applicable to a clustered distribution. In such a distribution the mean distance from a random point to the nearest individual will be greater, and the mean distance from that individual to its nearest neighbour will be less, than the corresponding means in a random distribution of the same density. Denoting the distance from a point to the nearest individual by r_a and the distance from that individual to its nearest neighbour by r_b, Batcheler and Bell apply a correction $\Sigma r_a / \Sigma r_b$ to obtain an estimator $0{\cdot}3183\, \Sigma r_b / (\Sigma r_a{}^2 \cdot \Sigma r_a / n)$*.

If only distances $\lessgtr R$ are measured, the correction depends not only on the inequality of the two means but also on the inequality of the observed and expected frequency of points at which both measurements are available. Combining these two corrections Batcheler and Bell obtain a combined correction $b^2 n\, \Sigma r_a / a^3\, \Sigma r_b$ where b is the number of points at which a measure of distance to nearest neighbour is obtained. This gives an estimator

$$0{\cdot}3183a \left/ \left\{ (\Sigma r_a{}^2 + (n-a)R^2)\left(\frac{b^2 n\Sigma r_a}{a^3 \Sigma r_b}\right) \right\} \right.$$

If two orders of clustering are present (i.e. clusters themselves arranged in clusters), a similar argument leads to an estimator

$$0{\cdot}3183a \left/ \left\{ (\Sigma r_a{}^2 + (n-a)R^2)\left(\frac{d^2 \cdot n\Sigma r_a}{a^2 b \Sigma r_d}\right) \right\} \right.$$

where d is the number of points at which a measure of distance of the nearest neighbour to its nearest neighbour is available and r_d is such a distance.

Batcheler and Bell found satisfactory agreement of the estimates with the known density in test populations. They point out that corresponding corrections could be derived for higher orders of clustering.

The derivation of density from distance measures assumes that the distances are measured to the centre of the plant. Unless care is taken to estimate the distance to the centre, rather than to the nearest point on the periphery of the stem, a constant bias will be introduced (Ashby, 1972)†. This is clearly important with trees.

If all individuals regardless of species are being recorded, the species and basal area (or other measure of size) of each individual may also be noted. From these data figures for relative density and relative basal area of each species are readily obtained, but the values obtained must be interpreted with caution because none of the methods gives a random sample of individuals, even if the

---

* The notation and form of presentation of Batcheler and Bell's estimators have been altered slightly to accord with those used above.

† Simberloff (1979) has examined the theoretical nearest neighbour relations of circles of known diameter.

individuals themselves are randomly distributed; more isolated individuals are more likely to be sampled. The resulting bias will be exacerbated if the distribution of individuals is not random, and especially if different species show different degrees or types of nonrandomness*.

Cottam and Curtis (1956) derived from the last three distance measures a measure of relative frequency (frequency recorded in samples as a proportion of frequency of all species). Like quadrat frequency, this is dependent partly on density and partly on type of distribution of the species concerned. It is, however, further affected by the density and pattern of other species present, and thus becomes an even more complex character of the vegetation and correspondingly difficult to interpret. It is true that relative density is similarly dependent on other species but there is the important difference that density is an absolute character, unlike frequency, and the corresponding absolute value can be obtained by multiplying by the estimated density.

Cottam and Curtis (1956) showed that the accuracy of the estimate of density, for the same number of points sampled, increases in the order: closest individual, nearest neighbour, random pairs and quarter method. The quarter method involves making a greater number of measurements to obtain a given accuracy than the random pairs method, but at considerably fewer points. They suggested its use, however, in preference to random pairs as occupying less time in the field and giving a greater amount of information on relative density and basal area.

Little use has been made of these methods in exact comparisons. Estimates of total density cannot readily be compared but instead the means of distance measurements may be compared by a *t* test or analysis of variance. Cottam *et al.* (1953) showed that the mean value of the four measurements at a point in the quarter method is normally distributed and hence this measure may be analysed directly. If random pairs are used square root transformation is advisable to reduce skewness. Relative density in different areas may be examined in a contingency table (cf. percentage contribution to the vegetation from point quadrats). Relative frequency, if it is thought useful, could be handled in the same way. The data for basal area are less tractable. Comparison of mean basal area per tree, either for total trees or for a species, may readily be made by a *t* test, but there is no ready means of comparing total basal area of trees per unit area, which involves the multiple of two values of which

* The use of an individual and its *n* nearest neighbours as a sample in the study of variability of composition *within* an area rather than average composition of the area, (Williams *et al.*, 1969) is a different aspect of nearest-neighbour relationships, representing the use of stands delimited by number of individuals rather than by area (p. 304).

estimates only are available. Shanks's (1954) calculation of estimates of density and mean basal area per tree from each point scarcely commends itself, as both estimates rest on quite inadequate samples and the resulting high standard error could only be reduced by using such large samples that the advantage of speed in the field would be lost.

If basal area is the measure of greatest interest, the method of 'variable-radius' sampling introduced by Bitterlich (1948; see Loetsch et al., 1973) is more useful. In this method a sighting gauge is used to count those trees which are within a distance from the sampling point not more than 33 times their trunk diameter, i.e. which subtend an angle of at least 1° 44' at the point. This ratio is so fixed that the count at a point multiplied by 10 corresponds directly to square feet of basal area per acre. The values obtained are approximately normally distributed and hence can be analysed directly. It must be emphasized that no other measures of any value are available from variable-radius sampling. The relative frequency figures, calculated by Rice and Penfound (1955), represent such a complex character, affected not only by density and pattern of individuals of the species concerned and of those of other species present, but also by the pattern of size classes of the different species, as to be virtually meaningless. Cooper (1957, 1963) applied the method to the assessment of canopy cover of shrubs and herbs, counting individual shrubs or herbs whose canopy subtended at least the minimum angle.

Lindsey et al. (1958) have made a careful assessment in forest of the efficiency of various areal and plotless sampling methods for both density and basal area. Their comparison is based on determination of the time necessary to sample sufficient units to reduce the standard error of the resultant estimate to an arbitrary level of 15%. They took into account not only time at sampling points but also time spent in moving between them. The last is clearly an important consideration but is often overlooked in discussions of efficiency of different methods. Their conclusion is that Bitterlich variable-radius sampling is to be preferred for basal area and that a 0·1 acre circular plot defined by rangefinder is to be preferred for density.

CHAPTER 3

Pattern

Ecology and plant geography are largely concerned with the causes of patterns of distributions, patterns of all scales from those of individuals within a small area to those of vegetation types and taxa over the surface of the world. One of the principal contributions from the use of numerical methods in ecology is the more exact detection and description of distribution patterns. Patterns of distribution differ not only in scale, but also in intensity, the degree of difference between different parts of the area under consideration. Very large scale patterns, such as those which interest the plant geographer, are outside the scope of this book*. Medium scale patterns, of varying intensity, are concerned mainly with assemblages of species, with vegetation rather than individual species, and are discussed in Chapters 6–9. This chapter is concerned with the patterns, generally of relatively small scale and low intensity, of individual species, the pattern or non-random distribution of individuals within what is generally considered a plant community.

The first work on non-randomness was naturally exploratory and able to contribute relatively little to ecological knowledge, but techniques of analysis of non-randomness now available are more useful. Before discussing technique the biological significance of departure from randomness must be examined.

A living plant and its various parts are subject to a multitude of variable factors, all affecting to greater or lesser extent its behaviour and performance, and even its continued existence. These factors include internal ones intrinsic to the plant and environmental ones. The external factors range from such obvious and measurable influences as temperature, humidity, and concentration of nutrients, through obvious but less readily described features such as soil texture to such little-understood effects as specific secretion from other plants. Intrinsic factors fall principally under the heads of effect of position of the part on the whole plant, and effect of age. Effective seed dispersal distance is also best considered an intrinsic factor. The effects of the numerous influencing factors are not independent but exhibit numerous and complex interactions of

* For the application of numerical methods to plant geographic problems see Pielou (1979).

which as yet we know comparatively little. Some few are well known as phenomena, though understanding of their mechanism is limited, e.g. the effect of concentration of one nutrient on the availability of another. Others have been singled out by plant physiologists for special study, e.g. the interaction of light, temperature, and carbon dioxide supply in photosynthesis. However, we lack more than the most superficial knowledge of many of the factors affecting the performance of plants in the field.

In any given area, of whatever size, some factors will be constant at any one time, while others will vary from point to point in the area. The smaller the area the greater in general will be the number of constant factors. The magnitudes of the ranges of difference in the varying factors may be considered in terms of their effects on the plant. If the effects of all factors on all species present are relatively small, it will be a matter of chance which species succeeds at any point, and the resulting distribution of individuals (or parts of individuals where such parts are largely independent) will be random. Such conditions of equality of effects of different factors will only hold when the range of values found is well within the limits of tolerance of all the species present. This follows from the much greater effect of small differences in an influencing factor when its value is near the limits of tolerance of a species. Now if one or few factors have a disproportionately great effect on performance or survival of a species, then the distribution of that species will tend to be determined by that particular factor or factors. If the values of the factor are themselves randomly distributed then the distribution of the species will also be random. Field experience shows that most environmental factors do not have a random distribution of different values (consider, for example, such factors as soil moisture and texture). We may thus put forward the hypothesis that departure from randomness of distribution of a species indicates that one or few factors are determining the performance or survival of the species. The converse does not necessarily apply. One factor may be overriding even if the species is randomly distributed. It is unlikely, however, in any community of more than a few species, that the factor will have a predominant influence on one species alone. If more than one species is concerned, whether the effect of the predominating factor is the same for different species or not, correlation between the occurrence or performance of species may be expected. (See Chapter 4.) Thus correlation gives an indication of predominance of one or few factors, even if individuals are randomly distributed.

Since study of the casual factors determining the distribution of plants and vegetation is a prime objective of ecology, any technique that can assist their detection is clearly of value. At the same time it must be emphasized that detection and analysis of non-randomness is a starting-point for further

investigation of the factors responsible and not an end in itself. Failure to
realize this is responsible for the apparently barren nature of some statistical
work in ecology.

   Though the pattern of distribution of a species within a community is a real
characteristic of that community, the appearance of non-randomness in a set of
samples is not an absolute characteristic but, like frequency, is dependent on
the size and, sometimes, the shape of sample area used. This may be illustrated
from Fig. 12, which shows a distribution of individuals with a general low
density but scattered patches in which the species has a high density. A
relatively small quadrat will clearly tend to be unoccupied more often in such a
community than in one with the same density of individuals randomly
distributed. Correspondingly, high values will occur more frequently also.
There will, in fact, be clear evidence of non-random distribution. With a very
small quadrat, however, so small that even when placed within the high-density
areas it is usually unoccupied, the position is rather different. The absolute
increase in number of unoccupied quadrats and of those containing more than
one individual will be so small that it will not be detected unless a very large
number of samples is taken. Thus the non-randomness will not be apparent. If
a large quadrat is used, such that it generally includes one or several high-
density patches, the grouping together of individuals will tend to affect only
their distribution within the quadrat and not the number it contains. The

Fig. 12. Patches of higher density imposed on a general distribution at lower
density, a type of distribution commonly found in the field.

distribution, judged by such samples, will thus tend to appear random.* These conclusions can readily be checked on artificial communities of counters of known distribution (cf. Greig-Smith, 1952a). Many types of non-random arrangement sampled by random quadrat throws will thus appear random with very small or large quadrats but non-random with intermediate sizes. It is also clear that the lower the density of a species within high-density patches, the greater is the minimum size or number of quadrats necessary to detect departure from randomness. It has commonly been stated that the rarer species in a community, unless propagated vegetatively, are randomly distributed, whereas the commoner species are usually not randomly distributed. This conclusion may result in part from this relationship between density and minimum quadrat size and number necessary to detect non-randomness and should not be accepted uncritically.

The distribution of individuals may be more regular than random expectation though this is very much less commonly found. If they are so distributed, very small quadrats will again fail to detect departure from randomness, but all larger quadrat sizes will do so and, in general, the larger the quadrat the more distinct the departure from random expectation will appear.

So far we have not questioned the appropriateness of the Poisson series as the basis of random expectation. The condition under which a Poisson distribution applies is that the chance of an event occurring is very small but if a sufficiently large sample is taken then some occurrences will be found. For quadrat data the relevant chance is that of a plant occurring at any point in the area under consideration. Provided this is low then a Poisson distribution may be expected if the area is sampled by any size of quadrat small relative to the whole area. In biological terms the expectation applies if the number of individuals of a species in the area is low relative to the possible number that could grow in that area. It should be noted carefully that the mean number of individuals per quadrat is irrelevant in deciding whether a Poisson expectation is appropriate. If the mean number per quadrat is, say, 10 or a hundred, Poisson expectation still applies if the maximum possible number that could occur in a quadrat is very much higher.

If the number of individuals actually occurring approaches the maximum possible, the Poisson expectation no longer applies. Instead, if individuals are randomly distributed, the frequencies of different numbers per quadrat will approximate to a binomial distribution obtained by expansion of $(p + q)^n$, where n is the maximum number possible per quadrat, p the chance of any one

---

* Whether the data from such a large quadrat fit random expectation exactly will depend on the pattern of the high-density patches (cf. Goodall, 1952a, p. 205).

of the possible 'places' in the quadrat being occupied (so that np is the mean number per quadrat) and $q = 1 - p$. The binomial series has $n + 1$ terms, giving the probability of 0, 1, 2, ... n individuals per quadrat, e.g. if the quadrat can contain a maximum of four individuals the expansion is

$$q^4 = 4pq^3 + 6p^2q^2 + 4p^3q + p^4.$$

Table 5 shows for a mean of 3 the probabilities of different numbers per quadrat given by the Poisson series and by binomial series corresponding to $\frac{1}{3}$, $\frac{1}{2}$, and $\frac{3}{4}$ of the maximum possible number of individuals. It will be seen that at these levels of relative density there is considerable discrepancy between the binomial and Poisson series, the discrepancy increasing as the density approaches the maximum possible. In each of the binomial series there is a greater proportion of values at and around the mean and fewer extreme values than in the Poisson series.

Table 5. Probabilities of finding 0, 1, 2, 3, etc. individuals in samples from various distributions with mean of 3

Number of individuals per sample	Distribution			
	Poisson	Binomial $(\frac{1}{3} + \frac{2}{3})^9$	Binomial $(\frac{1}{2} + \frac{1}{2})^6$	Binomial $(\frac{3}{4} + \frac{1}{4})^4$
0	0·0498	0·0260	0·0156	0·0039
1	0·1494	0·1171	0·0937	0·0469
2	0·2240	0·2341	0·2344	0·2109
3	0·2240	0·2731	0·3125	0·4219
4	0·1680	0·2048	0·2344	0·3164
5	0·1008	0·1024	0·0937	
6	0·0504	0·0341	0·0156	
7	0·0216	0·0073		
8	0·0081	0·0009		
9	0·0027	0·0001		
> 9	0·0012			
Variance	3·00	2·00	1·50	0·75

The density relative to maximum possible and, correspondingly, the maximum number of individuals that could occur in a quadrat of any specified size are theoretical concepts incapable of determination, though of some interest. A rough estimate may be possible from a knowledge of the morphology and behaviour of a species. What is important in the present connection is some indication of the relative density at which the Poisson series no longer gives a reasonable approximation to random expectation. Fortunately this may be easily obtained.

The variance of a Poisson series is equal to its mean. A measure of the degree of departure from Poisson expectation is therefore provided by the ratio of the variance to the mean. The departure of this ratio from unity has a standard error $\sqrt{(2/(N-1))}$, where N is the number of observations, and may be tested for significance by a t test.

Now a binomial series $(p+q)^n$ has a variance npq, which is always less than the mean np. The variance : mean ratio is thus

$$(npq)/(np) = q = 1 - p.$$

If this value is to be tested for departure from unity (on the assumption that the distribution is in fact Poissonian) we can put

$$1 - p = 1 - ts$$

and thus $p = ts$ gives the value of p, or proportion of maximum possible density at which difference between binomial and Poisson expectation would be just detectable at any desired level of probability. $s\left(= \sqrt{\dfrac{2}{N-1}}\right)$ is the standard error of the difference between the variance : mean ratio and unity, and t has the value (from table of t) corresponding to the number of observations and the desired level of probability.

$$
\begin{aligned}
&\text{For 100 observations } s = 0.1421\\
&\quad\quad 500 \text{ observations } s = 0.06331\\
&\quad 1000 \text{ observations } s = 0.04474\\
&\text{For } 5\% \text{ probability } \quad t = 2.0\\
&\quad\quad 1\% \text{ probability } \quad t = 2.6\\
&\quad 0.1\% \text{ probability } \quad t = 3.4 \text{ for } N = 100\\
&\quad\quad\quad\quad\quad\quad\quad\quad\quad\quad 3.3 \text{ for } N = 500 \text{ or } 1000
\end{aligned}
$$

Using these values we obtain the following approximate values of p:

N	Probability (%)		
	5	1	0·1
100	0·28	0·37	0·48
500	0·13	0·16	0·21
1000	0·09	0·12	0·15

If 5% probability is accepted as indicating significance, then with 100 throws significant departure from Poisson expectation would be expected to be detected in 50% of trials in a random population if density was equal to 28% of maximum possible or higher, but if 500 throws were used this would apply if density was only 13%. If the more extreme probability level of 1% was required before departure was accepted, density could be as high as 37% for 100 throws, or 16% for 500 throws, without departure being detected in 50% of trials. It will be noticed that this relationship between binomial and Poisson expectations is independent of quadrat size.

The expected distribution at densities approaching the maximum may be considered in another way. Since an individual has finite size there is always an area around its midpoint (at which it is conventionally regarded as situated) within which the chance of another individual occurring is depressed because two individuals cannot be superimposed. If the density relative to the maximum possible is low, the areas within which probability of occurrence is so reduced are small relative to the whole area and do not affect significantly the even distribution of probability over the area. If relative density is high this is no longer true; the distribution of probability of occurrence becomes markedly uneven and Poisson distribution no longer occurs.

If individuals vary widely in size the position is more complicated. At maximum density the relatively small spaces between large individuals may be occupied by relatively large numbers of small individuals, giving a pattern of individuals more uneven than the corresponding random distribution (Pielou, 1960). Though this might, in theory, complicate interpretation of non-random pattern, in practice it is unlikely to be important. It probably occurs in natural vegetation only rarely, principally in freely regenerating forest composed entirely or almost entirely of one species. Moreover, the practising ecologist is unlikely to treat individuals of such a wide range of sizes on the same basis.

There are two categories of departure from randomness, (1) that in which individuals tend to be clumped or aggregated together (Figs 2(c) and 12), sampling of which gives an excess of blanks and high values compared with random expectation, and (2) that in which individuals tend to be uniformly spaced (Fig. 2(a)), sampling of which gives a deficiency of blanks and high values and an excess of values around the mean. The former has been termed *overdispersed* (referring to the distribution curve obtained) and the latter *underdispersed*. Unfortunately these two terms have sometimes been used in the reverse sense (referring to the pattern of individuals on the ground). They have thus become a source of confusion and are better avoided. *Contagious* has been widely used for aggregated distributions generally, and this seems a suitable term, although originally applied by Pólya to a particular type of aggregation. The uniform distributions may be termed *regular*. Regular distributions have rarely been found in the field and attention has rightly been concentrated on contagious distributions.

Various tests and measures of departure from randomness have been proposed and the more important are detailed below. To demonstrate their use it will be convenient to consider their application to one set of data. The following results were obtained for number of shoots of *Carex flacca* per quadrat in 200 throws of a quadrat 10 cm square in limestone grassland.

Number of shoots per quadrat	0	1	2	3	4	5	6	7	8	9	10
Number of quadrats	134	34	12	8	8	—	1	1	1	—	1
Mean number per quadrat	0·725										

(1) χ^2 *test of goodness of fit.*   The Poisson series with the observed mean is calculated and the observed numbers of quadrats containing 0, 1, 2, etc., individuals compared with random expectation by a χ^2 test. This has been used by Blackman (1935) and many later workers.

For the *Carex flacca* data $m = 0·725$ and $e^{-m} = 0·48433$. (See Appendix Table 3.) Thus the expected number of empty quadrats is $200 \times 0·48433 = 96·866$. The remaining terms of the Poisson series are readily calculated in the manner shown below.

It is necessary to lump together all the figures for quadrats containing three individuals or more, so that all the expected values may be greater than 5. (This arbitrary level is that commonly used in calculations of χ^2 to eliminate the disproportionate effect of small differences from a low expectation.*) The total value is entered in the table of χ^2 with degrees of freedom 2 less than the number of values from which χ^2 is calculated, 1 degree of freedom being used in the determination of the mean. In this case there are 2 degrees of freedom available for the χ^2 test and the 0·1% point is 13·81. The probability of obtaining the observed distribution by chance is thus much less than 0·1% and the data show very clear indication of non-random distribution.

(2) *Variance : mean ratio* [Coefficient of dispersion (Blackman, 1942) or relative variance (Clapham, 1936)]. This test makes use of the equality of mean and variance of the Poisson distribution. If the ratio of variance to mean is less than one, a regular distribution is indicated, if greater than one, a contagious distribution.

---

* Cochran (1954), in a useful general discussion of χ^2, considers the generally recommended minimum expected value of 5 to be too conservative, resulting in a substantial loss of power in the test. (See also Everitt, 1977.)

Number per quadrat	Number of quadrats		Differ-ence	χ^2
	Expected	Observed		
0	$200e^{-m} =$ $200 \times 0\cdot48433 = 96\cdot866$	134	37·134	14·24
1	$200me^{-m} =$ $96\cdot866 \times 0\cdot725 = 70\cdot228$	34	36·228	18·69
2	$\dfrac{200m^2e^{-m}}{2!} =$ $70\cdot228 \times \dfrac{0\cdot725}{2} = 25\cdot458$	12	13·458	7·11
3	$\dfrac{200m^3e^{-m}}{3!} =$ $25\cdot458 \times \dfrac{0\cdot725}{3} = 6\cdot152$			
> 3	$\left. \begin{matrix} \\ 1\cdot296 \end{matrix} \right\} 7\cdot448$	20	12·552	21·15
	200	200		61·19

Two ways of testing the significance of the difference between the observed ratio and unity are available. The difference may be compared with its standard error by means of a t test. The standard error is independent of the density of individuals and depends only on the number of samples. Blackman (1942) used

$$s = \sqrt{[2N/(N-1)^2]}$$

where N is the number of samples. Bartlett (quoted in Greig-Smith, 1952a) has pointed out that the more correct value is $s = \sqrt{[2/(N-1)]}$. The difference is small for reasonably large values of N. A table of $\sqrt{[2/(N-1)]}$ for selected values of N is given in Appendix Table 4.

An alternative test of significance makes use of the index of dispersion $\dfrac{S(x-\bar{x})^2}{\bar{x}}$ i.e. $\dfrac{\text{variance}}{\text{mean}} \times$ (number of observations -1). Significance is assessed by reference to the table of χ^2, the observed value of the index of dispersion being entered in the table with degrees of freedom one less than the number of observations. This method of testing the variance : mean ratio is preferred by Skellam (1952) and David and Moore (1954).

For the *Carex flacca* data

$$Sx = 145, \; Sx^2 = 531, \; S(x-\bar{x})^2 = 531 - ((145)^2/200) = 425\text{·}875$$

Variance $425\text{·}875/199 = 2\text{·}140075$, variance/mean $= 2\text{·}9518$.
The standard error of the deviation of this value from unity is

$$\sqrt{(2/199)} = 0\text{·}1003.$$

Thus $t = (2\text{·}9518 - 1)/0\text{·}1003 = 19\text{·}46$, with 199 degrees of freedom and probability (from table of t) much less than $0\text{·}1\%$.

The index of dispersion is $I = 2\text{·}9518 \times 199 = 587\text{·}41$ with 199 degrees of freedom. The usual table of χ^2 includes values up to 30 degrees of freedom only. For higher degrees of freedom it may be assumed that $\sqrt{(2\chi^2)}$ is normally distributed about $\sqrt{(2N-1)}$ with unit variance, so that the value of $\sqrt{(2\chi^2)}$ $- \sqrt{(2N-1)}$ may be referred to the table of the normal deviate (Fisher, 1941 §20). The probability of χ^2 corresponds to that for a single tail of the normal curve so that the probability obtained from the table must be halved. Thus for the *Carex flacca* data

$$\sqrt{(2\chi^2)} - \sqrt{(2N-1)} = \sqrt{1174\text{·}82} - \sqrt{399} = 14\text{·}301.$$

From the table of the normal variate the corresponding probability is very low (χ^2 probability of $0\text{·}05\%$ corresponds to a value of $3\text{·}29$).

If deviation from random expectation is towards a regular distribution, the value of the index of dispersion is judged significant if the probability from the χ^2 table is usually high, e.g. probability of 95% corresponds to 5% for contagious distribution and 99% to 1%. If the procedure for a large number of degrees of freedom is followed, the value of $\sqrt{(2N-1)} - \sqrt{(2\chi^2)}$ is used and a low probability required for significance.

Numata's coefficient of homogeneity (Numata, 1949, 1954), which has been used in various investigations by Numata and others, is related to the variance:mean ratio. It is defined for a sample of n quadrats with density $\bar{x}$ and standard error s as

$$h = t \cdot (1/\sqrt{n}) \cdot (s/\bar{x}),$$

t being assigned the value corresponding to a desired probability. The value obtained clearly depends on the density and number of samples as well as on the departure from randomness. When expressed, as Numata suggests, as the ratio of the observed value to the value expected for a random distribution of the same density sampled by the same number of quadrats it reduces to $\sqrt{(\text{variance:mean ratio})}$.

(3) *Moore's ϕ test* (Moore, 1953). If the mean number of individuals per quadrat is high, or if the distribution is markedly contagious, some quadrats

will contain a large number of individuals which are both time-consuming to count and liable to counting errors. Moore has proposed a test dependent on the first three frequency classes (0, 1, 2 individuals per quadrat) only. He takes

$$\phi = (2n_0 n_2)/n_1^2$$

where n_0, n_1, n_2 are the numbers of quadrats containing 0, 1, 2 individuals respectively. For a Poisson distribution $\phi = 1$. For contagious distributions with an excess of empty quadrats and deficiency of quadrats containing one individual ϕ will generally, depending on the value of n_2, exceed unity. At least we can infer that if ϕ is significantly greater than unity, then the distribution is non-random. (Some regular distributions will also give $\phi > 1$.) The mean value of ϕ for samples from a Poisson distribution, and its standard error, which is dependent on the mean and number of quadrats, have been derived by Moore. He gives a table (Appendix Table 5) showing the value of ϕ plus twice its standard error, i.e. approximately the 5% point, for various values of N and the mean, m. Determination of m depends on complete enumeration of the quadrats. However, for each value of m there is a unique value of R, the percentage of quadrats falling in the first three classes, and the table may be entered according to the value of R instead of that of m. This is equivalent to estimating the mean from the first three classes only.

For the *Carex flacca* data $n_0 = 134$, $n_1 = 34$, $n_2 = 12$,

$$\phi = \frac{2 \times 134 \times 12}{34^2} = 2\cdot782, \quad R = \frac{134 + 34 + 12}{200} \times 100 = 90.$$

The 5% significance point for ϕ, from the table, is approximately 1·66 (nearest tabulated points $R = 92$, $\phi = 1\cdot66$ and $R = 81$, $\phi = 1\cdot68$) and hence non-random distribution is indicated.

(4) *Ashby and Stevens's test* (Ashby, 1935). This depends on random sampling by a quadrat subdivided into a number of smaller squares and comparison of the observed number of empty squares with the number expected from the density within the quadrat, which is $E = n(1 - n^{-1})^s$, where n is the number of squares and s is the total number of individuals in the quadrat. Although a comparison between observed and expected number of empty squares is available from a single quadrat sample, the distribution of the number of empty squares is so far from normal that data from a number of quadrats must be pooled before making a test of significance. The reader is referred to the original account for details of the test of significance.

(5) David and Moore (1957) have suggested a test based on division of an area into parallel strips and comparison of between strip and within strip variance of distance of individuals from the end. As this depends on knowledge

of the exact position of all individuals in terms of co-ordinates from two adjacent sides of the area taken as axes, its ecological utility is limited.

(6) Aberdeen (1958) has pointed out that if frequency data are available from several sizes of quadrat, any departure from linearity in a graph of log percentage absence against quadrat size indicates non-random distribution.

The above tests are concerned with detecting departure from randomness, though the variance: mean ratio is also a useful measure of degree of departure. A number of other measures have been proposed which give an indication of degree of departure rather than a test of its significance.

(7) *Index of clumping.* David and Moore (1954) later proposed the use of the value $\left(\dfrac{\text{Observed variance}}{\text{Observed mean}} - 1 \right)$, which they term *Index of Clumping* and which is zero for random distributions. They give a test for the significance of the difference between values obtained from two different samples, e.g. for the same species in two different habitats. It depends on the same number of quadrats being used in each case and takes the form of calculating

$$\omega = -\frac{1}{2}\ln\frac{v_1\lambda_2}{\lambda_1 v_2}$$

where λ_1, λ_2 are the observed means and v_1, v_2 are the observed variances of the two sets of data. If ω lies outside the range $\pm 2{\cdot}5/\sqrt{(N-1)}$, where N is the number of quadrats in each set, a significant difference in index of clumping is indicated.

If half or more of the total frequency is in the first three classes (0, 1, 2 individuals per quadrat), David and Moore recommend an alternative procedure based on the ϕ test of non-randomness. This alternative procedure, however, tests difference between the means and between the variances of the two sets jointly. Since the means are more likely than not to be different, it is of limited value only.

(8) *Observed density: calculated density ratio.* If a species is contagiously distributed, fewer quadrats will be occupied than in a random distribution of the same density. Thus, if density is calculated from the observed frequency according to the relationship $F = 100(1 - e^{-m})$, the value obtained will be less than the true one. The ratio of observed density to density calculated from frequency will therefore give a measure of the degree of non-randomness, and this will be greater than one for contagious distribution and less than one for regular distribution. A table of expected densities is given in Appendix Table 6. This measure was apparently first used by McGinnies (1934).

(9) Fracker and Brischle (1944) used a measure essentially similar to the last, calculating the ratio of the difference between the observed and calculated

densities to the square of the calculated density, i.e.

$$(m_{\text{obs.}} - m_{\text{calc.}})/(m_{\text{calc.}})^2.$$

This is zero for random, positive for contagious, and negative for regular distributions. Fracker and Brischle considered, apparently from empirical considerations, that values of from 0·0001 to 0·003 are to be expected for random distributions and that values above 0·02 indicate definite contagion.

(10) *Abundance:frequency ratio.* Whitford (1949) suggested the ratio of abundance (defined as mean density within occupied quadrats) to frequency as a measure of contagion. Abundance is related to density and frequency for

$$\text{Abundance } A = \frac{\text{Total number of individuals}}{\text{Number of occupied quadrats}}$$

$$\text{Density } D = \frac{\text{Total number of individuals}}{\text{Total number of quadrats}}$$

$$\text{Frequency } F = \frac{\text{Number of occupied quadrats}}{\text{Total number of quadrats}} \times 100$$

so that $A \times F = 100 D$. The same density may be produced by high frequency and low abundance (regular distribution) or low frequency and high abundance (contagious distribution). The ratio abundance:frequency is equal to $(100D)/F^2$.

This ratio has no fixed expectation for a random distribution, which severely limits its utility in making comparisons. Thus the expected value for $F = 1\%$ is 1·005 and for $F = 99\%$ is 0·047. Contagious distributions will give a value greater and regular distributions a value less than random expectation.

(11) Morisita (1959) has developed a measure of departure from randomness based on the measure of diversity proposed by Simpson (1949) rather than directly on Poisson distribution. (See also Morisita, 1962, 1971; Lloyd, 1967.) Simpson's measure is Σp^2, where $p_1, \ldots p_z$ are the proportions of individuals falling into each of Z groups (see p. 161). It ranges from 1 when all individuals are concentrated into one group to $1/Z$ when they are distributed equally between groups. Σp^2, the true population measure, is estimated by $\delta = \dfrac{\Sigma n(n-1)}{N(N-1)}$ where $n_1, \ldots n_q$ are the numbers of individuals observed in q groups and N is the total number of individuals observed. Morisita takes quadrats as the groups and sets up as a measure of dispersion

$$I_\delta = q\delta,$$

which will have the value 1 for random distribution, range from 1 to q for contagious distributions (or more accurately will tend to 1 for random distribution as N increases) and be below 1 for regular distributions.

Morisita proposes to test departure from randomness by referring

$$(I_\delta(N-1)+q-N)/(q-1)$$

to tables of F with $n_1 = q-1$, $n_2 = \infty$. The numerator of this fraction is, however, the index of dispersion, $S(x-\bar{x})^2/\bar{x}$, used in (2) above and may be referred directly to the table of χ^2.

For the *Carex flacca* data

$$N = 145, \ N(N-1) = 20880,$$

$$\Sigma n(n-1) = \Sigma n^2 - N = 531 - 145 = 386$$

$$\delta = \frac{\Sigma n(n-1)}{N(N-1)} = 0{\cdot}0184866$$

$$I_\delta = 3{\cdot}69732$$

$$I_\delta(N-1)+q-N = 587{\cdot}41, \text{ as before.}$$

A comparison of the average degree of non-randomness in two or more communities may sometimes be desirable. Measures of non-randomness for individual species might be combined in various ways. Curtis and McIntosh (1950) suggest that the ratio of the sum of observed densities to the sum of expected densities calculated from frequency should be used, i.e.

$$\Sigma m_{obs.}/\Sigma m_{calc.}$$

They point out that the measure is weighted in favour of the more plentiful species, which 'seems proper in a general statement about a plant assemblage'.

A number of tests and measures of departure from random expectation are available and it is important to consider their relative value. At first sight it might appear that the χ^2 test of goodness of fit with Poisson expectation would always be the most satisfactory test to use, but it has some disadvantages. Before calculating χ^2 those classes with a low expectation must be pooled; the usual, arbitrary, limit is a minimum expectation of five in each class. One of the commonest effects of non-randomness is the occurrence of a few quadrats with an unusually high number of individuals and, correspondingly, an increase in the number of quadrats with very few individuals or none. It is the quadrats with a high number of individuals and (if the mean number is high) those with few or none which have low expectations and have to be pooled. This may result in the discrepancy from Poisson expectation being largely obscured and the test failing to show any indication of non-randomness. Consider the following data for 100 throws:

Number per quadrat	Observed number of quadrats	Expected number of quadrats	
0	21	14·09	
1	27	27·61	
2	22	27·06	Mean
3	14	17·68	number
4	8	8·66	per quadrat
5	2	3·40	1·96
6	3	1·11	
7	1	0·31	
8	2	0·08	

This shows strong indication of contagion and gives a variance : mean ratio of 1·65 (probability from Index of Dispersion of much less than 0·1 %). Before applying the χ^2 test the table must be recast:

Number per quadrat	Observed number of quadrats	Expected number of quadrats
0	21	14·09
1	27	27·61
2	22	27·06
3	14	17·68
> 3	16	13·56

which gives a χ^2 of 5·55 with three degrees of freedom and probability 10–20 % not indicating departure from randomness. Another difficulty arises when the mean value is low and it is only possible to group the quadrats into two classes, empty and occupied, if all classes are to have an expectation greater than five and there is then no degree of freedom available for a χ^2 test, e.g. 100 throws with a mean of 0·29 have random expectation of 74·8 empty quadrats, 21·7 with one individual and 3·5 with two or more individuals so that the expected numbers must be grouped as 74·8 empty and 25·2 occupied quadrats. Both difficulties may be avoided by increasing the number of samples but it is frequently unpractical to do so without involving labour disproportionate to the information to be obtained.

The variance: mean ratio derives from one particular aspect of departure from Poisson expectation, the occurrence of abnormally high or low variance. Consider the following hypothetical case, for 101 throws with a mean of 1·00, quoted by Evans (1952):

Number per quadrat	Observed number of quadrats	Expected number of quadrats
0	20	37·16
1	76	37·16
2	—	18·58
3	—	6·19
4	—	1·55
5	5	0·31
> 5	—	0·05

In spite of the evident non-randomness, the variance is exactly equal to the mean, though the direct test against expectation shows a χ^2 of 68·29, with two degrees of freedom and probability much less than 0·1 %. Such a combination of generally regular distribution with occasional groups of individuals is unlikely in natural vegetation and would in any case be recognized without reference to a test. Much the commonest type of non-randomness is the occurrence of excessive number of empty quadrats and of those with high numbers of individuals, owing either to exclusion of the species from part of the area by unfavourable environmental factors or by the presence of competing species, or to aggregation brought about by inefficient propagule dispersal or by vegetative propagation. The variance : mean ratio test is normally sensitive to this type of discrepancy. The last two examples emphasize that either one of the χ^2 test of goodness of fit and the variance: mean ratio test may detect evident non-randomness when the other fails to do so. This has been demonstrated both for artificial 'communities' of discs with known pattern (Greig-Smith, 1952a) and for woody plants in forest (Greig-Smith, 1952b).

The variance : mean ratio as an indicator of non-randomness has been criticized by Skellam (1952) on the grounds that it is dependent on the size of quadrat used. It has been shown earlier in this chapter, however, that the appearance of non-randomness in discrete samples is always dependent on the size of sample unit used. The same supposed disadvantage applies to all the tests and measures proposed, as Curtis and McIntosh (1950) have shown for several of them. It is, indeed, the varying behaviour of non-random distributions with different sample sizes that enables information to be obtained on the scale at which non-randomness is operating. Jones (1955–6) has criticized the variance : mean ratio test on two grounds. He says that it appears to behave erratically when the mean is very small 'presumably because the distributions of deviations of the variance of a Poisson distribution from its mean is too strongly skewed'. It is true that results of the variance : mean ratio test must be interpreted with caution when the mean is very low, but it is under those conditions that a χ^2 test of goodness of fit can be applied only by greatly increasing the sample number. Moore's ϕ test, devised primarily for cases where the mean is relatively high, is also erratic with a very low mean, on account of the low expectation for quadrats containing two individuals. When the mean is very low we are thus dependent on the variance : mean ratio test for lack of a better one. Jones's second criticism is really a more general one, applying to all the tests and measures considered, viz. that it ceases to be applicable with the 'more abundant species' (more accurately those with a high density relative to the maximum possible), where the Poisson distribution itself ceases to be applicable. It has been shown that under these conditions a binomial type of expectation is more appropriate than a Poisson one. A binomial distribution is always more regular than a Poisson distribution of the same mean (cf Table 5). Thus, if the distribution of a species with density high relative to the maximum possible is shown to be contagious when compared with Poisson expectation, it can be assumed that it is contagious when compared with the true random expectation. In the absence of any means of setting up the true random expectation it is impossible to tell whether an observed distribution which is regular compared with Poisson expectation is also more regular than the true random expectation. However, in practice regular distributions are rare in the field and the difficulty is unlikely to occur often.

Moore's ϕ test, like the variance : mean ratio test, is sensitive only to certain types of departure from Poisson expectation but, like it, is usually affected by an undue proportion of empty quadrats, the commonest type of non-randomness. It takes cognizance, however, of less of the distribution than the variance : mean ratio and its principal advantage is in speed of enumeration and calculation.

Of the various measures of non-randomness proposed, the ratio of observed to calculated density, and the indices proposed by Fracker and Brischle and by Whitford all depend solely on the relative number of empty and occupied quadrats. They are thus affected by one aspect only of discrepancy, though an aspect that is frequently the most important in vegetation. Among them Whitford's index has the serious disadvantage that it has no fixed expectation so that only species of the same density *or* the same frequency can be compared directly. There seems no reason for preferring Fracker and Brischle's more complex formula to the straightforward ratio of observed to calculated density. There is no ready means of testing whether any difference between the values of these measures for two sets of data is significant, so that considerable caution must be used in comparing degrees of contagion of different species or in different areas. As a measure of non-randomness the variance : mean ratio is likely to be at least as sensitive as those dependent only on frequency, and in many cases more so. In the modified form of David and Moore's Index of Clumping its variance can be estimated so that the significance of any difference in degree of contagion can be assessed. In general, therefore, it is to be preferred as a measure.

The variance : mean ratio has a much wider range for contagiousness than for regularity. If, as may rarely happen, interest centres as much on degrees of regularity as on degrees of contagiousness, Lefkovitch's (1966) index is more appropriate. This takes the form

$$\frac{4}{\pi}\tan^{-1}\left(\frac{\text{variance}}{\text{mean}}\right) - 1$$

with expectation of zero for random distribution, -1 for perfect regularity and approaching $+1$ for maximum contagiousness.

In view of the interest in measures of point-to-plant and plant-to-plant distances as a means of estimating density, it is not surprising that attempts have been made to use a similar approach to detection of non-randomness (Skellam, 1952; Dice, 1952; Hopkins, 1954; Moore, 1954; Clark & Evans, 1954a; Thompson, 1956; Cottam *et al.*, 1957; Pielou, 1959, 1962a; Holgate, 1965a,b). Various suggestions have been made but almost all depend on the equivalence in a random distribution of mean distance from a random point to the nearest individual and mean distance from an individual to its nearest neighbour. One exception is Moore's suggestion that an area should be arbitrarily subdivided and the variances of nearest neighbour distance for the subdivisions compared, in order to test the uniformity of density over the area. Dice (1952) also avoids determination of density. He suggests examining the frequency distribution of the square roots of measurements to the nearest neighbour in each of the sextants around randomly selected individuals. The frequency curve obtained is normal for random distributions but skewed to the right for regular and to the left for contagious distributions. Clark and Evans (1954a) compare the mean nearest neighbour distance with that expected from density. Thompson (1956) has extended this comparison to the mean distance to the nth neighbour. Pielou (1962a) has considered the distribution of shorter nearest neighbour distances only, with the aim of detecting regularity, due to competition, within patches of high density. Pielou (1959) compares the mean distance from random points to nearest individual with that expected from density. (Mountford (1961) has pointed out that Pielou's procedure fails to take account of the sampling variability of the density estimate.) Skellam's (1952) procedure likewise depends on knowledge of the density. This is a serious objection in practice as the value of the distance measures is greatest in those circumstances where an enumeration of individuals to obtain the density is most difficult. Moore (1954) and Hopkins (1954) have both suggested using, in effect, an estimate of density obtained from point to nearest individual measurements instead. Hopkins uses as a *Coefficient of Aggregation* the ratio of the square of the mean distance between a random point and its nearest neighbouring individual (P) to the square of the mean distance between an individual and its nearest neighbouring individual (I), i.e. provided the same number of measurements are made between pairs of individuals and from points to individuals,

$$A = \Sigma P^2 / \Sigma I^2.$$

This coefficient is unity for random distributions, greater than one for contagious distributions, and less than one for regular distributions. The deviation of A from unity is tested for significance in the following manner. The parameter $x = A/(1 + A)$ has a mean value of 0.5 for random distributions and variance of $1/(4(2n + 1))$ (n = number of pairs of observations). The distribution of x tends rapidly to normality as n increases and if n is greater than 50, $(x - 0.5)$ may be treated as a normal deviate with zero mean and standard error $1/(2\sqrt{(2n+)})$, i.e. the expression $2(x-0.5)\sqrt{(2n+1)}$ may be referred to standard tables of the normal integral. If n is less than 50 reference must be made to tables of the incomplete beta function (Pearson, 1934). Hopkins gives a diagram (Fig. 13) showing the value of x corresponding to probabilities of 5, 1 and 0.1% and values of n from 20 to 50. There is one point of practical importance. Often the only practicable method of selecting a 'random' pair of individuals is to take the nearest individual to a random point and measure the distance to its nearest neighbour. As Cottam and Curtis (1956) have shown, this

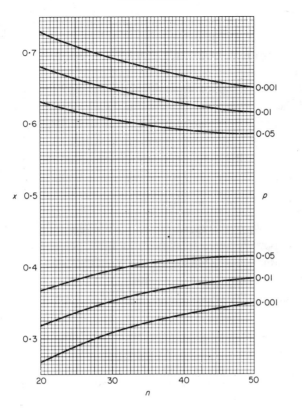

Fig. 13. Hopkins's method of detecting non-randomness from point-to-point and plant-to-plant distances; probability (p) of exceeding by chance the calculated value of x for n observations. (From Hopkins, 1954, by courtesy of *Annals of Botany*.)

results in a non-random sample giving a mean distance greater than the true value. Thus the expected values of A may be less than 1 and caution must be exercised in accepting a distribution as regular.

Holgate (1965a) has suggested two tests of randomness which depend only on measurements from a random sampling point to plants. Let X_s and X_t ($s < t$) be the distances from the point to sth nearest and tth nearest plants respectively. The ratio $Z_{st} = X_s^2/X_t^2$ has a mean value s/t and standard deviation $\left\{ \dfrac{s(t-s)}{t^2(t+1)} \right\}^{\frac{1}{2}}$ for a random distribution, i.e. a sample from n points of first and second nearest neighbours, Z_{12}, has an expected mean 0·5 and standard error of mean 0·2887 $n^{-\frac{1}{2}}$. An alternative test is based on the correlation between X_s and X_t, which has an expected value $r_{st} = (s/t)^{\frac{1}{2}}$ in a random distribution. The sampling standard deviation depends on a cumbersome formula; for first and second neighbours the expected mean is 0.7071 and the standard error of the mean is $0.7906n^{-\frac{1}{2}}$. Aggregation tends to increase both $\overline{Z}_{st}$ and r_{st}. The effect of regularity will depend on s and t, but both $\overline{Z}_{12}$ and r_{12}, the likeliest forms to be used in practice, will tend to decrease. The deviation of the $\overline{Z}_{st}$ from its expected value may be tested by reference to the normal distribution. The distribution of r_{st}, however, does not approximate to normality until very large sample sizes are reached, which limits its utility. Holgate conjectured, from limited practical tests, that the ratio test is more powerful in detecting contagiousness and the correlation test in detecting regularity, but this was not confirmed by Goodall and West (1979) in more extensive tests.

Eberhardt (1967) has suggested a test depending only on measurements from a random point to the nearest individual. This takes the form of the ratio of the mean of squared distances to the square of the mean distance, which has an expected value of 1.27 for random distribution and is greater for contagious and less for regular distributions.

Cottam et al. (1957) have attempted to interpret the meaning of indices of non-randomness derived from distance measures. They point out that, taking the simplest case of grouping of individuals into distinct clumps, aggregation can be described in terms of three variables, the clump mean area (total area of the population divided by the number of clumps), area of clumps (the mean value of the areas that would be obtained by drawing a planimeter round each clump) and the within-clump mean area (the mean value for the area of a clump divided by the number of individuals it contains). If any two of these factors are known for a fixed number of individuals in a fixed area, the third can be determined. Plant-to-plant distance is determined by within-clump mean area. It is this dependence on within-clump mean area that limits the value of indices based on plant-to-plant distances. Such indices will reflect mainly the nature of the clumps, primarily a result of reproductive behaviour and generally obvious on inspection, and scarcely at all the biologically more interesting pattern of the clumps themselves. An index based on point-to-plant distance is affected by the pattern of clumps as well, as Pielou (1959) has recognized, and is to that extent more satisfactory.

Clark and Evans (1954b) have shown that there is a fixed expectation for a random distribution of the proportion of individuals serving as nearest neighbour to 0, 1, 2, 3, 4, and 5 others. This proportion is affected by departure from randomness only if it involves direct effects between individuals. Clark (1956) has derived for a random distribution the proportion of individuals for which the relationship with the nth nearest neighbour is reflexive, i.e. each of the pair is the nth nearest neighbour of the other. Observed distributions may be compared with this random expectation. The labour involved in collecting data makes it unlikely that these approaches will have any great practical value. Bray (1962) has suggested another approach independent of distances measured and of any estimate of density. If n individuals are recorded at each sampling point, then, if a species is randomly distributed, the number of points at which 0, 1, 2, ... n individuals of that species are recorded will fall on a binomial distribution with p equal to the observed proportion of the species in all individuals recorded. A χ^2 test may be used to test the observed distribution against the expected binomial distribution.

The methods of detecting and measuring departure from randomness so far discussed are applicable only to species occurring as discrete, readily distinguished individuals, or those which present other countable units, e.g. rosettes or tillers. The latter usually show pronounced contagion owing to a number arising close together from the same individual and it is difficult to distinguish this from any non-randomness of individuals present. In many species even this modification of counting rosettes or tillers is impossible, e.g. many procumbent, rooting herbs. In many communities, therefore, it may be impossible by these techniques to analyse the distribution of all, or even of a majority, of the species present.

Species not amenable to counting, and many species that are, can conveniently be measured in terms of cover. If a set of points linked together in a constant manner, e.g. the frame of ten pins commonly used in making estimations of cover, is placed on vegetation in which the distribution of cover is random, the chance of a set of n points showing presence of a species at 0, 1, 2, 3, . . . n points will be given by expansion of the binomial series where p is the mean cover and $q = 1 - p$. If cover is contagiously distributed, i.e. if there are patches within which cover is greater than the mean for the whole area, then there will be a greater number of sets with no, or few, points occupied and with a high number occupied. Conversely, if cover is more uniformly distributed than random there will be a greater number of sets with about the mean number of points occupied.

The concepts of contagious and regular distribution in relation to cover should perhaps be examined more closely. In the density approach to pattern the convention is maintained that individuals occur at a point. It is because this is not true that the Poisson expectation for random distribution holds only within certain limits and we are forced to introduce the idea of density relative to the maximum possible. In considering cover no such simplifying convention is possible. Consider first a pair of points set a known distance apart as the sampling unit. The smallest unit of cover is a leaf or a length of stem. If the distance between the points is less than the average dimension of a leaf, the points will most commonly either both touch a leaf or neither touch a leaf and the data will show contagion when compared with random expectation. If the leaves are themselves uniformly distributed (cf. for example the circles in Fig. 14), and the two points are about the same distance apart as the mean distance between the leaves, an undue number of pairs will tend to fall one on a leaf and one off, i.e. a regular distribution will be indicated. If the points are at very much greater distance apart than this, then the observed data will tend to random expectation. In practice we are not likely to work with point spacings anywhere nearly as small as the dimensions of leaves, but the same consider-ations apply to spreading individuals or patches of species as apply to leaves, except that the indications will not be as clear-cut because there is not normally complete cover within the boundaries of one individual. To offset this the number of points within a set may be increased, the distance between the extreme members of the set bearing the same relation to the pattern of cover as the distance between the two points in the hypothetical case we have been discussing. It is evident from these considerations that, just as with density data, indications of non-randomness obtained will depend on the size of sampling unit used.

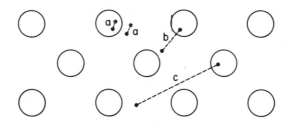

Fig. 14. Effect of spacing of sample points on indications of non-randomness of cover. (a) Spacing indicating contagious distribution. (b) Spacing indicating regular distribution. (c) Spacing indicating random distribution (see text).

The proportions of sets with various numbers of points occupied may be compared with the binomial expectation for the same mean by a χ^2 test of goodness of fit, just as density data are compared with Poisson expectation, e.g. 100 throws of a frame of four points arranged at the corners of a square of side 15 cm, were made on *Alchemilla vulgaris* agg. in a lawn. The mean cover was 14·75%. The observed number of frames having 0, 1, 2, 3, and 4 points, and the number expected from the expansion of $(0\cdot8525+0\cdot1475)^4$ were

Number of points occupied	Observed number of frames	Expected number of frames
0	64	52·82
1	20	36·56
2	10	9·49
3	5	1·09
4	1	0·05

After pooling values for 2, 3 and 4 points occupied, χ^2 is 12·58, with one degree of freedom and probability less than 0·1% indicating departure from randomness.

A test comparable to the variance : mean ratio used for density data may also be used. The variance of a binomial series $(p+q)^n$ is npq and the sampling variance of this variance is

$$\frac{2(npq)^2}{N-1} + \frac{npq(1-6pq)}{N}$$

where N is the number of throws (Fisher, 1941, §18). Variance of the observed data may thus be compared with that expected. For the example quoted

$$\Sigma x = 59,\ \Sigma x^2 = 121,\ \Sigma(x-\bar{x})^2 = 121 - (59^2/100) = 86\cdot19 \text{ and}$$
$$\text{Variance} = 86\cdot19/99 = 0\cdot8706.$$

The expected variance is $npq = 4 \times 0\cdot1475 \times 0\cdot8525 = 0\cdot5030$.
Difference between observed and expected variance is 0·3676.
Sampling variance of variance

$$\frac{2 \times 0\cdot5030^2}{99} + \frac{0\cdot5030(1-0\cdot7544)}{100} = 0\cdot006347.$$

Standard error of variance $\sqrt{0\cdot006347} = 0\cdot07967$.
The ratio of the difference to its standard error $0\cdot3676/0\cdot07967 = 4\cdot61$ with probability, from tables of the normal integral, of less than 0·01%. Alternatively, as with the Poisson distribution, the index of dispersion may be referred to the table of χ^2. The form of the index of dispersion appropriate to the binomial distribution is $\dfrac{\Sigma(x-\bar{x})^2}{\bar{x}q} = \dfrac{\Sigma(x-\bar{x})^2}{npq}$. For the example quoted this is

$86 \cdot 19 / 0 \cdot 5030 = 171 \cdot 35$, with 99 degrees of freedom and probability of less than $0 \cdot 01 \%$.

A useful approximate measure of departure from randomness, comparable to the ratio of observed to calculated density used for density data, is the ratio of the observed cover to cover calculated from frequency (percentage of frames with at least one point occupied). The expected proportion of frames of n points with no points occupied is $(1 - p)^n$. Hence the frequency $F = 100 - 100 (1 - p)^n$ and from this the cover corresponding to any given frequency can readily be calculated. In the present case the observed frequency is 36% and $100 (1 - p)^4 = 64$ from which $p = 0 \cdot 1056$ and the ratio of observed to calculated cover is

$$0 \cdot 1475 / 0 \cdot 1056 = 1 \cdot 40.$$

If the mean cover is very low only a limited number of possible distributions of occupied points can occur, e.g. for 100 throws of four points with a mean cover of 1%, only five arrangements are possible. Under these conditions the use of tests based on continuous distributions is not valid. (The same considerations would apply if percentage cover approached 100% but this is unlikely to occur in practice.) The difficulty can clearly be overcome by increasing the sample size, but this may not be practical. An alternative is to replace the sampling points by quadrats of suitable size, recording presence or absence only of the species concerned. The simplest method in most cases is to use a group of contiguous quadrats as the sampling unit. The mean frequency obtained in this way will be higher than the mean cover and, since the data may be treated in exactly the same manner as that from points, a suitable size of quadrat will permit tests of non-randomness on species too sparse to be tested by cover.

Another approach depending on frequency has been suggested by Jones (1955–56). If a number of quadrats are arranged in either a 'chessboard' pattern or a line, the occurrence of contiguous pairs or runs of quadrats containing a species may be examined. If aggregation of a species is on a scale larger than the size of quadrat used, the species will tend to occur in adjacent quadrats more frequently than would happen if distribution were random. If departure from randomness is on such a small scale that a unit of the resulting pattern is contained within one quadrat, then it will not be detected. On the other hand a widespread and abundant species may have such a high frequency in quadrats of the size used that departure from randomness may not be detected, although clearly demonstrable by density analysis. Formulae for calculating the expected number of contiguous pairs and runs under various conditions and the variances to which they are subject have been given by Krishna Iyer (1948,

1950).* The difference between the observed number of contiguous pairs or runs and that expected, compared with its standard deviation, is referred to tables of the normal curve. (See also Gounot, 1962, 1969; Cliff & Ord, 1971; Kooijman, 1976; Sokal & Oden, 1978.)

Pielou (1964, 1965, 1967, 1977) has developed methods of testing departure from randomness of mosaics of patches and gaps (for a single species) or patches of different species. A single species mosaic is examined by linked pairs of quadrats and is assumed to be random if the probability of a hit or a miss with one quadrat depends only on the result with the other, and these probabilities are constant over the area, i.e. the dimensions of patches and gaps are random. A multi-species mosaic is examined by a line of quadrats and assumed to be random if the probability of finding a species depends only on the result obtained in the previous quadrat and these probabilities are constant (as before) and, further, that the different species are randomly intermingled. This method was developed to examine the pattern of species having vegetatively produced patches rather than discrete individuals, but would be applicable to any vegetation having recognizable phases. Its limitation is that any one quadrat has to be assigned to one phase only; this is overcome for a single spacies by regarding a quadrat as within a patch if the species occurs at all in the quadrat, but is necessarily arbitrary for a multispecies mosaic.

So far it has been assumed that we are dealing with the pattern of individuals on an area of ground. Occasionally a linear arrangement may be involved, e.g. plants established along a crack in rock, trees along the bank of a river, and the question may be asked whether the spacing between individuals could have resulted from random placing on the line. For random placing, the distance, d, between successive individuals can be shown to follow the exponential distribution $f(d) = \lambda_e^{-\lambda d}$ where $\lambda = 1/\bar{d}$, the mean number of individuals per unit length. By integration of this function, the expected proportion of measured intervals between individuals in successive classes can be obtained. The observed frequency can thus be compared with that expected if individuals are randomly spaced. Hazen (1966) used this method to examine the distribution of epiphytes on the branches of trees. Branches were regarded as placed end to end and the spacing between individual epiphytes examined.

---

* The most generally useful formulae are those for a line of quadrats under conditions of 'non-free sampling', i.e. using the observed frequencies to calculate the expected values. The expected number of joins between quadrats with and without the species is $\dfrac{2\,n_1 n_2}{m}$ and its variance is $\dfrac{2\,n_1 n_2(2\,n_1 n_2 - m)}{m^2(m-1)}$, where n_1, n_2 are the number of quadrats respectively containing and not containing the species, and $m = n_1 + n_2$ is the total number of quadrats in the line.

Ryland (1972) has developed this approach to allow the detection of non-randomness from samples of different size. He was concerned with colonies of a bryozoan on fronds of the alga *Fucus serratus*. Fronds of *F. serratus* vary in size, but are the natural sampling unit. The area of each frond and the number of bryozoan colonies on it were recorded. For each frond the observed area was regarded for purposes of simulation as a linear measurement and the observed number of colonies were spaced along this length by a randomizing computer program. From the total data for all fronds, the mean interval was calculated and the observed frequency distribution compared with random expectation. Variation in the density of colonization on different fronds is apparent as an excess of small and large intervals and a deficiency of intermediate ones.

In the opening paragraphs of this chapter it was argued that the biological significance of any departure from randomness lies in the indication it provides that one or few factors are playing an overriding part in determining the survival of individuals. Departure from randomness is thus not of great interest in itself; its importance is as a guide to the factors controlling the survival and performance of individuals. Various attempts at a generalized interpretation of non-randomness have been made. It has frequently been pointed out that the commonest source of discrepancy between observed data and random expectation is the excessive number of blank quadrats, with the variance exceeding the mean. David and Moore (1954) state the issue succinctly: ' "Student" (1919) pointed out that the variance being greater than the mean is the usual cause of the Poisson hypothesis being inadequate for field data, and two mechanisms have since been put forward to "explain" this inadequacy. The first mechanism assumes that the Poisson parameter, λ, varies over the field. In the second mechanism the Poisson parameter is assumed to remain constant, but there is some form of dependence among the observations, a concept which Pólya (1930) called "contagion". . . . Feller (1943) has shown how it is impossible to distinguish between these two mechanisms on the basis of an observed set of data,* and probably the true state of affairs lies in between the two and may well be different for each species'.

It is evident that both mechanisms modifying the Poisson hypothesis are operative in natural vegetation, and the last phrase quoted is certainly an understatement. Vegetative spread and, for many species, seed dispersal will lead to the second mechanism being effective. Any botanist with field experience is well aware of the widespread occurrence of point-to-point

* More precisely, Feller showed it to be impossible *on the basis of an overall observed frequency distribution*.

variation within a plant community of such scale and degree that it is apparent by qualitative criteria alone. Watt (1947) has shown that some cases of such variation can be explained by the occurrence of cyclic changes in the composition of the vegetation, changes determined directly by changing behaviour of individuals with age, or indirectly by reaction of different species on their immediate environment. Other cases are dependent on predetermined differences in soil, topography or other environmental factors, varying in scale from the correlation of vegetation with catenas of soil types to the difference in composition of vegetation between ridge and furrow in grassland on old arable land. These are gross differences correlated with obvious environmental features, but it is evident that similar differences, of lesser degree, will occur, being expressed rather as differences in amounts of species and detectable only by quantitative means (Greig-Smith, 1979).

In spite of the abundant evidence that both mechanisms are likely to be operative in producing deviation from Poisson expectation, as was pointed out by Clapham (1932), attention was at first very largely focused on the presumed consequences of reproductive behaviour. Attempts to account for all non-randomness on this basis alone are doomed to failure. Nevertheless it is worth-while to examine various suggested distributions and consider their adequacy when point-to-point variation is absent.

(1) Archibald (1948) tested data for various salt-marsh species against Neyman's Contagious distribution and found a reasonably good agreement for species that did not fit the Poisson distribution. Fracker and Brischle (1944) on the other hand found that Neyman's distribution did not fit data for two species of *Ribes* in forest and suggested that there was a mixture of random original immigrants and contagiously distributed later generations. Neyman (1939), concerned with recently hatched insect larvae crawling away from egg clusters, assumed random distribution of larval clusters with the number per cluster also random, but with a limit to the distance to which larvae might crawl, i.e. an arbitrary limit to cluster size. His distribution is defined by two parameters, one proportional to the mean number of clusters per unit area and one to the mean number of units per cluster.

(2) Archibald (1950) tested a number of species from several communities (evidently principally from salt marsh and chalk grassland) against Thomas's Double Poisson distribution (Thomas, 1949) and again found that many species fitted it well. This distribution is based on similar assumptions to that of Neyman, clusters being assumed to be randomly distributed and the number of units, additional to the first, also being random.

Barnes and Stanbury (1951) found that data for early colonization of newly-exposed surfaces of china clay residues could be fitted to both Neyman's and Thomas's distributions. (Pielou (1957) has given reasons for expecting the two distributions to be fitted equally well by quadrat data.) Thomson (1952), on the other hand, found that only one species out of three tested in an 'old-field' community could be fitted to either distribution.

Evans (1953) has discussed the data of Archibald and Barnes and Stanbury in more detail and concludes that Neyman's distribution gives a satisfactory fit.

(3) Robinson (1954) suggested the use of the Negative Binomial distribution. If groups are randomly distributed and the number per group follows a logarithmic distribution then the resultant number of individuals per quadrat is given by the negative binomial distribution. Robinson fitted data of Steiger (1930) on prairie vegetation to this distribution and found good agreement. Evans (1953), however, considered the fit to the negative binomial not satisfactory. Unfortunately Steiger's data are scarcely adequate material for a test of the distribution, for, as Curtis (1955) has pointed out, Steiger deliberately selected the position of his quadrats to include as great a range of densities as possible and they cannot be considered unbiased samples for this purpose.

Various other distributions of a similar type have been suggested, but none of them, apparently, has been tested against field data.

Neyman's and Thomas's distributions are based on such similar assumptions that they may be considered together. Even if the primary assumption that reproductive behaviour is mainly responsible for non-randomness is accepted, the assumptions made in applying the distributions are not free from objections, which have been well put by Goodall (1952a): 'Both are open to the objection that the range of each postulated group is limited—in one case arbitrarily, in the other to the observational quadrat. A more natural assumption would be that the probability of finding a plant falls off as some continuous function of the distance from the centre of the group of which it is a member—the number of propagules being perhaps related to distance from mother plant by an inverse square law.' A further objection is the absence of any real evidence that the number of individuals per cluster is distributed randomly. Where spread of a species is vegetative this is probably often not so, as study of morphology indicates in many cases a tendency to production of a constant number of daughter individuals per individual per season. The successful fitting of either distribution provides, from the second parameter, an estimate of the number of individuals per clump. This is of some intrinsic interest but it is little help in disentangling the biological relations of the species of the community with one another and with environmental factors.

The use of the negative binomial is likewise dependent on an unproved assumption on the nature of clusters. The distribution is defined by the mean number per quadrat and by the exponent, which is dependent on the mean reproductive rate. Here again the utility of an estimate of the second parameter is doubtful. Robinson claims on theoretical grounds that the value of the exponent should be independent of external conditions and hence, for the same species, constant in different communities. He draws support from Steiger's data for this. It is not clear why reproductive behaviour should remain the same under different conditions; indeed, general knowledge of plants in the field would suggest that it would not do so. Little weight can be attached to Steiger's data in this connection for reasons already considered. Further, the thesis of a unique value of the exponent for any one species is not supported by Pidgeon and Ashby's (1940) demonstration that species may be randomly distributed in one area and markedly contagious in another.

In comparing the various suggested distributions, it must be remembered that there are so many possible theoretical distributions of this type that the satisfactory fit of data to one of them can scarcely be regarded as evidence for its validity in the absence of strong and independent biological evidence for the truth of its underlying assumptions.

The importance of point-to-point variation is emphasized by the fact that satisfactory fit to Neyman's and Thomas's distributions has only been obtained under conditions where such variation is minimized. Archibald's data are derived from samples of 100 or 500 *contiguous* quadrats each of area 20 cm² so that even with samples of 500 the total area sampled was only 1 m², small enough to avoid most if not all of any point-to-point variability present, particularly in the relatively uniform habitats of salt marsh and chalk grassland sampled. Barnes and Stanbury deliberately worked on an exceptionally uniform habitat (residues from china clay mining allowed to settle behind artificial dams) and were concerned with early stages of succession carried on by water- and wind-borne migrules.

The only practical mathematical approach to non-randomness based on the first mechanism, that of variability of the Poisson parameter over the field or point-to-point variability, appears to be that of Stevens (1937 and in Ashby, 1935)*. He derived the formula $E = n(1 - n^{-1})^s$ for the number of empty squares expected in a quadrat divided into n smaller squares and containing s individuals randomly distributed (p. 64). If the distribution is not random he suggested substituting

$$E = n(1 - n^{-1})^s[1 + s(s-1)c]$$

* There has been considerable interest in the sampling properties of various models of non-random distribution, but these have not led to practical approaches. (See, for example, various contributions in Patil *et al.*, 1971.)

where c is calculated to give the best fit to the data. c is then a measure of the variability of the Poisson parameter over the field. This was applied to *Salicornia europaea* agg. in salt marsh by Ashby (1935) and to *Bonnaya brachiata*, a weed species on arable land in India, by Singh and Das (1938). In neither case was any test made of goodness of fit given by the corrected formula for E, but in each case it clearly accounts for a considerable part of the departure from randomness. Although this approach is based on point-to-point variability, in both cases where it has been tried the sample quadrat used was of such size that only small-scale heterogeneity, due to clusters of the order of size produced by reproductive behaviour, would be likely to be detected. In other words, although the first mechanism (of point-to-point variability) was assumed, the second (of reproductive behaviour) was in fact investigated. Feller's (1943) conclusion that it is impossible to distinguish the effects of the two mechanisms is once more emphasized. Theoretically there is no reason why the sampling quadrat should not be made very much larger to take account of point-to-point variability, but, except for species of very low density, this would involve so many squares within the quadrat to allow a reasonable proportion of empty squares that enumeration of a sufficient number of quadrats would be impractical. There is also a limitation in that the method applies only when c is small relative to the mean probability of occurrence of an individual. Finally it is difficult to interpret the value of c either in terms of the plant or of the habitat.

Erickson and Stehn (1945) suggested that a given habitat could be divided into two parts, one favourable to a species and the other unfavourable or at least not occupied, and that observed data could be divided into two sets, one derived from the favourable and one from the unfavourable portions of the habitat. To obtain such a division they plotted $\log(x!y)$ against x (y being the number of quadrats containing x individuals). A Poisson distribution plotted in this way gives a straight line

$$\log(x!y) = \log(Ne^{-m}) + x\log m$$

(from $y = (Ne^{-m}m^{x})/x!$, for N quadrats with mean m). They fitted a straight line by eye to the points corresponding to higher values of x and from it derived the number of, and mean density within, quadrats to be assigned to the favourable portion of the area. The distribution for quadrats assigned to the unfavourable portion was then obtained by difference and, in the majority of cases tested, was found to give a good fit to Poisson distribution. This approach, which has been largely overlooked, is realistic in concentrating attention on variability in the Poisson parameter. Its value is, however, limited by the assumption that the Poisson parameter shows discontinuous variability; this is unlikely in many circumstances. It will be invalidated if clumping due to reproductive behaviour is present. If there is reason to think that a number of quadrats which cannot be occupied by the species have been included in the data, it may be useful to derive the Poisson distribution corresponding to the occupied quadrats only. The density within occupied quadrats is $m/(1 - e^{-m})$, where m is the mean of the corresponding Poisson distribution. From the observed density within occupied quadrats m can readily be calculated by successive approximation and the appropriate distribution derived.

Kemp and Kemp (1956) have considered the expected distribution when cover is recorded by frames of pins. Following Robinson's (1954) suggestion that cover might be expected to follow the Beta distribution,* they derive a distribution based on the assumption that the probability of a pin hitting the species is constant within a frame but has a Beta distribution between frames. The resulting distribution may be defined in terms of the mean and variance of the Beta distribution. The variance of the Beta distribution is a measure of the variability between frames, i.e. of point-to-point variability. They obtained a satisfactory fit to cover data taken from Goodall (1952b). Once again the basic assumptions may be questioned. No real evidence has been presented for the appropriateness of the Beta distribution. Further it is unlikely, for the usual spacing of pins, that the probability within a frame is constant, though this objection may be overcome by using a smaller frame. Lastly, as with Stevens's approach, the second parameter, though measuring point-to-point variability, is difficult to interpret in biological terms.

Summing up the results of attempts to use modifications of the Poisson or binomial distributions to account for discrepancies found in field data, it appears that they result either in a statement of the obvious, viz. that a species is tending to occur in compact clumps of several individuals, or in definition of the pattern of distribution in terms of constants to which no precise biological meaning can be attached. It seems likely that a more empirical approach, while inevitably unattractive to the mathematical mind, will be more rewarding to the ecologist, who is aware of the very large number of variables which he has to consider in all but the most favourable circumstances.

Two conclusions that emerged from the earlier investigations of pattern were that species are relatively rarely randomly distributed even within small and apparently homogeneous areas, and that contagious is very much more common than regular distribution (Ashby, 1948). The rarity of regular distribution is at first sight surprising, as it might be expected to occur as the

---

* Defined by two parameters l and m. The probability of a fractional area x occurring in a quadrat is given by $y = \dfrac{\Gamma(l + m + 2)}{\Gamma(l + 1)\,\Gamma(m + 1)}\, x^l (1 - x)^m$.

result of competition between individuals. Absence of reports of regular distributions might perhaps be attributed to many workers having selected the easier species in a community, i.e. normally the less dense, to investigate. However, in forest communities in Trinidad, Greig-Smith (1952b) considering all woody species, found only very slight indications of regular distribution even when individuals of different species were grouped under size classes. Similarly, Jones (1955–6) working in forest in Nigeria makes no mention of the occurrence of regular distributions. The only case so far reported where a major component has been shown to have even a slight tendency to regularity is for *Shorea albida* in lowland swamp forest in Borneo (Greig-Smith, 1979).

If distribution is contagious then, whatever the exact nature of the contagion, the actual occurrence of individuals on the ground is, in qualitative terms, patchy; either there are patches in which a species is present and patches in which it is absent or patches in which it occurs more abundantly and patches of less abundance. If we are to discover the causes of this patchiness information about the size of patches is desirable and may be essential. It will at least permit the elimination of factors which do not exhibit patchiness of a similar scale as possible causes. The distribution of individuals may be more complex than the occurrence of patches of two types, e.g. a species might show groups of plants each originally derived from a single individual by vegetative reproduction, with the groups themselves being more numerous in some patches than in others. In other words heterogeneity may be present on several different scales concurrently and it is desirable to detect the existence of the different scales and their approximate dimensions.

Some indication of scale of heterogeneity may be obtained by varying the size of sampling unit in the tests already described and noting the size at which indications of non-randomness disappear or decrease markedly, but this is a laborious and insensitive method. It is in any case of little practical use for the larger scales generally produced by variation in environmental factors. If the non-randomness is very marked, systematic sampling (e.g. by a grid of contiguous quadrats) and subsequent insertion on a plan of *isonomes* (i.e. lines joining samples of equal density or other measure) (Pidgeon & Ashby, 1942), may be sufficient to indicate relatively large-scale heterogeneity. Greig-Smith (1952a) suggested a more discriminating technique based on systematic sampling. A grid of contiguous quadrats is used, each side of the grid having a number of grid units which is a power of 2, e.g. 16×16, 32×64, and the number of individuals per quadrat or grid unit is counted in the normal way. The total variance between grid units can be apportioned by the usual technique of analysis of variance to differences between units within blocks of 2, blocks of 2 within blocks of 4, blocks of 4 within blocks of 8, etc., the blocks being square for even powers of two and oblong for uneven. If the arrangement is perfectly random over the whole area, the mean square for all block sizes should be the same and equal to the mean number of individuals per grid unit, i.e. the variance : mean ratio should be unity whatever sample size is used. If the distribution is contagious it is to be expected (M. S. Bartlett quoted in Greig-Smith, 1952a), and has been found by trial on known patterns of discs, that the variance will rise up to a block size equivalent to the area of patches. If the patches are themselves random or contagious the variance will be maintained at this level with increasing block size. If the patches are regular, and this is necessarily the case when alternating low and high density areas of similar size are involved, the variance will fall as block size is increased further. If more than one scale of heterogeneity is present the behaviour will be repeated as block size reaches this scale. After analysis of variance of the data obtained from the grid in this way the mean square is plotted against block size and the position of sharp rises and peaks noted.

In the method as originally proposed and described above, the peaks correspond to the average *area* of different phases of a scale of pattern. In most circumstances it is more useful to associate the units in the same direction throughout, i.e. to sum them along a transect rather than into the squares and rectangles of a grid, so that the peaks correspond to the average *linear dimensions* of the phases (Kershaw, 1957a). Sampling may then be by spaced transects instead of a complete grid, allowing larger scales of pattern to be investigated by a smaller number of units and eliminating difficulties due to patches of the different phases being elongated in a constant direction. If there is reason to suspect that patches are not isodiametric, e.g. on sloping ground, separate analyses may be made in two directions at right angles. As Kershaw (1957a) pointed out, the variance of the largest block size (the full length of the transects) must be ignored as it includes an element dependent on the spacing between them.

Either grid or transect sampling can be used with frequency as the measure of species representation by subdividing the grid unit into a number of smaller units in each of which presence or absence is recorded, i.e. local frequency (e.g. Greig-Smith, 1961a), Kershaw (1957a) extended transect sampling to cover by using line transects and recording presence or absence at points equally spaced along the line. The basic unit is then a number of consecutive points. Kershaw (1958), working in grassland, found five points at a spacing of 1 cm apart satisfactory. If frequency or cover is used as the measure, the expected variance for a random distribution is that of the corresponding binomial distribution npq (n, number of subunits or points; p, mean frequency or cover; $q = 1 - p$). Pattern may be analysed in the same way as with density data for other measures of species representation, e.g. area of shrub canopy (Greig-Smith & Chadwick, 1965), yield of grassland (Morton, 1974a), or properties of individual plants, e.g. growth increment over a fixed period (Lowe, 1967), or the levels of environmental variables, e.g. soil depth (Kershaw, 1958), soil level (Pemadasa *et al.*, 1974), soil pH (Anderson, 1965), concentration of exchangeable cations in the soil (Morton, 1974b). With such continuous measures there is no 'expected' variance available for random distribution of the variable concerned.

The interpretation of these analyses may be made clearer by two examples. Kershaw (1958, 1959) using cover in basic units of 5 cm found consistent peaks at block size 64 (*c*. 3 m) for several species in a series of upland reseeded pastures of different ages. Figure 15 shows a typical curve, for *Agrostis tenuis*.* A similar transect analysis of depth of soil at successive points, using a basic unit of 10 cm, gave a peak at block size 32, indicating the same scale of heterogeneity (Fig. 16). This suggested a correlation of depth of soil and abundance of the species not previously suspected. On following up this indication the relationship was confirmed.

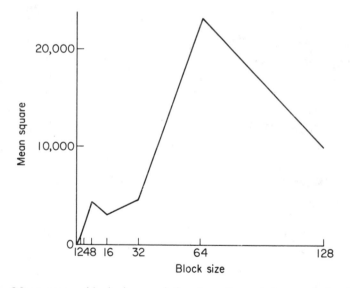

Fig. 15.   Mean square: block size graph for *Agrostis tenuis* in reseeded upland pasture, 7 years after sowing. Transformed cover data from transects with basic unit of 5 cm (data from Kershaw, 1957b).

---

* The peak at block size 8 corresponds to smaller patches, due to vegetative spread, within the larger ones. Such 'morphological peaks' commonly appear in analyses, but their nature is generally readily recognized.

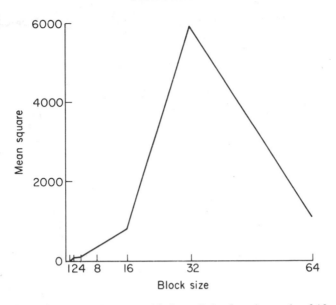

Fig. 16. Mean square: block size graph for soil depth at intervals of 10 cm along transects in reseeded upland pasture (data from Kershaw, 1957b).

Phillips (1953, 1954a) used the method to obtain information on morphological response of *Eriophorum angustifolium* otherwise available only with great difficulty. Rhizomes arise from the base of upright shoots of *E. angustifolium* in three vertical planes only, owing to the one third phyllotaxy of the species. These may be arranged, relative to the direction of the rhizome of which the parent shoot is the terminal portion, as two forward and one back or one forward and two back, i.e. there are six possible positions which daughter rhizomes may occupy. Those in the three forward positions are longer than backward ones on the same plant. The daughter rhizomes turn up to form new aerial shoots and the same pattern is repeated. The aerial shoots flower only after several, commonly three, years, after which they, and the rhizomes from which they arise, die. The plant thus consists of a system of limited extent of shoots linked by rhizomes. These facts could readily be established by excavation in suitable open habitats such as mud pools, but the difficulty of extracting the rhizome systems intact from more compact soils prevented direct study of varying length and production of rhizomes in different habitats. A typical grid analysis (Fig. 17) from a favourable habitat shows a 'primary double peak' (block sizes 2 and 8) corresponding to the grouping around the parent shoot of shoots from the shorter backward and longer forward rhizomes respectively, and a 'secondary peak' (block size 64) due to the

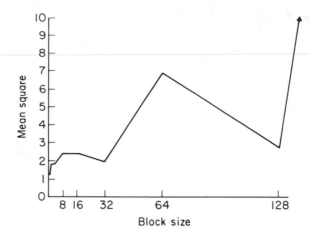

Fig. 17. Typical mean square: block size graph for *Eriophorum angustifolium*. (From Phillips, 1954a, by courtesy of *Journal of Ecology*.)

grouping of shoots linked in one system by rhizomes. Figure 18 shows two grid analyses from a habitat unfavourable to *E. angustifolium* and illustrates the type of information obtained. Here both peaks occur farther to the left, indicating that rhizomes are shorter, and the primary peak is much lower and not distinguishable into two parts, showing that fewer rhizomes are produced from each shoot.

Two aspects of the variance: block size graph are important in interpreting the pattern present, the position of peaks, which reflect the scale or scales

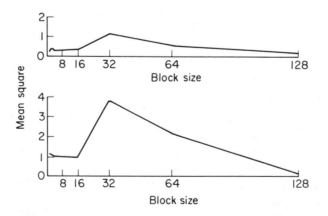

Fig. 18. Two mean square: block size graphs for *Eriophorum angustifolium* in *Calluna vulgaris-Erica cinerea* communities. (From Phillips, 1954a, by courtesy of *Journal of Ecology*.)

present and the heights of peaks, which reflect the intensity of pattern, or contrast between phases. The accuracy with which the technique detects the scale of pattern present has been the subject of several empirical investigations (Kershaw, 1957a; Usher, 1969; Errington, 1973). In the most straightforward case in which, on the average, phases of high and low abundance of equal dimensions alternate (or clumps are separated by unoccupied areas of equal dimension), a peak will indicate the true scale of pattern provided that the transect starts at a boundary between the two phases (Fig. 19(a)). If, however, the transect starts in the middle of either phase (Fig. 19(b)), the peak will be at the next smaller block size. Intermediate starting points (Fig. 19(c)) will give increased variance at both block sizes, which is greater depending on the exact position of the starting point. In patterns of this type there is thus likely to be a 'drift' of peaks to the left (Usher, 1969).

Kershaw (1957a) (see also Errington, 1973) showed that if there are random clumps or patches of high abundance (Fig. 19(d)), and the overall abundance is relatively low, there is liable to be a drift of peaks to the right, a peak occurring at the next block size above the one corresponding to the clump size. This results from the occasional occurrence of two clumps or parts of clumps within one block of the next size up (as in the first block of four units in Fig. 19(d)), reinforced by some reduction of the variance at the true clump size due to clumps overlapping blocks of their own size as in Fig. 19(b). Increase in the sample size (more or longer transects) will correct this drift to the right.

If patches of high abundance alternate with patches of low abundance of markedly different dimensions (or clumps are more or less regularly arranged at relatively low overall abundance), unless either abundance phase coincides with a block size, a peak will tend to occur at half the total unit of pattern (Fig. 19(e)) (Errington, 1973).

The limitation of block sizes to a power of 2 times the dimension of the basic unit allows an approximate estimate only of the true scale of pattern present, apart from the effects discussed above. A final difficulty in interpreting the position of peaks is that, because the analysis depends on variance, a pattern and its reverse, i.e. one in which areas of high abundance are assigned low abundance and vice versa, will give essentially the same graph, as Pielou (1977) has clearly demonstrated.

Usher (1969) suggested allowing for the effects of starting point of the transect by trying various starting points and accepting the maximum block size obtained for a peak as the correct value. Hill (1973a) suggested the simpler approach of taking the average value of mean square corresponding to all starting points. If the observations along a transect are $x_1, x_2, \ldots \ldots, x_n$, the

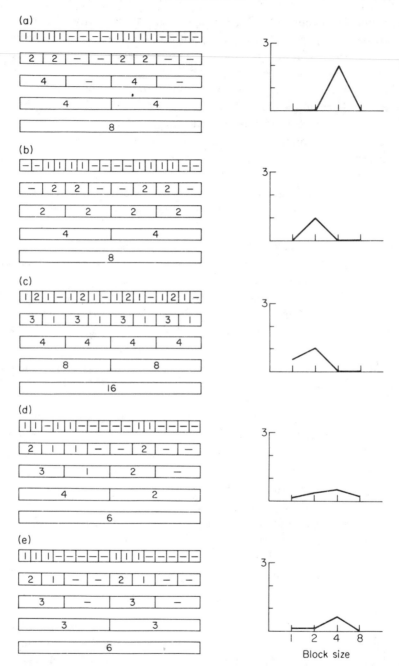

Fig. 19. Short transects with totals for different block sizes and corresponding mean square:block size graphs (see text).

mean square at block size 2 is

average of $(\frac{1}{2}(x_1 - x_2)^2, \frac{1}{2}(x_3 - x_4)^2, \ldots \ldots, (x_{n-1} - x_n)^2)$

and at block size 2 is

average of $(\frac{1}{4}(x_1 + x_2 - x_3 - x_4)^2, \frac{1}{4}(x_5 + x_6 - x_7 - x_8)^2, \text{etc.})$

and so on for larger block sizes. These expressions clearly omit a number of terms of the same form, corresponding to different starting points, e.g. $\frac{1}{2}(x_2 - x_3)^2, \frac{1}{4}(x_2 + x_3 - x_4 - x_5)^2$. An estimate of variance at block size 1, largely independent of starting point, 'two-term local variance', is

average of $(\frac{1}{2}(x_1 - x_2)^2, \frac{1}{2}(x_2 - x_3)^3, \frac{1}{2}(x_3 - x_4)^2 \text{ etc.})$.

Similarly, for block size 2, the two-term local variance is

average of $(\frac{1}{4}(x_1 + x_2 - x_3 - x_4)^2, \frac{1}{4}(x_2 + x_3 - x_4 - x_5)^2, \frac{1}{4}(x_3 + x_4 - x_5 - x_6)^2,$ etc.),

and so on for larger block sizes. This eliminates the need for n to be an exact power of 2 and also allows an estimate of variance to be made for any block size less than $\frac{1}{2}n$. Thus the two-term local variance for block size 3 is the average of all terms of the form $\frac{1}{6}(x_1 + x_2 + x_3 - x_4 - x_5 - x_6)^2$. If there is an exact periodicity in the data, a peak at block size p will result in peaks of diminishing height at $3p, 5p, \ldots$. (Usher, 1975). Field data will rarely show such precise periodicity but the effect should be kept in mind when interpreting analyses by Hill's method.

There are thus considerable difficulties in determining the precise scale or scales of pattern present in a set of data. This is not, however, as great a disadvantage in the practical use of the technique as might appear. Only if pattern is due to the morphology of the species concerned is the precise size of patches likely to be of interest and in most species they are evident on inspection in the field, and corresponding peaks in pattern graphs are readily recognized even if not at exactly the expected scale. Only if such field recognition is difficult are any more elaborate techniques of pattern analysis required, e.g. *Eriophorum angustifolium* (Phillips, 1954a), *Nardus stricta* (Chadwick, 1960). Much more commonly analysis of pattern is applied in an exploratory way to seek possible relationships between the pattern of different species or between the pattern of a species and patterns of environmental variables, as in Kershaw's (1958) study of *Agrostis tenuis*, quoted above. Then the interest is not in the exact scale of pattern but in whether different species have corresponding patterns or which environmental variables have a similar pattern to that of a species and are therefore worth investigating further as possible controlling factors.

The intensity of pattern, or contrast between phases of high and low abundance reflected in the height of a peak, is of interest because a high peak suggests a strong influence of the controlling factor on the abundance of a species, and a low peak that the species is less sensitive to the controlling factor. As Hill (1973a) pointed out, there is a natural definition of intensity; under random thinning a measure of intensity should remain roughly constant. The ratio of observed variance at a particular block size to that expected for random distribution (e.g. Greig-Smith, 1961b; Anderson, 1965) does not satisfy this requirement, but decreases with decreasing mean. For density data Hill suggests the ratio

$$(\text{Variance} - \text{mean})/(\text{mean})^2,$$

the expected value of which does remain roughly unaltered.

Kershaw (1970) has suggested using the average difference between counts in the two blocks of an adjacent pair as a measure of intensity. Like the (observed variance)/(expected variance) ratio this is not independent of the mean and, further, has no defined value for random distribution. If divided by the mean it remains constant under thinning but still has no defined value for random distribution. It has thus no advantages over measures based on variance and additionally has to be separately calculated*.

For environmental variables and measures of species representation for which there is no 'expected' variance, the ratio variance/(mean)^2 is an appropriate measure of intensity of pattern; it will retain roughly the same value if the same proportionate decrease is applied to all values recorded. For frequency and cover data there is no simple measure of intensity which remains constant under decreasing probability of occurrence in each subunit.

The testing of significance of peaks or troughs in the mean square: block size graph presents difficulty. The ratio of the mean square at each block size to the expected variance may be tested by referring the sum of squares to the table of χ^2 with the appropriate number of degrees of freedom. In this way significance bands for departure from randomness at any desired level can be derived†. Thompson (1958) provided a brief table of 95% limits and a fuller table from Greig-Smith (1961b) is reproduced in Appendix Table 7. If no observed value falls outside the significance bands, the null hypothesis of random distribution can be accepted. However, species are rarely randomly distributed and interest centres on which peaks are significant. Once non-randomness has been demonstrated, the tests for individual block sizes are no longer valid. A similar objection applies to Greig-Smith's (1952a) use of the F

* Kershaw suggests that it may be calculated from the variance, but this is true only if all the successive differences are equal.

† As departure can be either towards regularity or contagiousness, a two-tailed test is needed, i.e. for 95% probability reference is made to the 0·0275 and 0·0975 values of χ^2.

test to compare the mean square at higher block sizes with that for individual units and to F tests between successive block sizes. Thompson (1955, 1958) pointed this out, but showed that if some likely model of non-random pattern can be set up, the expected mean square and its standard error can be calculated for each block size; if the observed mean squares do not deviate significantly, the model can be accepted as a satisfactory description of the pattern. However, pattern analysis is generally used in an exploratory way and it is not usually possible to set up an appropriate model. In the absence of adequate tests of individual peaks, assessment has commonly been made subjectively. This is materially assisted by their consistency in a series of analysis; if a peak recurs at the same block size or shows a regular drift in a series of related communities, there can be little doubt of its validity (Kershaw, 1957a; Thompson, 1958)*.

More recently Mead (1974) has devised a method of testing the significance of individual peaks which is not dependent on a model of the pattern present. The scores for successive block sizes can be represented as a hierarchy. Mead suggests that the data can be viewed as a divisive or as an agglomerative process, and alternative questions asked: 'Is the division of each block total into two half totals random?' or 'Given the scores at any particular scale in the hierarchy are the pair totals compatible with random pairing?'. After considering several possible tests, he concludes that an agglomerative approach, 'the 2 within 4 randomisation test', is the most satisfactory, providing a fully valid test. This is based on the absolute difference between the two pair totals in a run of four blocks and may be illustrated from a small set of data, quoted by Mead.

Sixteen successive values, forming four sets of pairs are (0, 2, 2, 0), (0, 0, 1, 10), (11, 1, 0, 2), (5, 9, 4, 10). The three possible differences for each set are (0, 0, 4), (11, 9, 9), (10, 8, 12) and (0, 10, 2) respectively. The total of differences observed is $0 + 11 + 10 + 0 = 21$. In a series of random pairings the average differences for the four sets would be 4/3, 29/3, 10 and 4 respectively, totalling 25. The observed total is less than 25 and we require the probability of obtaining a value of 21 or less by chance.

The probability of obtaining a total of 17, $(0 + 9 + 8 + 0)$, the smallest possible, is clearly $\frac{2}{3} \times \frac{2}{3} \times \frac{1}{3} \times \frac{1}{3} = \frac{4}{81}$. The next smallest value possible is 19, which could be obtained in three different ways $(0 + 11 + 8 + 0, 0 + 9 + 10 + 0, 0 + 9 + 8 + 2)$, the probabilities of which are 2/81, 4/81 and 4/81 respectively, totalling 10/81. There are five different ways of obtaining the next smallest value of 21, with a total probability of 14/81. The required cumulative probability is thus $(4 + 10 + 14)/81 = 28/81$, indicating that the observed deviation towards regularity is not unlikely to have occurred by chance.

* Mead (1974) has pointed out that suggested approaches to determining the number of peaks in a graph (Goodall, 1961; Greig-Smith et al., 1963) are statistically unsound (see also Goodall, 1963).

If the observed total of difference is greater than that expected, indicating a deviation towards contagiousness, the probabilities are summed in the opposite direction. Thus if the order was $(2, 2, 0, 0)$, $(0, 1, 0, 10)$, $(11, 2, 0, 1)$ and $(5, 4, 9, 10)$, giving a total of differences of 35, the required probabilities are for totals of 37 $(1/81)$ and 35 $(3/81)$, with a cumulative probability of $4/81$. As this approach is dependent on the difference between pairs, no test is available for individual units.

A disadvantage of this test is that it is not independent of the scale of variation within each set of four scores, and a set which shows a large range will tend to dominate the randomization distribution. Mead therefore suggests a distribution-free version. Each set of differences within a set takes one of three forms (a, a, b), (a, b, c) or (a, b, b) where $a < b < c$ and a, b, c are all odd or all even. These three forms are replaced by $(0, 0, 2)$, $(0, 1, 2)$ and $(0, 2, 2)$ which gives approximately equal variance for the three forms, while retaining the simple nature of the combined randomization. (Exactly equal variances would be given by using $(0, 0, \sqrt{3})$, $(0, 1, 2)$ and $(0, \sqrt{3}, \sqrt{3})$).

For both the original randomization test and the distribution-free version a test based on the normal distribution may be expected to give a good approximation to the exact test for reasonably large sets of data. This is based on the mean difference between pairs and the mean variances of the sets of four. If there are n_{10} observations a and n_{12} observations b from n_1 distributions (a, a, b), n_{20} a's, n_{21} b's and n_{22} c's from n_{22} distributions (a, b, c) and n_{30} a's and n_{32} b's from n_3 distributions (a, b, b), the mean difference is

$$\frac{n_{21} + 2(n_{12} + n_{22} + n_{32})}{n_1 + n_2 + n_3},$$

the expected mean difference is

$$\frac{2n_1 + 3n_2 + 4n_3}{3(n_1 + n_2 + n_3)}$$

and the variance of the mean difference is

$$\frac{\frac{8}{9}(n_1 + n_3) + \frac{2}{3}n_2}{(n_1 + n_2 + n_3)^2} *.$$

$$Z = \frac{n_{21} + 2(n_{12} + n_{22} + n_{32}) - \frac{1}{3}(2n_1 + 3n_2 + 4n_3)}{\{\frac{8}{9}(n_1 + n_3) + \frac{2}{3}n_2\}^{\frac{1}{2}}}$$

is then referred to the table of the normal distribution.

* Mead (personal communication) has suggested that this value, which assumes that the mean is fixed by hypothesis, is more correct than that given initially ($\{\frac{4}{3}(n_1 + n_3) + n_2\}/(n_1 + n_2 + n_3)^2$), which is based on two degrees of freedom for each set, i.e. it assumes that the mean is calculated from the data.

The use of the forms with exactly equal variances makes the randomization distribution cumbersome, but there is not the same objection with the normal approximation, when

$$Z = \frac{n_{21} + \sqrt{3}(n_{12} + n_{32}) + 2n_{22} - \frac{1}{3}(\sqrt{3n_1} + 3n_2 + 2\sqrt{3n_3})}{\{\frac{2}{3}(n_1 + n_2 + n_3)\}^{\frac{1}{2}}}.$$

The difference for most sets of data is slight, but it may be worth using the second formula in critical cases.

It should be noted that a set of four identical scores do not contribute to the 2 within 4 randomization test, though they will have provided evidence of regularity at the smaller scale at which scores became identical. This contrasts with examination of mean squares where once regularity is established, it is reflected in low mean squares for all scales at which successive scores are identical.

Zahl (1974, 1977) has used Scheffé's method of comparing variances to assess significance of variances for different block size. His procedure was developed for a grid, and is not readily adaptable to transects (Ripley, 1978), and it is concerned with identification of a single scale of pattern only. These limitations make his test of little practical value.

Analysis of pattern by this method is most useful in vegetation that is, apart from relatively small scale clumping, apparently homogeneous. It is sometimes necessary, however, to analyse vegetation showing an evident overall trend in abundance of the species along the length of the transects, or a less pronounced trend may not be recognized initially, though it will become evident if the totals for the larger block sizes are examined. Such an overall trend results in a steady rise in mean square at the larger block sizes, which may mask scales of pattern present (Fig. 20).

The effect of the trend on the mean square: block size graph may be reduced sufficiently to expose smaller scales of pattern by deducting terms for covariance with position from the sums of squares (Greig-Smith, 1961a,b). The object of this correction for position is to refer the variation between blocks to an average regression line instead of to the mean value, rather than to allow for as much as possible of the effect of position. The correction for covariance is therefore based on a regression calculated for the total data rather than on the sums of squares and cross products for each block size separately. This can sometimes result in the correction being greater than the sum of squares from which it is subtracted, i.e. in a negative mean square. This implies that the deviations of the block means concerned from the regression line are in the opposite sense to their deviations from the overall mean (Fig. 21). The negative sign is ignored in presenting the analysis as a graph. It should also be noted that

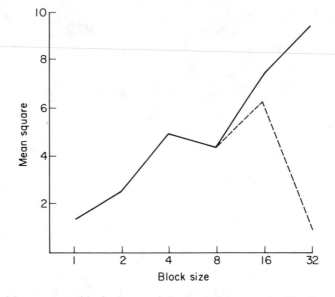

Fig. 20. Mean square: block size graph for local frequency in 10 × 5 cm units of *Agrostis stolonifera* in a dune slack with (broken line) and without (continuous line) correction for covariance with position. (From Greig-Smith, 1961b, by courtesy of *Journal of Ecology*.)

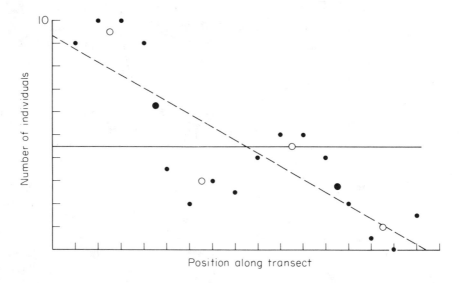

Fig. 21. Numbers of individuals in successive units along a transect, with (——) mean value and (- - -) regression of number on position. O, means for blocks of four; ●, means for blocks of eight.

the degree of freedom for the covariance is not orthogonal to any of the sum of squares from which it is subtracted. Thus the corrected sum of squares for which only one degree of freedom was initially available is valueless.

In spite of this undesirable feature of producing negative sums of squares under some circumstances, this procedure of allowing for an overall trend is usually satisfactory in practice. A more precise treatment is to calculate the expected value for each unit from the regression equation and analyse the residuals directly (Goodall, 1963). Hill (1973a) has suggested a trend free 'three-term local variance', analogous to the two-term local variance described above. This is the average of

$$\tfrac{1}{6}(x_1 - 2x_2 + x_3)^2, \tfrac{1}{6}(x_2 - 2x_3 + x_4)^2 \text{ etc. for block size 1,}$$

$\tfrac{1}{12}(x_1 + x_2 - 2x_3 - 2x_4 + x_5 + x_6)^2, \tfrac{1}{12}(x_2 + x_3 - 2x_4 - 2x_5 + x_6 + x_7)^2$ etc. for block size 2 and the corresponding expressions for larger block sizes. He later (Hill, unpublished) suggested an alternative 'compact trend-free variance', using terms of the form $\tfrac{1}{6}(x_1 - 2x_2 + x_3)^2$ for block size 1, $\tfrac{1}{4}(x_2 + x_3 - x_1 - x_4)^2$ for block size 2, $\tfrac{1}{84}(4(x_3 + x_4 + x_5) - 3(x_1 + x_2 + x_6 + x_7))^2$ for block size 3, $\tfrac{1}{8}(x_3 + x_4 + x_5 + x_6 - x_1 - x_2 - x_7 - x_8)^2$ for block size 4 etc.

The nested block analysis of variance described above has been modified and adapted in various ways.

Usher (1975) has made the interesting suggestion that it is useful to analyse *amplitude pattern* as well as the more usual *abundance pattern*. Consider the following sequences of values along a transect:

(a) 1212121212121212
(b) 1212121203030303.

They have the same mean, 1·5, and will show pattern at block size 1 (and no other pattern); with this very artificial example (a) has a mean-square/mean ratio for units within blocks of two of 0·33, indicating some tendency to regularity, and (b) has a ratio of 1·67 indicating a tendency to clumping. Transect (b) differs from (a) in that the contrast between high and low density phases is greater in the second half of the transect but this feature is not detected by the normal analysis. Usher suggests transforming the data to their absolute deviation from the overall mean, i.e. $x_i' = |x_i - \bar{x}|$, and analysing the transformed data in the usual way. For transect (a) the amplitude is constant but in transect (b) the amplitude shows pattern at block size 8. If an overall trend is present, the data can be transformed to absolute deviation from the value expected from the regression line, i.e. $x_i'' = |x_i - E(x_i)|$. Amplitude pattern can only exist if abundance pattern on a smaller scale is present. Usher quotes field data for *Lotus corniculatus* in which there were indications of amplitude pattern, but it is not clear how general it is or how helpful its detection may be in elucidating the structure of vegetation.

Goodall (1961) suggested that the difference in variances at different block sizes should be examined by calculating polynomial regressions of log variance on log spacing rather than be interpreted in terms of peaks and found that up to quartic regressions were necessary to fit field data. There has been disagreement over which approach involves fewer assumptions (Greig-Smith *et al.*, 1963; Goodall, 1963; see also Mead, 1974), but Goodall's approach appears to be less fruitful of hypotheses about the control of species distribution.

Goodall (1974) pointed out two disadvantages of the method. Variances for larger block sizes are based on fewer degrees of freedom than for smaller block sizes, and the estimates of variance for different block sizes are not independent. Viewing the analysis as concerned with variance between samples of the same size at different spacings, he suggested using random pairs of units from transects or grids, e.g. for a sample at a spacing of 4 units, one unit is selected randomly and from those units at 4 units distance from it, a second one is selected randomly to obtain the appropriate comparison (of the form $\frac{1}{2}(x_1 - x_2)^2$). Once a unit has been used it is deleted and is not available for further comparisons. Thus variances at different spacings are independent and the number of degrees of freedom on which they are based can be made equal, or approximately so, by choosing the appropriate number of pairs at each spacing. Since the variances are independent they may be compared by the F test, or the homogeneity of the whole set examined by Bartlett's test (p. 36). Further, Goodall suggested that an initial analysis should be performed on half the total data only and indications of significant spacings then examined independently in the remainder of the data. This technique certainly overcomes an important limitation of the original technique, the testing of significance of peaks. How important this is depends on how far the analysis is regarded as an exploration of the data preparatory to further study and how far as an end in itself. Zahl (1977) examined its performance on artificial data showing clustering at a single scale only and concluded that it was less satisfactory than the conventional analysis, but it has apparently been little used on field data. As Goodall pointed out, it does allow incomplete data, with some units not recorded, to be analysed, but this situation is unlikely to arise in practice. Ludwig and Goodall (1979) extended this spacing approach to the examination of all pairs of samples at a given spacing. As with Hill's (1973a) use of all comparisons of adjacent blocks, the estimates of variance are then not independent and no test of significance can be made.

Orloci (1971) has suggested using an information measure as the measure of heterogeneity instead of variance. This can be partitioned into quantities reflecting comparisons between blocks of different sizes in the same way as variance. The appropriate measure of the information carried by a set of samples is $2I = 2\sum x \ln \dfrac{x}{\bar{x}}$ (cf. p. 113).

This is calculated for the blocks of each size. (Note that $\bar{x}$ is the mean for the block size, not the mean for grid or transect units.) The information referring to contrasts between blocks of one size within blocks of the next size is obtained by subtraction in the same way as for sums of squares. For example, part of the analysis of Orloci's data for *Andropogon scoparius* is:

Information between blocks of 32   333·83
Information between blocks of 64   238·88
Information between blocks of 32
      within blocks of 64   333·83 − 238·88 = 94·95.

The method is applicable only to categorical data (density counts, local frequency, cover from point quadrats) and not to measurements (biomass, environmental data etc.) but where it can be used involves less computation. The information content measures departure from complete uniformity of values in all units, i.e. complete regularity. Orloci obtained confidence limits for departure from random expectation by analysis of repeated computer simulation of random distribution.

Morisita (1959) has suggested an approach essentially similar to nested block analysis of variance. His measure of dispersion, I_δ, is calculated for a series of quadrat sizes and the ratio $I_{\delta(s)}/I_{\delta(2s)}$ plotted against quadrat size (taken as $2s$). The resulting graph shows peaks at quadrat sizes corresponding to the scales of pattern present. Morisita has applied this method to the data of Evans (1952) and a more extensive trial in various types of forest has been made by Ogawa *et al.* (1961).

Various other approaches to the analysis of pattern (as opposed to the detection of non-randomness or the fitting of observations to theoretical models) have been proposed.

Yarranton (1969a) suggested that regressions of the amount of a species in a sample on the amount of the species in samples at different specified distances away should be calculated, and applied the method to data on cover of several species of lichen from transects on a cliff face. The spacing, or spacings, giving large regression coefficients are accepted as indicating scales of pattern. Like Hill's two-term local variance and Goodall's determination of variance at different spacings, it is not limited to scales at 2^n times the basic unit. Yarranton applied the original form of nested block analysis of variance to the same data and concluded that the two methods gave much the same indications for larger scales of pattern, but that the regression method was more informative at the smallest scales.

Spectral analysis is a technique widely used in describing a series of events in time, but it is also applicable to a spatial series and several authors have considered its possible use in pattern analysis (Bartlett, 1963, 1964; Hill, 1973a; Ripley, 1978; Usher, 1975). The observed values in quadrats along a transect are regarded as determined by a number of functions each showing a wave form of variation along the transect and differing in wavelength from two quadrats to the complete length of the transect, so that the observed value x_i in a quadrat is

$$x_i = c_0 f_0(i) + c_1 f_1(i) + c_2 f_2(i) \ldots \ldots$$

where the $f(i)$'s are the values of the functions for quadrat i and the c's are constants calculated for the particular set of data. If functions of the form of 'square waves' (i.e. capable of taking the values 1, 0, -1 only) are used, the analysis is closely related to block-size analysis of variance (Ripley, 1978). Spectral analysis is normally in terms of sine and cosine waves, which have the advantage that the starting point is not important.

If the values observed along the transect are $x_1 \ldots \ldots x_i \ldots \ldots x_m$, then

$$x_i = c_0 + \sum_{j=1}^{m/2} \left\{ c_j \cos\left(\frac{2\pi i j}{m}\right) + s_j \sin\left(\frac{2\pi i j}{m}\right) \right\}.$$

With $j = 1$, the curves have wavelength equal to transect length, with $j = m/2$, the successive values of $(2\pi i j)/m$ are $\pi, 2\pi, 3\pi \ldots \ldots$, for which sine is always zero and cosine alternates between 1 and -1 (Fig. 22). Thus the value of x_i may alternatively be expressed (Ripley, 1978) as

$$x_i = c_0 + \sum_{j=1}^{m/2-1} \left\{ c_j \cos\left(\frac{2\pi i j}{m}\right) + s_j \sin\left(\frac{2\pi i j}{m}\right) \right\} + c_{m/2}(-1)^i.$$

The constants are calculated:

$$c_0 = \bar{x}_i$$

$$c_j = \left\{ \sum_{i=1}^{m} x_i \cos\left(\frac{2\pi i j}{m}\right) \right\} \frac{2}{m}$$

$$s_j = \left\{ \sum_{i=1}^{m} x_i \sin\left(\frac{\pi i j}{m}\right) \right\} \frac{2}{m}.$$

Including c_0, as many constants are calculated as the number of quadrats in the transect and any set of quadrat values can be described by this analysis.

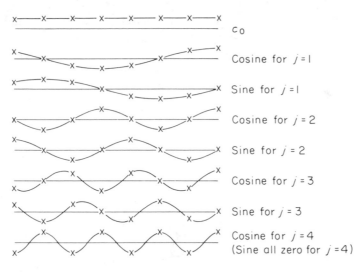

Fig. 22. A plot of the eight waves which form the basis of spectral analysis for eight quadrats at the positions indicated along a transect. A particular pattern of counts in quadrats may be described by some combination of these waves (see text). (From Ripley, 1978, by courtesy of *Journal of Ecology*.)

The *periodogram*, $I_j = (c_j^2 + s_j^2)m/8\pi$, is proportional to the reduction in the sum of squares obtained by fitting sine and cosine waves of period m/j, and interpretation is made in terms of the calculated values of the periodogram for different scales. A smoothed function of the periodogram is plotted against the period of the waves; peaks then correspond to scales of pattern. As with Hill's method, pattern at a particular scale is likely to result in harmonics in the analysis at larger scales. Spectral analysis has had only limited trial with field data, but it appears doubtful whether it is appreciably more useful than the simpler methods.

The value of information on pattern lies very largely in the relation between the patterns of different species or the pattern of species and that of environmental factors. The examination of the relationship between patterns is discussed in Chapter 4.

The methods described in this chapter are mostly so sensitive that they can profitably be used only within areas that are apparently homogeneous or nearly so. More marked pattern is often present even within the limits of a community as accepted by most ecologists. Such pattern may usefully be examined by means of ordination techniques, discussed in Chapter 8 (cf. Anderson, 1963; Whittaker *et al.*, 1979).

A number of other methods of analysing pattern have been proposed from theoretical considerations but are not discussed here because they are unrealistic in relation to the field situation either in assuming only a single scale of pattern to be present or only the smallest scale to be of interest, or in concentrating on tests for departure from randomness. These were features of the pioneer work but it has since become clear that such approaches are rarely profitable. (For an account of some of such recent theoretical developments, see contributions to Cormack & Ord (1979).)

CHAPTER 4

Association between Species

If one or few factors have a predominating influence on the occurrence of individuals of a species, the effect on the spatial distribution of individuals within the community will, as we have seen, depend on the distribution of different levels of the influencing factor. Suppose, for example, the occurrence or non-occurrence of a species in a particular area is dependent on percentage moisture within the soil. If the distribution of values of percentage moisture in the soil were quite random over the area, so that the value at any point were independent of the values at other nearby points, the distribution of individuals of the species would also be random, and no pattern due to its control by one factor would be apparent. In fact, as in this case, it is difficult to imagine most influencing factors having random distributions of levels, and the very widespread occurrence of non-randomness is in itself evidence that this normally is not so. The possibility must, however, be taken into consideration. Such a state of affairs can clearly occur in the special case of the influencing factor being the presence of an individual of another species, which may itself be randomly distributed. Thus, in the simplest example of a host-specific parasite, if the host is randomly distributed, the parasite will also be random unless the incidence is so high that more than one individual of the parasite may be dependent on a host individual. Such a clear-cut relationship between species is evident on inspection but less precise interrelationships occur, which are not so readily detected, e.g. the association between desert shrubs and annuals in California shown by Went (1942), and the effect of *Agropyron repens* on other species (Osvald, 1947; Hamilton & Buchholtz, 1955). These are attributable to such causes as the local altering of levels of moisture and mineral nutrients, critical to one species, around individuals of another species, or, in some cases, to the effect of root exudates.

When the factor influencing one species is the presence of individuals of another species, samples of suitable size will show association between the two species. This may be apparent either by correlation between quantitative measures of the two species or, if the effect is a more pronounced one, by the species occurring together in samples either more or less frequently than chance expectation, according as the influencing species favours or is antagonistic towards the influenced species. In the more general case of other influencing factors varying randomly over the area, indications of their importance may still appear from study of interspecific association. In any community of more than a few species it is unlikely that an influencing factor will influence one species only, and concurrent influence on several species will result in association between them.

Information on association between species is valuable apart from the detection of overriding influence on distribution where direct study of pattern fails. If several species in a community exhibit pattern, indicating control by influencing factors, study of association between them will provide evidence of any grouping of the species into assemblages of like response to the influencing factor or factors. Consideration of such species assemblages is similar to classification of plant communities. Indeed the distinction between species groupings in the present context of variation within an accepted plant community and species grouping into units accepted as different plant communities is largely one of magnitude of difference. Techniques of examination are similar and the distinction between aspects considered here and those postponed to the discussion of communities in later chapters is rather arbitrary.

In assessment of association, as of pattern, provided that no measure of the degree of association is required, samples need not be random.* The null hypothesis tested is that there is no correlation between the occurrence of one species and another, and if this is true all samples are statistically independent of one another in this respect. The simplest procedure in the field is, generally, to use systematic sampling. There is a slight risk that equally spaced samples might give data from one phase only of a periodic variation in the vegetation. This risk can, in any case, be avoided by using contiguous quadrats, e.g. a grid or parallel lines of quadrats arranged in two sets perpendicular to one another. The contiguous quadrats are also advantageous in permitting grouping to give larger sizes. If a measure of degree of association is to be derived from the data, then the samples must be random, as otherwise no confidence limits can be assigned to the measure and no valid comparisons can be made between the measures for different pairs of species.

According as the species are present in all, or nearly all, samples or not, association between them may be tested in terms of correlation either of some quantitative measure of the species or of their presence or absence in the samples. There is rarely any choice of type of data to use for a particular case.

Presence or absence data for pairs of species can readily be tested in a 2×2

---

* There is one minor exception to this when the two species being considered differ widely in frequency (see p. 117).

contingency table. Suppose the numbers of samples containing both species A and B, species A alone, species B alone, and neither species are as follows

		Species A		
		+	−	
Species B	+	a	b	$a+b$
	−	c	d	$c+d$
		$a+c$	$b+d$	$a+b+c+d = n$

If the occurrence of species B is quite independent of that of species A, then of the $(a+c)$ samples containing species A a proportion $(a+b)/n$ would be expected to contain species B, i.e. the expected number of samples containing both species is $((a+b)(a+c))/n$. The expected number for the other cells of the table can similarly be calculated or obtained by subtraction from the marginal totals. The observed numbers can be compared with this expectation based on independence by calculating χ^2 for the table, or by the exact solution, as appropriate (p. 38).

Pielou (1969, 1977) pointed out that this use of a 2×2 contingency table assumes that the marginal totals are fixed, i.e. the test applies to the particular set of samples examined whereas the question at issue is whether the species are associated in the whole population (the area from which the samples have been taken). To make the more general test, the marginal totals themselves should be regarded as random variates giving estimates only of the probability of a species being found in any one sample. Pielou gives an alternative procedure which assumes this. The usual χ^2 test is conservative, giving a greater probability of an observed deviation, and the difference is normally small. The χ^2 test is thus usually satisfactory, at least for isolated comparisons, though the discrepancy may become more important if a large number of comparisons are being made.

Unless interest centres on a few species only, it is generally worth while to make complete species lists for each sample, so that presence and absence data are available for a considerable number of pairs of species in a community. If a number of comparisons are made, the results of a few of them will individually correspond to a low probability by chance alone. Thus, if one hundred pairs of species are examined, the contingency tables for five of them will have an individual probability of 5 % or less and similarly one out of a hundred may be expected to show a probability of 1 % or less. No importance can therefore be attached to isolated cases of apparent association when large numbers of comparisons are being made. This point has often been overlooked.

It is sometimes worth while to examine the joint occurrences of more than two species to clarify particular points. This may be illustrated from an example. Greig-Smith (1952b) found in a set of samples from secondary forest in Trinidad some evidence of association between three species, *Amaioua corymbosa*, *Lacistema aggregatum*, and *Alibertia acuminata*, which occurred in 58, 54, and 42 samples respectively out of 100. The observed and expected numbers of joint occurrences and probability levels of chance occurrence were:

	Observed	Expected	P
Amaioua–Lacistema	37	31·32	<0·05
Amaioua–Alibertia	29	24·36	0·05–0·1
Lacistema–Alibertia	29	22·68	<0·05

The suggestion that these three species formed an ecological grouping was supported by each showing negative association with two other species, themselves negatively associated. The expected numbers of joint occurrence of all three species in the various possible combinations are readily calculated, e.g. that for *Amaioua* with *Lacistema* but without *Alibertia* is $0·58 \times 0·54 \times (1 - 0·42) \times 100 = 18·17$. The observed and expected numbers in each of the eight possible classes are

	Am. Lac. Alib.	Am. Lac. —	Am. — Alib.	— Lac. Alib.	Am. — —	— Lac. —	— — Alib.	— — —
Observed	23	14	6	6	15	11	7	18
Expected	13·15	18·17	11·21	9·53	15·47	13·15	8·11	11·21
Deviation	+9·85	−4·17	−5·21	−3·53	−0·47	−2·15	−1·11	+6·79
χ^2	7·38	0·96	2·42	1·31	0·01	0·35	0·15	4·11

The total χ^2 is 16·69 with four degrees of freedom, corresponding to a probability of less than 1%. This tests the hypotheses that all three species are completely independent in their occurrence. As Pielou (1977) and Goodall (1973) pointed out, other hypotheses may usefully be tested when more than two species are involved. Everitt (1977) gives a full discussion of the range of hypotheses relating to a three-way table, but some of these are scarcely relevant in this context. The most informative tests are likely to be whether there is an excess of samples containing all three species or an excess of samples containing none of them or both. As Pielou notes, the conclusions to be drawn from these two results are not identical. Excess of 'empty' samples suggests that parts of the area are unfavourable to all three species, excess of samples containing all three that there is overall positive association between them.

The number of samples, m, containing none of i species is approximately normally distributed with variance

$$V_{\mathrm{m}} = E(m) - \{E(m)\}^2 + N(N-1)\Pi_i \frac{(N-n_i)(N-n_i-1)}{N(N-1)}$$

where $E(m)$ is the expected value and species i occurs in n_i samples out of N (Pielou, 1969). Thus $c = (m - E(m) - 0.5)/\sqrt{V_m}$ may be referred to tables of the normal distribution. The subtraction of 0.5 is a continuity correction. The number of samples containing all species may be tested in the same way, substituting n_i for $(N - n_i)$. For the Trinidad data $c = 2.62$ $(P < 0.01)$ for samples containing none of the species, $c = 3.68$ $(P < 0.001)$ for samples containing all three.

This approach could be applied to the interactions of a number of species but it becomes impractical for more than three or four species as the number of possible groupings increases geometrically, e.g. there are sixty-four possible groupings of six species, and the expected number of occurrences in each group becomes low. In comparisons of joint occurrences of three or more species it may be necessary to group some classes so that no expected value is below 5, the usual arbitrary level for χ^2 calculations. There is not the same need to correct for continuity as, with the increased number of degrees of freedom, the number of possible tables for any set of 'marginal totals' becomes much greater and the probability distribution curve much more nearly continuous. The number of possible sets of species to be tested if they are taken in sets of more than two also increases rapidly with the number in a set, so that the labour involved in analysis becomes heavy. It is not generally worth while, therefore, to examine associations between more than two species unless there is strong indication from the 2×2 tables of interesting relationships involving more than two species. Alternative treatments are more appropriate to segregation of communities and will be considered in that context.

Comparison of observed with expected numbers of occurrences of species in samples is a very useful and flexible technique, which can be applied to a variety of situations. Where interest centres on certain categories of association only, records may not be available of samples containing no species. This situation may be illustrated from Went's (1942) data, already mentioned, on the association between shrubs and annuals in desert vegetation in California. Data were obtained along a belt transect of the number of shrubs of different species, the species of annual occurring round the base of each shrub and also of the number of occurrences of each annual species away from any shrub. There is thus no category of samples containing neither shrubs nor annuals; indeed there are no defined sample areas in the ordinary sense at all. The data are presented in the form of percentages but it is possible to reconstruct the data from the totals for different species. Table 6 shows the reconstructed data for four of the annual species, with the less numerous shrub species grouped together as 'other species'. Three principal questions may be asked. (1) Do the annuals under investigation tend to occur with shrubs, rather than away from them, more or less frequently than chance expectation? (2) Among the annual individuals found with shrubs is there evidence that they tend to occur more frequently with some shrub species than others? (3) If so, is the relationship the same for different annuals? In the absence of records of blank samples no direct answer can be given to the first question. An affirmative answer to the second, while certainly indicating association between annuals and individual species of shrubs, might arise from positive association with some species and negative with others. This would make the first question irrelevant, as the relationship with shrubs as a class would then depend on the relative numbers of different species. (In the present case a tendency for annuals to occur with shrubs was obvious on inspection and this justified the incomplete type of sampling used.)

If shrub species do not have differential effects on the occurrence of annuals, the number of occurrences of an annual species would be expected to be distributed among the different shrub species in proportion to the numbers of the latter. The expected number of occurrences calculated in this way are shown in Table 6 beneath each observed number. From each cell of the table an item can be calculated contributing to the total χ^2, which has 16 degrees of freedom. If the five items for each annual species are summed, a χ^2 for each, having four degrees of freedom, is obtained. Each is highly significant, indicating that no annual species is randomly distributed among the shrub species. This could be due either to a general effect of the different shrub species on all annual species, or to specific effects between particular shrub and annual species. The total χ^2 for the table, 181·38, may be partitioned to test this. From the total observed and total expected number of occurrences of annuals for each shrub species (bottom line of the table) a χ^2 with four degrees of freedom can be obtained testing agreement of total annual occurrences with expectation. This is found to be 104·69. If this is subtracted from the total, the remaining portion is due to heterogeneity, or varying behaviour of different annual species in relation to the different shrub species. Thus we have

	χ^2	Degrees of freedom	P
Deviation	104·69	4	< 0.001
Heterogeneity	76·69	12	< 0·001
Total	181·38	16	

If there is a significant heterogeneity item, as in this case, it is important to notice that no importance can necessarily be attached to a significant χ^2 for

Table 6. Association between annuals and shrubs in a desert in Southern California. Observed (above) and expected (below) occurrences. Data reconstructed from Went (1942)

	Shrubs								No connection with any shrub
	Encelia farinosa (living)	Encelia farinosa (dead)	Franseria dumosa	Hymenoclea salsola	Other species	Total	χ^2	P	
Number of shrubs in transect	248	84	172	136	133	773			
Annuals									
Phacelia distans	74	49	88	92	49	352	34·45	<0·001	—
	112·93	38·25	78·32	61·93	60·56				
Malacothrix californica	12	29	46	21	17	125	50·07	<0·001	—
	40·10	13·58	27·81	21·99	21·51				
Emmenanthe penduliflora	10	30	6	27	24	97	70·49	<0·001	2
	31·12	10·54	21·58	17·07	16·69				
Rafinesquia neomexicana	3	9	21	6	5	44	26·37	<0·001	1
	14·12	4·78	9·79	7·74	7·57				
Total	99	117	161	146	95	618	104·69		
	198·27	67·16	137·51	108·73	106·33				

deviation as, if the different annuals are associated in opposite senses with shrub species, the deviations of the total annual occurrences for different shrub species will depend on the relative numbers of different species of annuals. The analysis of χ^2 thus establishes that the annual species do not occur randomly with different shrub species, and that they differ in their interactions with different shrub species.

Analysis can often be carried further. Here, for instance, examination of the data suggests that a large part of the total deviation is due to annuals tending to occur with dead rather than living *Encelia farinosa*. All four annual species occur less frequently with living and more frequently with dead *Encelia* than chance expectation. Biological considerations suggest that dead shrubs are likely to be different in their effects on their immediate environment from those of living shrubs, regardless of species. In Table 7 the data for dead *Encelia* and for six* cases of occurrence with dead shrubs among 'other species' have been excluded and the marginal totals and expected values recalculated. The total χ^2 for each annual species is still highly significant as is the heterogeneity χ^2 for the whole table, confirming that none of the four annuals is randomly distributed amongst the shrubs and that the former differ among themselves in their interaction with the shrubs.

The data for *Malacothrix californica* and *Rafinesquia neomexicana* in Table 7 indicate similar responses to the different shrubs. If a similar table is drawn up with these two annuals only, the analysis of χ^2 is

	χ^2	Degrees of freedom	P
Deviation	57·62	3	< 0·001
Heterogeneity	2·05	3	0·5–0·7
Total	59·67	6	

The heterogeneity χ^2 is here not significant, confirming that the interaction between these two species and the shrubs is, as far as the observations indicate, the same†.

* It has been assumed that the total number of dead shrubs is the minimum indicated by recorded cases of joint occurrence with annuals, i.e. the possibility of some dead shrubs having had no associated annuals is ignored.

† An alternative to χ^2 in the analysis of contingency tables is the *likelihood ratio criterion*, variously symbolized as G and χ^2_L, which is considered to be preferable (e.g. Sokal & Rohlf, 1981, Everitt, 1977). It takes the form

The examples quoted by no means exhaust the possible information obtainable from the data, but they are sufficient to show the flexibility of the contingency table approach.

If two species are present in all or nearly all the samples, association may be evident as a relationship between the abundance of the species. An obvious approach is to calculate a correlation coefficient between the abundance values for the two species. As an example, consider the following data for cover of *Cirsium acaule* and *Festuca ovina* in adjacent 10 ft square plots on chalk grassland, each sampled by 112 points (seven frames of sixteen points). The figures are number of points out of 112 hitting the species, and have been arranged in order of cover of *F. ovina.*

Festuca ovina	*Cirsium acaule*
99	10
95	4
83	22
82	13
68	35
64	26
62	21
49	36
46	37

(With more extensive data the values for one species may be plotted against those for the other to see if there is any indication of relationship before calculating the correlation coefficient.)

There is some evidence that the cover of *Cirsium acaule* increases with decreasing cover of *Festuca ovina*, though the relationship is not very exact and cannot be accepted without testing. Denoting cover of *Festuca* by y and of

---

$$2(\Sigma \text{ observed} \times \ln \text{ observed} - \Sigma \text{ observed} \times \ln \text{ expected})$$

and is referred to the table of χ^2. χ^2 is an approximation to G for large samples, so that for many tables the values will be similar. For Table 6, the analysis of G is

	G	Degrees of freedom	P
Deviation	107·82	4	< 0·001
Heterogeneity	71·72	12	< 0·001
Total	179·54	16	

Table 7. As Table 6, but dead shrubs omitted

| | Shrubs | | | | | | |
	Encelia farinosa	Franseria dumosa	Hymenoclea salsola	Other species	Total	χ^2	P
Number of shrubs in transect	248	172	136	127	683		
Annuals							
Phacelia distans	74	88	92	49	303	31·14	< 0·001
	110·02	76·30	60·33	56·34			
Malacothrix californica	12	46	21	17	96	34·92	< 0·001
	34·86	24·18	19·12	17·85			
Emmenanthe penduliflora	10	6	27	20	63	35·90	< 0·001
	22·88	15·86	12·55	11·71			
Rafinesquia neomexicana	3	21	6	5	35	24·75	< 0·001
	12·71	8·81	6·97	6·51			
Total	99	161	146	91	497	69·42	
	180·46	125·16	98·96	92·41			

	χ^2	Degrees of freedom	P
Deviation	69·42	3	< 0·001
Heterogeneity	57·29	9	< 0·001
Total	126·71		

Cirsium by x the necessary calculations are:

$(n = 9)$

$\Sigma x = 204$, $\Sigma x^2 = 5\,776$, $(\Sigma x)^2/n = 4\,624$, $\Sigma (x - \bar{x})^2 = 5\,776 - 4\,624 = 1\,152$

$\Sigma y = 648$, $\Sigma y^2 = 49{,}520$, $(\Sigma y)^2/n = 46{,}656$, $\Sigma (y - \bar{y})^2 = 49{,}520 - 46{,}656$

$\quad = 2\,864$

$\Sigma (xy) = 13{,}074$, $(\Sigma x \Sigma y)/n = 14{,}688$, $\Sigma [(x - \bar{x})(y - \bar{y})] = 13{,}074 - 14{,}688$

$\quad = -1\,614$

Correlation coefficient

$$ r = \frac{\Sigma [x - \bar{x})(y - \bar{y})]}{\sqrt{[\Sigma (x - \bar{x})^2 \, \Sigma (y - \bar{y})^2]}} = \frac{-1614}{\sqrt{(1152 \times 2864)}} = -0{\cdot}8886. $$

Variance of r $V_r = \dfrac{1 - r^2}{n - 2} = \dfrac{1 - 0{\cdot}8886^2}{7} = 0{\cdot}0300557.$

Standard error of r $\sqrt{0{\cdot}0300557} = 0{\cdot}1734.$

The value of t testing the departure of r from zero (the expected value if the values of x and y are independent) is $0{\cdot}8886/0{\cdot}1734 = 5{\cdot}13$. This value of t has seven degrees of freedom, one having been used in determination of r, and the corresponding probability is less than 1%.[*] The indication is thus confirmed that the cover of *Festuca ovina* is depressed where that of *Cirsium acaule* is greater, as indeed might be expected from the morphology of the latter.

Any quantitative measure may be used in tests of association and, indeed, different measures may be used for the two species. For example, to test the relationship between an abundant grass species and a forb of low cover per individual, it would be appropriate to use cover for the grass and density for the forb.

An alternative to the correlation coefficient is the use of a rank correlation coefficient, which has the advantages of greater ease of computation and of not giving undue weight to extreme values. Of those available the most suitable is Kendall's *tau*. (See, for example, Kendall, 1948). To estimate this coefficient the data are arranged so that one variate is in natural order from the greatest to the smallest value and the order of ranks of the other variate is noted. The coefficient is then obtained by examining all possible pairs of values of the second variate, noting which are in natural and which in reverse order and expressing the excess as a proportion of the total comparisons. Thus, for the

---

[*] The values of r corresponding to certain probability levels for a range of values of n have been tabulated directly, e.g. Fisher and Yates (1943). The calculation of t is thus generally unnecessary.

data discussed above, the rankings are:

Festuca ovina	1	2	3	4	5	6	7	8	9	
Cirsium acaule	8	9	5	7	3	4	6	2	1	
P	1	0	2	0	2	1	0	0	—	+ 6
Q	7	7	4	5	2	2	2	1	—	− 30
S	−6	−7	−2	−5	0	−1	−2	−1	—	−24

The score, S, of excess of positive or negative comparisons, is most readily calculated in the way shown. For each entry for *Cirsium acaule* the number of entries to the right of it which are greater is noted as P and the number which are less as Q. S is equal to $P - Q$. The total number of comparisons made is $\frac{1}{2}n(n-1) = 36$. Thus $\tau = -24/36 = -0.67$. It will be seen that for complete concordance in ranking of the two variates all comparisons will be positive and for completely reversed rankings all comparisons will be negative. Thus τ ranges from $+1$ for perfect positive association to -1 for perfect negative association.

For $n \geqslant 10$, S is approximately normally distributed and the significance of τ may be tested by comparing with its standard deviation and referring the ratio to the table of the normal distribution. The observed value of S should first be reduced by unity to correct for continuity. The variance of S is $\frac{1}{18}n(n-1)(2n+5)$ (Kendall, 1948). Kendall provides a table, which is reproduced in Appendix Table 8, for values of n up to 10. From this the probability in the present case is approximately 0.6%.

If there are ties in the ranking of one or both species, the tied ranks are averaged, e.g. if the second and third highest values are equal they are both given the rank of $2\frac{1}{2}$. The number of possible comparisons is reduced and the coefficient then takes the form

$$\frac{S}{\sqrt{[(\frac{1}{2}n(n-1)-T)(\frac{1}{2}n(n-1)-U)]}}$$

where T is $\frac{1}{2}\Sigma t(t-1)$, the sum of the expressions $t(t-1)$ for each group of t tied ranks in one variate, and U is, similarly, $\frac{1}{2}\Sigma u(u-1)$, the sum of the expressions $u(u-1)$ for each group of u tied ranks in the other variate. Under these conditions the variance of S is:

$$\frac{1}{18}\left\{ n(n-1)(2n+5) - \Sigma t(t-1)(2t+5) - \Sigma u(u-1)(2u+5) \right\}$$

$$+ \frac{1}{9n(n-1)(n-2)}\left\{ \Sigma t(t-1)(t-2) \right\}\left\{ \Sigma u(u-1)(u-2) \right\}$$

$$+ \frac{1}{2n(n-1)}\left\{ \Sigma t(t-1) \right\}\left\{ \Sigma u(u-1) \right\}.$$

There remains one type of association to be considered, viz. where one species is present in all or nearly all the sample areas and the other is absent from many sample areas, and has a low measure in samples where it is present. Here interest centres on the relation between the amount present of one species and the presence or absence of the other. The only test possible is to classify the samples into those with and those without the second species and calculate the mean measure of the first species for the two classes. A t test may then be applied to the difference between the means. If this is to be done, sampling must be random, contrary to the general condition for tests of association, so that a valid estimate of the standard error of the two means may be obtained.

As in sampling for pattern determination, so in sampling to test association, care should be taken not to include obviously different communities in the same set of samples. If different communities are included, strong indication of association will be obtained, positive between species belonging to the same communities, negative between those belonging to different communities. It is true that these associations reflect the influence of controlling factors, but it is an influence that is apparent without quantitative examination, and the resulting associations may obscure less obvious relationships between the species. The effect of sampling an obviously heterogeneous area may be illustrated by an example. If two species have frequencies of 30 % and 40 %, and 14 joint occurrences (against random expectation of 12) are found in 100 samples, the corrected χ^2 is 0·45. If 100 samples containing neither species are added, the expected number of joint occurrences falls to 6, and the corrected χ^2 is 13·79 with probability of about 0·1 %. Bray (1956) has given a clear example of this effect in the field.

The effect of sample size on indications of association has been largely overlooked. Not only does it affect the indication obtained but useful information can be derived from the size of sample at which indication of association changes. There are four causes of association which may be distinguished by the behaviour of indications of association in relation to changing sample size.

(1) If the quadrat used is of the same order of size as individuals, negative associations will appear which mean no more than that two individuals cannot occupy the same place. As quadrat size increases this association will disappear.

(2) If some species present have individuals much larger than those of other species, spatial exclusion by the former may impose positive associations among the latter. Thus, if a species A has individuals large enough to exclude, from a quadrat of the size used, species B and C, having individuals small

enough to occur together, these two species will show positive association. This effect is not uncommon, e.g. in grassland containing both tussocks of grasses, perhaps several thousand square centimetres in area, and individuals of small forbs. Indication of association will disappear when the quadrat size is increased above the size of the individuals of the larger species.

(3) If two species respond similarly to a controlling factor which has a defined pattern of values, they will show positive association up to the size of quadrat corresponding to the scale of heterogeneity of the controlling factor. With further increase in size of quadrat, the indications of association will disappear. Conversely, opposite responses to the controlling factor by two species will result in negative association, disappearing at a quadrat size above that of the scale of heterogeneity of the controlling factor. If several controlling factors are operating, giving pattern with several scales of heterogeneity for the species, indications of association may change at two quadrat sizes. Figure 23 shows two species which are both much more abundant in certain parts of an area; within those parts the species are favoured by different levels of a second controlling factor, giving pattern on a second, smaller scale. Here, as quadrat size is increased, indications of association will at first be negative, then positive, and finally disappear. The positive association of two species imposed by spatial exclusion by a third, described above, is really a special case of this

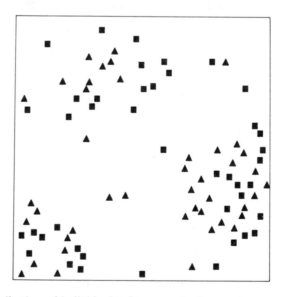

Fig. 23. Distribution of individuals of two species (see text).

similar response to a controlling factor. That it is a simple spatial exclusion effect will generally be apparent on considering, in relation to the growth form of the species present, the quadrat size at which is disappears.

(4) Direct effects between species, e.g. by root excretions or local modification of the soil factors round an individual, will, if causing positive association, give indications of association at all quadrat sizes large enough to include individuals of both species. If, on the other hand, the effect is one of negative association, indications of association will disappear when the quadrat becomes larger than the average area of influence of an individual. The result will be like that of spatial exclusion, but with indication of association not disappearing until a quadrat size greater than would be expected from simple spatial exclusion.

It is thus clear that the information derived from association data based on a single size of sample will be very incomplete and difficult to interpret. If circumstances allow, it is well worth while making observations with a number of sample sizes. This strongly reinforces the argument for using a scheme of systematic sampling, which will normally allow data for larger samples to be obtained by grouping adjacent samples. If circumstances permit only a single size, it may be advantageous with quantitative data to calculate partial correlation coefficients, taking account of the amount of other abundant species, as Dawson (1951) suggested. This will eliminate spatial exclusion effects, but, as Goodall (1952a) pointed out, may obscure relationships between groups of species.

One difficulty in interpretation of association data derived from a series of increasing sample sizes may arise from the change-over from presence or absence records to quantitative correlation. Comparison of quantities is the more sensitive test. Consider two species both more abundant in limited parts of an area only, and competing actively within those parts (cf. Fig. 23). From the nature of the two tests it is almost always impossible validly to apply them both to the same set of data. Suppose, however, that examination of presence and absence at a sample size where both species have nearly 100 % frequency shows positive association, resulting from their common restriction to parts of the area. If the sampling unit is then increased slightly so that calculation of a correlation coefficient is the appropriate test, negative association may appear, owing to the strong competition effects between the two species where they are abundant being sufficient to outweigh their joint abundance in certain samples only. The negative association at the larger sample size represents a 'carry-over' of the spatial exclusion effect, made evident by the greater sensitivity of comparison of quantitative measures.

This reversal of indication at change to the correlation coefficient test may be illustrated from data given by Kershaw (1956). Pattern analysis of upland reseeded pasture by transects of 5 cm basic units (of five points for cover determination), had shown heterogeneity of *Agrostis tenuis*, *Lolium perenne*, and *Dactylis glomerata* at a block size of 64 units, determined by predominance of *Agrostis* on patches of shallow soil and of *Lolium* and *Dactylis* on deeper soil (see p. 88). Association tests of data from single units showed negative association, interpreted as spatial exclusion, for all three pairs of species. With blocks of 8 units *Agrostis–Lolium* and *Agrostis–Dactylis* showed negative and *Lolium–Dactylis* positive association, reflecting similar response by *Lolium* and *Dactylis* to the controlling factor. For blocks of 16 units it was necessary to change to calculation of a correlation coefficient. Most areas examined showed no association between *Lolium* and *Dactylis* but where their joint cover was high, i.e. where direct competition might be expected, there was evidence of negative association, though scarcely significant.

There has been considerable interest, especially among animal ecologists, in measures of degree of association as opposed to tests of association. Where a correlation coefficient can legitimately be used as a test its value can be used as a measure of association, provided the samples on which it is based are random, and that the correlated measures are normally distributed. Vegetation measures are frequently not normally distributed and if the correlation coefficient is to be used as a measure of association it will generally be necessary to transform the data first (see Chapter 2). Non-normality of the data does not affect the validity of the use of the correlation coefficient as a test of the existence of association. For data of the contingency type a wide range of measures has been proposed, many of which are unsatisfactory in some respect or other. The subject has been critically reviewed by Goodall (1973). Only when the main interest is the relationship between the same pair of species in a number of different communities does it seem likely that a measure of association will be informative. Generally, the existence of association and the scale on which it appears are more important.

Pielou (1961) introduced a concept of *segregation*, related to that of association, referring to the degree to which two species are intermingled, regardless of scales of pattern (Fig. 24). This is approached in terms of nearest neighbour relationships. In a two species population four classes of nearest neighbour relationships may be distinguished.
(1) Individuals of A whose nearest neighbour is an individual of A
(2) Individuals of A whose nearest neighbour is an individual of B
(3) Individuals of B whose nearest neighbour is an individual of A
(4) Individuals of B whose nearest neighbour is an individual of B.

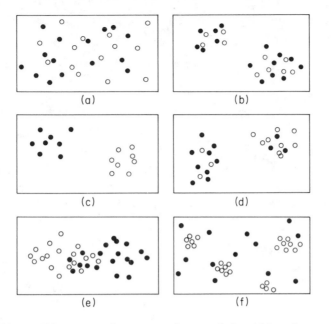

Fig. 24. Six possible patterns for two-species populations: (a) and (b) unsegregated; (c) fully segregated; (d), (e) and (f) partly segregated. (From Pielou, 1961, by courtesy of *Journal of Ecology*.)

Denoting the numbers in these classes by f_{AA}, f_{AB}, f_{BA} and f_{BB} respectively a contingency table may be drawn up:

		Base plant		
		Species A	Species B	
Nearest neighbour $\{$	Species A	f_{AA}	f_{BA}	Na'
	Species B	f_{AB}	f_{BB}	Nb'
		Na	Nb	N

(N, total number of relationships examined; a, b and a', b', proportions of base plants and of nearest neighbours which are A and B respectively). This may be tested in the usual way by a χ^2 with one degree of freedom. Segregation will be indicated by the observed numbers of AB and BA relationships being less than random expectation. Pielou defines as a *coefficient of segregation*

$$S = 1 - \frac{\text{Observed number of AB and BA relationships}}{\text{Expected number of AB and BA relationships}}$$

$$= 1 - \frac{f_{AB} + f_{BA}}{N(a'b + ab')}.$$

This can have negative values when AB and BA relationships are more numerous than random expectation. It ranges from -1 for isolated pairs made up of one A and one B to $+1$ for complete absence of AB and BA relationships.

The connection between segregation and association can be clarified by considering the cases illustrated and discussed:

Segregation	Indication of association with increasing quadrat size
Negatively segregated (isolated pairs AB)	positive at all sizes
Unsegregated Fig. 24(a)	nil
Fig. 24(b)	positive → nil
Fully segregated Fig. 24(c)	negative → nil
Partly segregated Fig. 24(d)	negative → nil
Fig. 24(e)	negative → positive → nil
Fig. 24(f)	negative → nil

Segregation is affected more by small scale effects than by large, just as indications of pattern obtained from plant-to-plant distances are more sensitive to small scale effects.

In a further consideration of segregation Pielou (1962b) proposed fitting data for runs of different species in transects to various modifications of random expectation. This is analogous to the fitting of modified Poisson distributions to data for single species and is subject to the same objections (see p. 83).

Goodall (1965) emphasized that segregation involves both the pattern of the species concerned and the association between them and suggested two approaches based on interplant distances that would be sensitive only to association.

If the nearest individual of a different species to a plant is identified and its nearest individual of a different species is the starting plant, the two may be described as 'mutually nearest neighbours'. Goodall identified all such pairs in a sample of vegetation. He derived an expression for the exact probability of the observed number of pairs of different composition, but this is laborious to calculate and he derived the required probability by repeated random sampling from a population with the observed number of individuals of different species involved in mutually nearest neighbour pairs.

This test shows whether individuals of two species tend to occur closer together than species in general, but it does not provide a complete test. 'It would be possible for species A to occur only in close proximity to species B, yet for this not to be displayed in the results because species C, a more common one, was even more closely associated with B'. A test which is independent of other species is provided by comparing the mean distance from species A to the nearest individual of species B, with the mean distance from random points to the nearest individual of species B. If the species are independently distributed the expected difference is zero. A smaller mean distance from A to B indicates positive association and a greater one negative association. Goodall used a square-root transformation on the distance measures and compared them by a t-test. Another approach to segregation based on interplant distance (Peterson, 1976) is discussed in Chapter 6 (p. 166).

Kershaw (1960, 1961) considered the comparison of the patterns shown by different species in the same set of samples. The association between species at different block sizes may be examined in terms of presence or absence, or of correlation coefficient applied to the totals for the block size in question, as in the data quoted above. Kershaw pointed out that the total covariance between two species may itself be analysed in terms of portions appropriate to different block sizes. If the representation of two species A and B in each unit of a grid or transect are added, an analysis of pattern of (A + B) may be prepared. If there is no association between the species, the resulting variance at any block size will equal the sum of the variances of the two species individually. If there is association then

$$V_{A+B} = V_A + V_B + 2C_{AB}$$

where C_{AB} represents the covariance at that block size. The covariance can then be plotted against block size (Fig. 25). This approach has the advantage that the covariance isolated is that appropriate to the relationship of the species considered for blocks of x units within blocks of $2x$ units; the direct calculation of a covariance from the totals for blocks is liable to be affected by relationships at larger block sizes. The corresponding correlation coefficient for relationship between species within each block size, $r = C_{AB}/\sqrt{(V_A V_B)}$, may be used in a test of significance. Alternatively the correlation coefficient

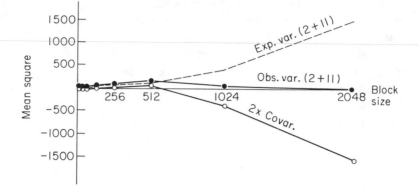

Fig. 25. Expected and observed variance of combined data and covariance, plotted against block size, for two species. Data for cover along a transect extending from acidic into basic grassland. Species 2, *Festuca rubra*; species 11, *Holcus lanatus*. (From Kershaw, 1961, by courtesy of *Journal of Ecology*.)

itself may be plotted against block size. Austin (1968a) pointed out that there may be discrepancies between the positions of peaks in covariance and correlation graphs for the same data; a small covariance value may give a high correlation coefficient if the associated variances are also small. For example, the following data

Species A	12	1	16	1	15	2	14	0
Species B	1	20	1	10	2	11	1	25

have covariance values of $-99{\cdot}625$ and $-9{\cdot}875$ at block sizes 1 and 2 respectively, and the corresponding correlation coefficients are $-0{\cdot}9016$ and $-0{\cdot}9633$.

Hill's (1973a) approach of averaging all terms of the appropriate form to avoid the effect of a particular starting point and permit estimates for intermediate block sizes may also be applied to the covariance between species. The covariance between species X and Y for units within blocks of two is the average of the terms

$$\tfrac{1}{2}(x_1 - x_2)(y_1 - y_2), \tfrac{1}{2}(x_3 - x_4)(y_3 - y_4) \ldots \ldots \text{ etc.}$$

The corresponding 'two-term local covariance' is the average of

$$\tfrac{1}{2}(x_1 - x_2)(y_1 - y_2), \tfrac{1}{2}(x_2 - x_3)(y_2 - y_2) \ldots \ldots \text{ etc.}$$

Similarly the 'two-term local covariance' for blocks of two is the average of

$$\tfrac{1}{4}(x_1 + x_2 - x_3 - x_4)(y_1 + y_2 - y_3 - y_4), \tfrac{1}{4}(x_2 + x_3 - x_4 - x_5)(y_2 + y_3 - y_4 - y_5)$$
$$\ldots\ldots \text{etc.}$$

If either species has an overall trend these procedures fail to detect the relationships at different scales. If the block totals are compared, the effect of overall trend may be eliminated by calculating partial correlation coefficients allowing for the effect of position (Greig-Smith, 1961a), but again these are liable to be affected by relationships at larger block sizes. Calculation of covariances from analyses corrected for trend of the two species separately and the sum of the two species is not possible because the variances and covariances are no longer additive. Appropriate estimates can, however, be obtained by using a 'three-term local covariance' analogous to Hill's 'three-term local variance'. This takes the form of the average of

$$\tfrac{1}{6}(x_1 - 2x_2 + x_3)(y_1 - 2y_2 + y_3), \tfrac{1}{6}(x_2 - 2x_3 + x_4)(y_2 - 2y_3 + y_4) \ldots\ldots \text{etc.}$$

for units within blocks of two, and correspondingly for larger scales.

The failure of other procedures is illustrated by the following artificial data for two species showing trends in opposite senses, positive relationship at block-size 4 and some degree of negative relationship at block-size 1.

Species A 1   5   6   5   4   3   6   6 10 13 14 14 11 13 12 14
Species B 18 16 15 14 11 10   7   9   9   8   7   5   4   0   1   0

	1	2	4
Correlation coefficient between blocks	−0·80	−0·82	−0·81
Partial correlation coefficient between blocks	+0·52	+0·89	+0·94
Analysis of covariance (no correction for trend)	−1·19	−5·06	−1·75
Analysis of correlation (no correction for trend)	−0·57	−0·99	−0·35
Two term local covariance	−0·83	+1·13	−22·92
Three term local covariance	−0·85	+1·98	+12·56

Only the three term analysis correctly indicates the known features of the data.

Analysis of covariance allows exploration of the relationship between two species only. Noy-Meir and Anderson (1971; Walker *et al.*, 1972) have described an approach, 'multiple pattern analysis', which gives an estimate of the degree of relationship between all species at different scales. This involves preparing variance/covariance matrices for each block size and adding them to give a combined variance/covariance matrix, which is then subjected to principal component analysis (p. 242). Each component represents a set of relationships between species. The variance represented by each component can then be partitioned among block sizes and a variance: block size graph drawn.

If interest centres on relationships between individual plants of different species, the procedures so far discussed are of limited value. With quadrat sampling, spatial exclusion (p. 117) will tend to produce negative associations, which will override other effects as quadrat size is reduced. Cover is less sensitive to small-scale associations because the shoot of a single plant may spread over a considerable area (Stowe & Wade, 1979).

Yarranton (1966) designed a sampling technique to detect contacts between individuals of different species. This technique, which is particularly suited to unistratal vegetation such as grassland (Turkington *et al.*, 1977; Turkington & Harper, 1979) and the saxicolous bryophyte community which Yarranton examined, involves sampling by points, either systematically or randomly positioned, and recording the species present and the species which touches the first species nearest to the point. A 'pseudospecies', 'no contact', is necessarily involved in the collection of data to record cases where the point does not hit a plant and where a plant is growing in isolation and hence has no contact with any other species; 'no contact' can be treated as a species or ignored in subsequent analysis. The numbers of joint occurrences of pairs of species can be examined for evidence of association between them, but the usual 2×2 contingency table used by Yarranton as one approach and by Turkington is not valid (de Jong *et al.*, 1980). The expected number of joint occurrences cannot be calculated in the manner used with 2×2 contingency tables for data from a set of quadrats, where the null hypothesis assumes that occurrence of different species is independent, i.e. any or all of the full species complement may occur in any one quadrat. With contact samples only two species occur and there are necessarily two species present.

De Jong *et al.* (1980, 1983) outline the appropriate procedures for calculating expected values. To simplify calculation, they discard any sample where there is no contact with any other species, i.e. 'no contact' is not accepted as a 'pseudospecies'. If the nearest contact is with another individual of the same species the contact is recorded for the nearest point at which a different species is in contact. On this basis they give a method for deriving numerically maximum likelihood values for expected numbers of different species pairs. If samples for which the nearest contact is with an individual of the same species are discarded a rather simple method of deriving expected values, 'iterative proportional fitting' (Denning & Stephan, 1940; see, for example, Everitt, 1977), is available. The limitation of this is that it may be very difficult in the field to distinguish two individuals of the same species in contact from a vegetatively derived patch. In their later paper, however, De Jong *et al.* (1983) show that if the distinction between contacts of B with A and A with B is retained for the purposes of analysis, a more straightforward procedure is available for the sampling method which accepts the nearest contact with another species, thus eliminating the difficulty of having to distinguish individuals of the same species in the field.

Stowe and Wade (1979) suggested sampling by equidistant points, d units apart, along transects. If no plant occurs within a distance $r (r < \frac{1}{2}d)$ a blank is recorded; presence may refer to whatever part of a plant is considered appropriate, e.g. stem, tiller, leaf, any part of shoot etc. The values of d and r are fixed with reference to the type of vegetation being examined. The results take the form of a string of species occurrences or blanks e.g.

AACCCA-BBB-NNCAAB-D for species A, B, C, ... N.

Stowe and Wade used two alternative techniques of examining such data. One, species juxtaposition, derives from Pielou's (1967) method of testing departure from random expectation in the sequence of species along a transect through a vegetation mosaic. The sequence is 'collapsed' by replacing runs by single occurrences, i.e. the sequence above becomes

ACA-B-NCAB-D.

On the assumption that the sequence of species is random, the expected number of transitions, m_{ij}, from species I to species J can be calculated by an iterative procedure, either that used by Pielou (1967) or by iterative proportional fitting referred to above. An overall test of the observed number of transitions, n_{ij}, against expected can be made by χ^2 or G (Pielou, 1967). Stowe and Wade examined individual species pairs by an index of association

$$J_{ij} = \frac{1}{2} \ln \frac{n_{ij}}{m_{ij}} \cdot \frac{n_{ji}}{m_{ji}},$$

using the geometric mean of the transitions from I to J and from J to I. (Clearly n_{ij} and n_{ji} are interchangeable, according to the direction in which the transect is recorded). J_{ij} is normally distributed with mean zero and variance which may be calculated from the data (see Stowe & Wade, 1979); it is positive for positive association and negative for negative association.

The second technique divides the transect into 'species regions'. If the sequence is again collapsed, the 'region' of a species is regarded as including entries on either side of its entries, e.g. in the above example the regions of A are

ACA- and CAB. The occurrences of another species, e.g. B, within and outside
the regions of A can now be counted and entered in a 2×2 table:

	Occurrences of B	Occurrences of species other than A or B
Within region of A	a	b
Outside region of A	c	d

The χ^2 value (or exact test) for this table then tests whether an individual's
chance of being species B is independent of whether or not it occurs in regions
of A. For a given pair of species there are two tests, depending on which is the
'primary' species, the regions of which are examined. The approach thus
distinguishes between B being more likely to occur in regions of A than outside
them and the converse. This represents a possible biological difference—A may
be indifferent to the occurrence of B, but B be dependent on the occurrence
of A.

The examination of species-regions may be extended to the uncollapsed
sequence and occurrence within regions of different radius determined. Stowe
and Wade (1979) suggested that this is most effective if annuli, rather than radii,
surrounding the individuals of the primary species are considered, e.g. only
individuals, say, three points distant from the primary species, are considered
to be within the species-region. Change or disappearance of association with
increasing distance can then be detected.

CHAPTER 5

Correlation of Species Distribution with Habitat Factors

THE determination of gross differences in vegetation by environmental factors, as between the successive stages of a typical hydrosere, is accepted as self-evident; in such cases the factors or complex of factors responsible can generally be detected with reasonable certainty. On the other hand, it is uncertain how far the smallest differences between samples of vegetation are environmentally determined and how far they are matters of chance as to which species, of a number fitted to do so, happened to establish first. Intermediate degrees of difference between vegetation can commonly be related, with increasing certainty, to differences in environmental factors. At all levels, however, correlation between present environmental differences and present vegetation may break down owing to past changes in environment and vegetation, so that the determining factors are primarily historical ones. Compare, for instance, the tree growth resulting from secondary succession in areas completely cleared of forest at some previous time, with that occurring near enough to uncleared forest for the original species to return.

In so far as vegetational differences are determined by environmental differences, they may be expected to correlate with them. If the vegetational or environmental differences, or both, are small, then correlation will not be readily detected by subjective examination, and only objective assessment of suitable data will reveal the relationship. This applies especially where several factors, themselves perhaps not independent, are jointly responsible. Evidence for determination by chance factors and, to some extent, by historical factors is negative. Their importance can only be assumed from apparent absence of determination by environmental factors. Important as the study of correlation between vegetation and the level of particular environmental factors thus is, it must be emphasized that mere correlation between the two variables is no proof of causal relationship either direct or indirect. Both may be determined by some other factor, e.g. Conway (1938) pointed out that performance of *Cladium mariscus* is correlated with high temperature but the favouring factor is incidence of bright sunlight, which is responsible also for increased temperature. Additional information and general ecological knowledge may lend such support to a supposition of causal relationship that it is

129

reasonable to accept a correlation as reflecting such relationship but final proof can generally be obtained only by experiment in which all other factors are held constant. Unfortunately, the appropriate experiment may not be possible because some other factors cannot be held constant, or may be impractically complex or lengthy to carry out.

We are here concerned with the occurrence of a single species, or a particular state of a species (e.g. plants in fruit) in relation to environmental factors. If clearly distinct vegetation types are being examined, similar considerations apply, but the discussion of the more general problem of relating variation in composition of vegetation to variation in environmental factors is deferred to Chapter 9.

As in all numerical ecological work, a suitable type of sampling* must be adopted if the maximum information is to be obtained on correlation between vegetation and environment. As sampling requirements vary to some extent according to the type of data being collected, sampling procedure will be considered in relation to the different categories of data discussed below. However, one misleading procedure, which has been used in some studies of the relationship between the distribution of species and such factors as soil acidity, must be mentioned first. Samples of the soil are taken from around the roots of a number of individuals of the species and the pH (or other factor) determined for each. If the number of occurrences at each pH value is plotted against the pH, a peaked curve, often approximately normal in form, but sometimes bimodal, commonly results. Some workers, e.g. Salisbury (1925), have interpreted the peak or peaks as optimal pH values. Volk (1931) and Emmett and Ashby (1934) have pointed out that the shape of the curves is affected by the frequency of soils of different pH values available to the species, and that the frequency of different pH values around individuals of the species must be compared with the frequency of different pH values in the habitat examined, regardless of whether the species in question is growing there or not. As Ashby (1936) later pointed out, by such an erroneous procedure telegraph poles would show an 'optimal' pH value. Such data, drawn only from soil around the species, are adequate only to indicate the range of pH (or other factor) within which a species occurs and can provide no information on the behaviour in relation to varying values of the factor within the range tolerated.

Both plant and environmental data may be recorded either quantitatively or qualitatively. As in studies of association, the abundance and distribution

---

* Sampling here refers to the positioning of samples. Environmental factors themselves should be determined in as efficient a manner as possible, but these aspects are outside our present scope and are, in any case, amply dealt with elsewhere.

of the species concerned will determine the form of the plant data. If the species is present in all, or nearly all, samples, then a quantitative measure will be necessary. If it is sparse, not only may recording merely of presence or absence be adequate to detect any relationship with the environmental factor, but quantitative measures may have so small a range of values, or so intractable a distribution curve, as to be impractical. Whether the environmental factor should be recorded quantitatively or not depends on the nature of the factor as well as the degree of precision desired. Some factors such as pH, light intensity and humidity can scarcely be classified other than by measurement. The degree of accuracy to which they are measured can of course be varied, and should be related to the magnitude of differences expected. On the other hand features such as soil texture, amount of litter, etc., can often be satisfactorily graded on inspection into a few categories, which may be sufficient to reveal relationships without the time-consuming physical or chemical analysis necessary for quantitative measures. Even such factors as the level of mineral nutrients may be graded as high, medium or low by approximate methods of analysis. Other factors, such as the presence and amount of undecomposed litter, where the contrast between presence and absence implies a much greater difference in environment than does a variation in amount of litter, can scarcely be treated other than qualitatively without confusing effects on two quite different scales. The form which the data for environmental factors should take thus requires careful consideration before the field work is begun.

Since both plant and environmental data may be either qualitative or quantitative, there are four categories of data to be considered.

(1) Both plant data and environmental factors qualitative.
(2) Plant data qualitative, environmental factors quantitative.
(3) Plant data quantitative, environmental factors qualitative.
(4) Both plant data and environmental factors quantitative.

The first category, with all records qualitative, is comparable to association data with both species recorded as present or absent only, and the same considerations apply. It need not be considered in detail; it is sufficient to repeat that sampling may be either systematic or random, and that the data are examined in a contingency table. More than one environmental factor may be graded for each sample and the data treated in the same way as that for association between more than two species, but the results may be difficult to interpret. In general, examination of correlation jointly with a number of qualitatively recorded factors is not very profitable, though it may be useful where a species grows only where a number of rather narrowly defined environmental conditions are satisfied.

If the plant data are qualitative, and the environmental data quantitative, and a single environmental factor only is being considered, the test of relationship is straightforward. The mean values of the environmental factor for the plant classes (commonly 'with' and 'without' a species) are determined and compared by a t test. If several plant classes have been distinguished, a single comparison of their differences may be made in the form of an analysis of variance, comparing between-class differences with within-class differences, or the means may be compared in pairs, as seems appropriate. Data should not be combined in one analysis of variance if the variances for the different plant classes differ greatly from one another, as a significant difference may then be due to the differing variances rather than differing means of the classes. Widely differing variances may be allowed for in t tests in the way described earlier (p. 33).

Data of this type are recorded in two rather different circumstances, viz (1) examination of the general difference in level of an environmental factor in two or more different communities spatially separate from one another; (2) investigation of the relationship between point-to-point variation in the environmental factor within a community and the distribution of individuals of species within the community. The two cases, though differing only in scale of differences involved, do present rather different practical problems of sampling. In the former case the several vegetation types must each be sampled randomly (with some restriction of randomization if deemed desirable) for estimates of the environmental factor. In the latter case those parts of the community with and without the species in question must likewise be sampled randomly. In practice the two types 'with' and 'without' the species are often so intermingled that it is more convenient to allow each sample not only to be randomly selected in relation to the environmental factor, but also to be allotted randomly to the 'with' and 'without' classes. Samples taken at intervals along transects through the community have been used, e.g. Emmett and Ashby (1934) working on pH and occurrence of *Pteridium aquilinum* and *Vaccinium myrtillus*, Jowett and Scurfield (1949b) working on various soil factors and occurrence of *Holcus mollis* and *Deschampsia flexuosa*. Emmett and Ashby, working on large areas, drew arbitrary transects on a map and then followed them by compass, sampling at 50-yard intervals. Jowett and Scurfield, working in defined woodlands, drew transects selected 'at a fixed equal distance apart, the distance being based on the size of the woodland and selected arbitrarily beforehand as that giving the appropriate number of samples. At arbitrarily fixed equal intervals (also selected previously) along each transect, soil samples were taken. . . .' The latter procedure is in fact a completely systematic scheme of sampling, in which the position of the first sample determines that of the remainder, and hence is not a satisfactory basis for statistical analysis. Emmett and Ashby's procedure results in random sampling between transects (every point in the area has an equal chance of being represented) but the data from any one transect are linked in a systematic manner and, strictly, the data from one transect should be treated as one unit associated with 'presence' or 'absence' of the species. Clearly this cannot be done, so that this procedure is not entirely satisfactory either.

The transect method can be modified to give random samples. Equally-spaced transects are laid down in the way proposed by Jowett and Scurfield and a number of samples taken from each proportionate to its length. The actual position of each sample is fixed by two random numbers, one indicating distance along the transect and the other distance at right angles to it, the latter having a maximum value equal to the distance between transects. This results in a satisfactory form of restricted randomization, and one that is reasonably quick in the field. Alternatively, the common procedure may be adopted of using two lines at right angles as axes of random co-ordinates to give unrestricted randomization. It will generally be necessary to fix some arbitrary area around the sampling point within which the species is recorded as present or absent, e.g. Emmett and Ashby used a quadrat of area 10 square feet around the point at which the soil sample was taken. If the species being investigated is relatively sparse in the community, it may be necessary, in order to obtain sufficient data for the 'species present' class, to take the nearest individual to the random point selected and sample the environmental factor around or beneath it. This is unobjectionable, provided that no difference is made in the manner of taking the soil or other environmental sample from that used in the 'species absent' class, as, for example, by sampling a different soil layer.

Bieleski (1959) described a case where it was desired to test the correlation with light intensity of tree seedlings of very low density. The usual approach would be to record the number of individuals in small quadrats together with the light value at the centres of the quadrats. The low density made this impracticable as a single light reading from a quadrat large enough to include a reasonable number of individuals would have little meaning. Bieleski therefore used a relatively large quadrat and recorded light intensity at each of a grid of points and also at all seedlings within the quadrat. The distribution of light intensities in the two sets of data were then compared. The sampling procedure is not entirely satisfactory as the data on the total range of light values are based on a systematic sample. Bias is unlikely to be serious as at the scale of the grid used (1 m apart within 6 m square quadrats in scrub) there is unlikely to be any correlation between points; random samples within the quadrat could equally well be used.

The testing of the observed difference in the values of environmental factors for different plant classes has given rise to some confusion. Bieleski pointed out that difference in variance between 'species present' and 'species absent' class is equally indicative of correlation between species occurence and level of the environmental factor. Any difference in variance should therefore be tested before proceeding to a test of differences of means. A t test is, as we have seen, the appropriate test of difference of means in straightforward cases of two categories only, but the approach used by Emmett and Ashby (1934) and a criticism of it by Jowett and Scurfield (1949a) must be mentioned. As shown above, the sampling procedure used by Emmett and Ashby is not entirely satisfactory, but we may, for the present purpose, ignore this and consider only the handling of the data; data obtained by a satisfactory sampling procedure would have the same form. Their data are presented in the form shown in Table 8, which includes those for *Pteridium* and *Vaccinium* from one of the localities considered. They calculated the expected frequency of samples with and without *Pteridium* (or *Vaccinium*) in each pH class on the assumption of no correlation and, after grouping pH classes 4·8–5·1 and 5·8–6·1, tested the observed frequencies against the expected by a χ^2 test. The expected frequency adopted for *Pteridium* is 51·772 which is the mean of the observed frequencies, omitting the first three classes, which contained no samples with *Pteridium*. This method of calculating the expected frequency gives a biased estimate, weighted in favour of pH classes containing a large number of samples. An unbiased estimate is obtained from the total number of samples, 212, and the number containing *Pteridium*, 108, and is 50·943%. There is room for legitimate difference of opinion whether the results from the first three classes should be included in the calculation of expected frequencies. It might be said that, since no *Pteridium* was found below pH 5·1, this represented the lower limit of tolerance. The number of samples with lower pH is, however, small and in this case it is perhaps wiser to attribute to chance the absence of *Pteridium* from them. If these classes are omitted in calculating expected frequency, the expected frequency should not be applied to them nor should their deviations be included in the χ^2 as Emmett and Ashby did. There is also an error, as Jowett and Scurfield pointed out, in calculation of the χ^2 in that the contribution of the deviations in the *Pteridium* absent class is omitted.

If the primary intention in making these observations was, as Emmett and Ashby imply, to test whether pH was correlated with the presence or absence of a single species, the appropriate test is comparison of mean pH for samples with and without that species. If this is done the values of t are 2·02 for

Table 8. Frequency distributions of occurrence of *Pteridium aquilinum* and *Vaccinium myrtillus* in classes of pH (From Emmett & Ashby, 1934, by courtesy of *Annals of Botany*.)

pH class mean	Number of samples	Frequency of		Percentage occurrence	
		Pteridium	*Vaccinium*	*Pteridium*	*Vaccinium*
4·8	2	—	2	—	100
4·9	2	—	2	—	100
5·0	2	—	2	—	100
5·1	5	1	4	20·00	80·00
5·2	13	6	8	46·15	61·54
5·3	17	8	11	47·07	64·71
5·4	7	4	3	57·14	42·86
5·5	44	27	28	61·36	63·64
5·6	78	41	54	52·56	69·23
5·7	16	5	12	31·25	75·00
5·8	7	5	1	71·43	14·29
5·9	9	4	3	44·44	33·33
6·0	7	5	1	71·43	14·29
6·1	3	2	—	66·67	—

Pteridium (probability slightly less than 5%) and 3·40 for *Vaccinium* (probability *c.* 0·1%), indicating that there is some correlation in this habitat between pH and the occurrence of the two species, contrary to Emmett and Ashby's conclusions.* If a χ^2 test is used, based on an expected occurrence of *Pteridium* of 50·943% in each pH class, a total χ^2 of 13·685 with seven degrees of freedom is obtained, corresponding to a probability of between 5 and 10%. A substantially similar conclusion is drawn but the χ^2 test is evidently less sensitive, owing to grouping of the extreme classes necessary to bring all expectations above 5. Jowett and Scurfield apparently assume that the object of the observations was to determine whether *Pteridium* and *Vaccinium* had different pH 'preferences'. They compared the mean pH for samples containing *Pteridium* but not *Vaccinium*, with that for samples containing *Vaccinium* but not *Pteridium*, arguing that if neither species is influenced by pH, the two categories should show no difference in mean pH; the difference corresponded to a t of 2·85 (probability 2%). This method of testing provides information of interest but it brings into consideration not only the influence of pH on the

* If, however, the first three pH classes are omitted in the *Pteridium* calculations, providing a more stringent test, t is 0·957, with probability 30–40%.

occurrence of *Pteridium* but also the influence of the presence of *Vaccinium* on occurrence of *Pteridium* and vice versa. The situation may occur in which two species separately are not correlated with pH, but their relative competitive strength is. In this case significant difference would be shown by Jowett and Scurfield's test. This test is further open to objection that part of the data, those samples containing both species or neither, is ignored.

If it is desired to consider the correlation of the occurrence of the two species with one another as well as with pH, this may be done by an analysis of variance including the interaction between the species. Table 9 shows this analysis for Emmett and Ashby's data.* It will be seen that both the primary differences for presence and absence of *Pteridium*, and of *Vaccinium*, and the interaction between them, are significant. Too much reliance cannot be placed on the conclusions in this case, owing to the unsatisfactory features in the original sampling already discussed. If the data were accepted they would indicate not only definite pH 'preferences' by the two species but also differential effects on one another at different pH levels. There is, however, no direct information from the analysis as to whether the mean pH value for samples with *Pteridium* differs from that for samples with *Vaccinium*. It could be obtained by comparing the mean pH of all samples containing *Pteridium* with that of all samples containing *Vaccinium*, but would have little ecological meaning in view of the interaction between the species in relation to pH.

This contingency approach used by Emmett and Ashby may sometimes be useful. If the environmental factor has been measured with such a low degree

Table 9. Analysis of variance of pH in samples with and without *Pteridium aquilinum* and *Vaccinium myrtillus* (data of Table 8)

	Degrees of freedom	Mean square	F	P
Pteridium present v. *Pteridium* absent	1	0·36136	7·79	0·001–0·01
Vaccinium present v. *Vaccinium* absent	1	1·00564	21·68	< 0·001
Interaction	1	0·65100	14·03	< 0·001
Error	208	0·04386		

* Note that here, as is likely to be the case with most data of this type, the numbers of samples in different subclasses are not proportionate and the preparation of an analysis of variance is not so straightforward as for the usual orthogonal data. See, for example, Snedecor (1946, § 11.10).

of accuracy relative to the range observed that the number of different observed values is small, and this sometimes applies to pH, the distribution is so discontinuous that it is better to regard the values as grades of the environmental factor and test by contingency χ^2. Another difficult situation may arise if the distribution curve of the environmental factor in all samples together departs widely from the normal, and cannot be brought even approximately to normal form by transformation, so that the t test is inappropriate. This would apply if the curve of pH values were markedly bimodal, indicating perhaps that two quite different environments had been included in the samples. Here again the contingency approach is appropriate, though it would be better to repeat the observations, and to separate the supposedly different environments, if they can be identified. This may not always be possible, and the non-normal distribution curve may sometimes prove to be a real feature of a single environment. Finally, it may be noted that the contingency approach will take account of difference of variance as well as difference of mean.

It is sometimes useful to take several measurements of the environmental factor at each sampling site. This applies especially to soil characteristics which may vary up and down the soil profile. Such data are conveniently compared by an analysis of variance classifying the values for the environmental factor not only according to the plant class but also by position in the soil profile. This may be illustrated by the data in Table 10, showing values for percentage calcium carbonate at the surface and depths of 6 and 12 inches in five different vegetation zones from the foreshore community (stage A) to fixed dune pasture (stage E) (Gimingham *et al.*, 1948). Replication of samples was provided by laying down three transects at right angles to the vegetation zones and taking one sample at random within each vegetation zone on each transect. The regular pattern of the vegetation zones as strips parallel to the shore permitted this linking together of samples from different stages as units. Where, as more commonly happens, the vegetation pattern is irregular, replication would take the usual form of random sampling within each vegetation type and a transect item would not appear in the analysis of variance. Observations had suggested that different positions along the shore received differing amounts of wind-blown sand. This prompted the decision to remove an item in the analysis representing the sum of squares for transects, which decision was thus an ecological rather than a statistical one. The highly significant mean square for transects confirmed the correctness of doing so. We are not concerned here with the detailed conclusions to be drawn from the data; broadly they indicate little difference in total carbonate in the different vegetation stages (stages not significant), different concentrations at different depths (depths significant), and varying relation between carbonate and depth in different stages (interaction significant), probably owing to differing balance between leaching and accretion of sand.

There remains the case of concurrent measurements of several environmental factors, with qualitative plant recordings. Each environmental factor may be tested in turn for correlation with species occurrence, but it is clearly desirable to compare the total environmental information with the presence or absence of a species or vegetation type. General ecological experience suggests that sometimes an extreme value of one environmental factor, which would otherwise be unfavourable to the survival of a species or type of vegetation, may be compensated by an unusually high or low value of another factor, itself perhaps detrimental in other conditions. (Compare, for example, the occurrence of *Calluna vulgaris*, normally calcifuge, on calcareous soils under extreme climatic conditions.) Hughes and Lindley (1955) have pointed out the value of the generalized distance, D^2, developed by Mahalanobis and by Rao (see Rao, 1952; Blackith & Reyment, 1971; Sneath & Sokal, 1973) for this kind of data. D^2 is calculated from the means, variances and covariances of the various factors measured in replicate samples of two or more different groups. If only one factor is measured D^2 reduces to the t test. Like t it represents a measure, in terms of the factors measured, of the amount of difference between groups of samples. The probability of getting a given D^2 by chance can be determined, thus giving an indication of the significance of difference between groups. Its use may be illustrated from an example quoted by Hughes and Lindley. Determination of exchangeable calcium, available phosphate, exchangeable potash and pH were available for a number of samples of each of six soil types in Snowdonia. D^2 was calculated for each pair of soil types, giving the values shown in Table 11. These indicate that the soil types A' and B' are not significantly different from one another in terms of the four properties measured, nor are the soil types A and C. The six soil types may thus be grouped into four classes, which are shown in Fig. 26 with the lines between them proportional in each case to the amount of difference between the groups. This example refers to sets of samples defined by the soil type but they might equally well have been defined by the vegetation they bore.

The third category of data, with qualitative information on the environmental factor and quantitative plant data, can be dealt with more briefly, as considerations in handling the data are essentially similar to those in the last category. The basic technique is the comparison of means of the vegetation measures for the different environmental classes, either by t test for pairs of environmental classes or by analysis of variance for a number of classes considered together. More complex situations may be dealt with by methods analogous to those discussed in connection with the last category. When the environmental classes are spatially well distinct from one another the problem resolves itself into comparison of a vegetation feature in two distinct

Table 10. Percentage calcium carbonate in dune sand (From Gimingham *et al.*, 1948, by courtesy of *Transactions of the Botanical Society of Edinburgh*.)

Vegetation	Depth	Transect			mean	Stage mean
		I	II	III		
Stage A	Surface	68	ʹ57	56	60·3	
	6 in.	68	70	61	66·3	60·7
	12 in.	60	52	54	55·3	
Stage B	Surface	59	50	51	53·3	
	6 in.	62	56	52	56·7	57·7
	12 in.	67	67	55	63·0	
Stage C	Surface	54	50	57	53·7	
	6 in.	60	69	58	62·3	59·1
	12 in.	69	64	51	61·3	
Stage D	Surface	64	62	57	61·0	
	6 in.	62	53	58	57·7	58·8
	12 in.	59	53	61	57·7	
Stage E	Surface	52	48	48	49·3	
	6 in.	68	62	65	65·0	55·8
	12 in.	61	53	45	53·0	

Depth mean: Surface = 55·6; 6 in. = 61·6; 12 in. = 58·1.

Analysis of variance

	Sum of squares	Degrees of freedom	Mean square	F	P
Stages	118·80	4	29·700	1·49	> 0·2
Depths	278·53	2	139·267	7·01	0·001–0·01
Interaction	608·13	8	76·017	3·82	0·001–0·01
Transects	370·53	2	185·267	9·32	< 0·001
Error	516·80	26*	19·877		
Total	1892·80	42*			

* Total and error degrees of freedom are reduced by two because two samples were lost and their values estimated. (See Snedecor, 1946, § 11.6 for discussion of estimation of missing data.)

Table 11. D^2 analysis of Snowdonian soil series (From Hughes & Lindley, 1955, by courtesy of *Nature, London*). Figures are the values of D^2 for the comparisons indicated. (See Fig. 26 for key to soil types)

	A′	B	B′	C	C′
A	2·66***	3·66***	4·46***	0·26	3·07***
A′		1·91***	0·30	2·18***	1·44***
B			2·44***	2·21***	2·48***
B′				3·76***	1·78*
C					2·09**

Probability * 0·01–0·05, ** 0·001–0·01, *** < 0·001.

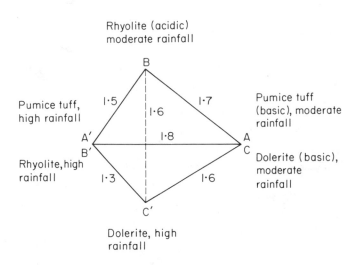

Fig. 26. Relationships of six Snowdonian soil series, based on D^2 analyses. Figures are the values of D between series. (From Hughes & Lindley, 1955, by courtesy of *Nature, London*.)

communities, already discussed in Chapter 2. The requirements of sampling will be apparent by analogy with those for the last category. Samples may be selected for the environmental factor, but within the selected environments the vegetation must be sampled randomly. It may, however, be more convenient to allow a sample to be randomly assigned to environmental classes if the latter are closely intermingled.

The last category of data includes those cases with quantitative measures of both plant and environmental factor. The sampling procedure is straightforward. Since the null hypothesis is that the two measures are unrelated, there is no need for either to be sampled randomly. Practical considerations will normally dictate the generally quicker and easier method of systematic sampling. The data obtained bear an evident resemblance to data on association between species when both species are recorded quantitatively, where, as we have seen, the appropriate test is the calculation of a correlation coefficient. In this case, however, there is an important difference in the nature of the data. When testing association between species it can be assumed initially, in the absence of any indications to the contrary, that the part played by the two species in the relationship is similar. The results of association tests may be used to postulate influence on one species by another, but generally the investigator is concerned to make a general test only of interrelationship in occurrence and, if any relationship is suspected, it is commonly control of both species by an environmental factor that is in mind. In testing correlation of vegetation with an environmental factor, however the two variables can normally be assumed to have different rôles in the relationship. In the absence of evidence to the contrary it is reasonable to assume that the level of the environmental factor is determining directly or indirectly the level of abundance of the plant. It may be that both are controlled by some other environmental factor, in which case the correlation coefficient is perhaps more appropriate, or even that the environmental factor is controlled by the level of the species, as might apply, for example, to atmospheric humidity, but the initial assumption of the controlling part played by the environmental factor is nearly always justifiable ecologically. In these circumstances the use of regression analysis is more appropriate.

The reason for preferring regression analysis to the correlation coefficient, which is concerned only with the degree of interdependence, will be clearer if the derivation of a regression line is considered. If concomitant observations are made on any two variables, e.g. plant density and soil moisture content, the results may be plotted on a graph with the y and x axes representing plant density and soil moisture content respectively. If there is a relationship between the two, the points will not be scattered over the whole area of the graph, but will approximate more or less closely to a line, according to the exactness of the relationship. If the fit to a line is reasonably good, the appropriate line may be sufficiently clear on inspection. The fitting of a regression line is the calculation of a line (straight or of other defined form) which best fits the data. If we calculate the linear regression of y (the *dependent variable*) upon x (the *independent variable*), the line fitted is $y = a + bx$, in which the constants a and b are so fixed that the sum of squares of deviations of all points from the line, measured in terms of y, are minimized.* Conversely, the regression line for x upon y, generally written as $x = a + by$, minimizes the deviations measured in terms of x (Fig. 27). Clearly the two lines will not necessarily be the same. In one case the fitting depends primarily on values of y, in the other primarily on values of x. Put another way, the regression of y upon x measures the expected change in y for a given change in x, and the regression of x upon y the expected change in x for a given change in y. In the example quoted, if the soil moisture level is controlling the density of the plant, it is logical and justifiable to estimate the increase or decrease in density for a given change in soil moisture content, but the reverse process is ecologically meaningless. In the present context we are concerned less to evaluate a relationship than to show that it exists. The existence of a relationship is shown by the line making an angle with the axes, i.e. by the coefficient b having a value differing significantly from zero. Like any other statistic calculated from data, b is subject to random variability and appropriate tests are available to determine the probability of chance occurrence of a value of b as great as that actually obtained, and hence of deciding the significance of an indicated relationship.

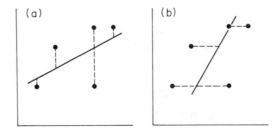

Fig. 27. Regression lines of (a) y upon x and (b) x upon y for the same points. The broken lines show the deviations minimized by the regressions.

We have so far considered the relationship only in terms of a straight line. Within the limits of a particular set of observations this may approximate to the truth. Equally, however, the relationship may be expressed on the graph by a curve. It may be perfectly possible to derive an equation connecting x and y in some other way, e.g. $y = a + bx^2$, $\log y = a + bx$, $\log y = a + b \log x$, etc. The likelihood of such a line fitting the data should be judged, before embarking on

---

* The line is actually calculated as $y = a' + b(x - \bar{x})$.

$$a' = \bar{y}, b = \frac{\Sigma(x - \bar{x})(y - \bar{y})}{\Sigma(x - \bar{x})^2}.$$

the calculations, by plotting the transformed values of x and/or y and seeing if the relationship then approximates to a straight line. If no straight line relationship is obtainable in this way, a satisfactory fit can often be obtained to an equation with several terms in x, e.g. $y = a + bx + cx^2 + dx^3$, etc. To do so, however, from the point of view of establishing a relationship with environmental factors, is a dubious procedure, as there is usually no possibility of placing a biological interpretation on such a complex relationship. Before trying any such relationship careful thought should be given to the meaning of the type of equation being tested.* That is not to say that such relationships should never be considered. If, for example, a measure of abundance was being correlated with altitude, the relationship might really be with various meteorological factors of which some varied directly as altitude and others as the square root of altitude, in which case a relationship of the type $y = a + bx + c\sqrt{x}$ would be suitable.

If several environmental factors have been measured, the joint relationship of the plant measure with them cannot be plotted as a simple two-dimensional graph, but the same type of treatment may be applied to the data, e.g. if two environmental factors, represented by x_1 and x_2, have been measured, then a *multiple regression* $y = a + bx_1 + cx_2$ may be calculated. The significance of the whole regression and of the coefficients of each independent variable may be tested. The mode of calculation takes into account any correlation between the independent variables.

Sometimes some environmental factors may be recorded qualitatively and others quantitatively. They may be included in the same regression analysis if the qualitative data are binary, i.e. distinguish between two states only. For example, one of a number of sets of observations by Blackman and Rutter (1946) on the relationship between density of *Endymion non-scriptus* and light intensity was in a mixed plantation of oak and larch. It was suspected that the nature of the canopy, oak or larch, might be affecting the density of *Endymion*, apart from the differing light intensities beneath the two species. They included in the regression a binary variable taking the value 1 if a sample was beneath oak canopy and zero if it was not, and obtained the multiple regression $y = 35\cdot11x - 3\cdot33z - 4\cdot26$, where y = square root of density, x = light intensity as fraction of full daylight, $z = 1$ if beneath oak canopy, 0 if not. From this regression equation it follows that at a given light intensity the density of *Endymion* is less beneath oak canopy than beneath larch.

A difficulty may arise, even with a simple regression on a single independent variable, if the data include any marked discontinuity. The

* It must also be remembered that perfect fit to a set of n observations can always be obtained by an expression with $n - 1$ terms in x.

relationship between the two variables may be different in the two subsets, so that when they are combined no relationship is apparent (Fig. 28) (Williams & Lance, 1968). If there is reason to suspect the occurrence of this effect, the data can be examined by a suitable technique (Chapter 7) to identify any marked discontinuity and relationships examined in the subsets separately.

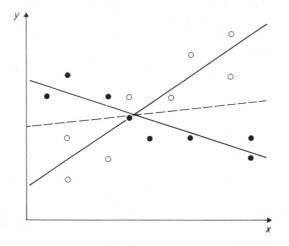

Fig. 28.   The effect of discontinuity on regression. Two groups of samples (O, ●) have different relationships between the dependent and independent variables (——), but when combined show no significant relationship (- - -). (From Williams & Lance, 1968, by courtesy of *Statistician*.)

The use of regression is considered further in Chapter 9.

There is one exception to the use of systematic sampling for the study of relationships with quantitative data. Sometimes interest centres on the nature of relationships rather than on their existence, e.g. information may be wanted whether the difference in abundance of a species for a given difference in an environmental factor is the same in two different communities. This is normally obtained by comparing the regression coefficients obtained from the data from the two communities. If the sampling is systematic there is no means of testing the difference between the coefficients as no unbiased estimate of their variances is available. Just as in the comparison of means the sampling on which they are based must be random if their variances and so the variances of the difference between them is to be estimated, so, if two regression coefficients are to be compared, the values of the dependent variates must be obtained from random samples. This should be borne in mind in planning observations if the need to compare coefficients with one another is likely to arise.

No mention has been made of the number of samples necessary in observations on vegetation and environmental factors. No single guiding principle can be given. For data of the first category it is necessary that no expected class of observations should be so low that the most extreme deviation has a probability greater than that adopted as the limit for significance. If a mean is being determined, as in the second and third categories, the approximate methods of deciding sample size discussed in Chapter 2 may be applied, but the number of samples necessary for a stringent test clearly depends on the magnitude of the observed difference. It may be helpful to plot successive values of the difference of means, rather than the means, and continue sampling until the difference has reached a more or less steady value. For the fourth category little guidance is available except previous experience with similar observations, particularly if multiple regressions are involved. As such observations are generally time-consuming it is well not to collect too extensive data until a preliminary examination of a relatively small number of observations has shown whether the approach is a promising one.

The four categories differ in their convenience of sampling and analysis and, to some extent, in their sensitivity. There is, however, rarely much choice of category for a particular investigation. Whether plant data are treated quantitatively or qualitatively depends on the number of blank entries and the range of values recorded. Environmental data likewise can generally be recorded only in one form or the other, unless an arbitrary boundary in quantitative data is drawn between 'high' and 'low' values.

CHAPTER 6

Plant Communities—I. Description and Comparison

In a sense the preceding chapters have all dealt with plants as they occur in communities. We have been concerned with the information on abundance and spatial distribution that may be obtained from samples taken within a defined area, the vegetation of which may, in the broadest sense, be termed a plant community. The emphasis has, however, been more on the differential behaviour of individual species both within and between such communities. In this and the next three chapters we are concerned with the characterization and comparison of the whole assemblages of species constituting the communities. Consideration of these leads on to problems of classification of communities, whether any natural classification, as opposed to *ad hoc* classification of a particular set of samples, is possible and, if so, how it may be achieved. This is fundamentally a separate subject, but it is so bound up with characterization of communities that the two can scarcely be considered apart.

Any discussion of plant communities as units raises the question whether or not they are more than abstractions made by ecologists from vegetation, the variation of which is continuous in space and, perhaps, in time. On the one hand there is the view that there are no discontinuities in natural vegetation, except where there are discontinuities in the physical environment, as at the junction between the outcrops of different geological strata. This view, of a continuum of vegetation, implicit in the outlook of many ecologists, was clearly stated by Gleason (1926) and adopted and developed especially by Matuszkiewicz (1948), by Curtis and his associates (see especially Curtis & McIntosh, 1951; Brown & Curtis, 1952; Curtis, 1959) and by Whittaker (1952, 1956, 1960). At the other extreme there is the view of the community as an organism (Clements, 1916) or quasi-organism (Tansley, 1920), postulating that the individuals and species within a community so interact as to increase one another's potentiality of survival. This concept necessarily implies more or less sharply defined boundaries between one community and another. Conversely, if communities do have sharply defined boundaries it follows that they must have some degree of reality as units.

Rejection of the organismal concept of the community, it may be noted, does not necessarily exclude the existence of separate communities as units, quite apart from the effects of environmental discontinuities. If species had ranges of tolerance in relation to environmental differences that tended to coincide, so that the total number of species in a region could be arranged in a considerably smaller number of groups, the members of each having approximately the same limits of tolerance, then distinctive communities, with more or less well-defined boundaries, would be expected, each corresponding to, and composed of, one of the groups of species of similar tolerance. The existence of such a phenomenon seems to be implied in the approach of both the Zürich–Montpellier and Uppsala schools of plant sociology, both of which erect a hierarchical system of classification of communities, without accepting any organismal or quasi-organismal view of the community. No extensive examination of the limits of tolerance of a geographical group of species in relation to all environmental factors has apparently been made; indeed, it is probably impossible except as a theoretical ideal. The work of Curtis and his associates, quoted above, points against any general coincidence of limits of tolerance as does most work on individual environmental factors, e.g. Hora (1947), Grime and Lloyd (1973), Hansen and Jensen (1974).

Reference has been made to formal systems of classification because their erection carries implications of the existence of discrete communities on the ground. A clear distinction must be made between the abstract concept of vegetational units of different grades, embodied in a system of classification, and the actual existence in the field of communities with defined boundaries between them. The real existence of the classificatory units implies the real existence of communities in the field. The converse is not necessarily true; communities might be clearly distinguishable in the field, but the communities so distinguished might form a continuum not susceptible of classification except by drawing arbitrary boundaries. The distinction between the units of classification and communities is analogous to that between species and individual plants.

Goodall (1954a) attempted an objective assessment of the reality of plant communities. He argued that a community, if it has real existence, should show homogeneity of composition within its boundaries and cited evidence that homogeneity is not found on any of three criteria, (a) spatial distribution of individuals, i.e. agreement with Poisson expectation, (b) lack of correlation between quantities of species in replicate samples, (c) constancy of variance with increasing quadrat spacing. He rightly suggested that, although complete homogeneity may not exist, there may be a much greater degree of homogeneity within one stand (community in the sense used here) than between different stands. This difference in level of homogeneity may be sufficient to allow delineation of stands. It follows from this that, if samples are drawn from more than one stand, of pairs of samples those near together are more likely to be drawn from the same stand than those far apart. Hence variance between adjacent samples should be less than between those farther apart. He quoted data showing that in several areas examined this was not so. Unfortunately, in each case the samples appear to have been taken entirely within an area the vegetation of which would be regarded on most subjective criteria as a single community, and the investigations refer rather to pattern of the type discussed in Chapter 3 than to the differences between communities. The approach is an interesting and original one, but needs to be applied to areas including more diverse vegetation before it can assist greatly in assessing the reality of communities.

Transects which include a transition between clearly distinct types of vegetation commonly illustrate the difficulties of delineating the boundaries of communities. If the occurrence of species along such a transect is noted, it is rarely found that the points at which they appear or disappear from the record coincide even approximately, unless there is a very pronounced environmental discontinuity. There is in fact great difficulty in drawing any boundary between the two types, although there may be a length of the transect in which change in composition with position is very much more rapid then elsewhere, i.e. there is a more or less well-defined zone of transition or ecotone. The question has been well summarized by Webb (1954). 'The fact is that the pattern of variation shown by the distribution of species among quadrats of the earth's surface chosen at random hovers in a tantalizing manner between the continuous and the discontinuous'.

The reality or otherwise of distinctive units of vegetation in the field remains to some extent an open question and one likely to remain a matter of considerable interest for its bearing on differing concepts of the plant community*. It has, however, relatively little importance in many fields of ecological interest. Empirical description of vegetation cannot wait for clarification of theoretical concepts and ecologists are well aware that satisfactory approximate boundaries can often be drawn between different vegetation types. Even if no approximate boundaries can be fixed, it is still possible to delimit arbitrarily areas away from doubtful transition zones, areas which may be studied in as exact a manner as possible. Such a procedure is quite justifiable provided it is remembered that the information obtained applies only to the area delimited, and that when it is compared with that from

---

* Differing views on the continuity or otherwise of variation in composition of vegetation have been reviewed by McIntosh (1967b; see also the responses to this review in Dansereau, 1968).

another area and found different, the two areas may represent the end points of a continuous series the intermediates of which have been rejected as transitional. Though theoretically this is a serious limitation, for many practical purposes the disadvantage is lessened because the very fact of an area being recognized subjectively as distinctive and not transitional is evidence that it represents a significant proportion of the total vegetation of a region.

However the boundaries of areas selected for description are fixed, the technique of characterization and comparison of the vegetation of areas is the same. Moreover, they apply whether the objectives of investigations are contributions to ecological theory, empirical cataloguing of vegetation, or information of immediate economic importance in agriculture or forestry. We now turn to consideration of these techniques. It will be convenient in doing so to use the term 'stand' to denote any area the vegetation of which has been treated as a unit for purposes of description. This will help to emphasize that we are dealing with concrete samples of vegetation and not with abstractions of classification.

No two stands can be identical, if sufficient detail is taken into account. Moreover, they may differ in so many characters that it is impossible to take all of these into consideration. A subjective selection must, therefore, be made of criteria to be used in characterization and comparison of stands. This selection will be made in the light of the kind of information desired and the scale of differences to be examined.

The principal criteria commonly used may be grouped under the following heads.

(1) Floristic composition—the species present.

(2) Measures of abundance of species.

(3) Performance of individuals of species. This may be assessed either qualitatively, e.g. setting viable seed or not, or quantitatively, e.g. proportions of tillers flowering, mean height of plants, etc.

(4) Growth and life form. These concepts are familiar enough, though difficult of precise definition. They represent attempts to classify individuals on the basis not of their taxonomic affinities, but of vegetative morphology.* Indeed, they may cut right across taxonomic affinity, as in the placing of succulent species of *Euphorbia* and members of the Cactaceae in the same class. Growth or life form grouping in relation to vegetation description may be comprehensive, assigning every species to its appropriate class in some such

---

* *Life form* and *growth form* have slightly different connotations. Life form is generally used where the classification is believed to have an adaptational significance, as in Raunkiaer's system, and growth form when no such significance is intended, but different authors vary in their use of the terms (cf. Du Rietz, 1931).

system as that of Raunkiaer (1934), or it may be concerned with the occurrence of certain distinctive features of vegetative form, e.g. thorniness, formation of stilt roots, etc.

(5) Physiognomy. The physiognomy of vegetation is likewise difficult or precise definition; it refers to the appearance of the stand as a whole and is closely connected with growth form, being largely influenced by the proportions of individuals of different growth forms; additional features depend on the performance and arrangement of individuals. It is the criterion least susceptible to exact description, though it is a most useful characteristic to an experienced ecologist. The terms for vegetation types used in common speech, such as woodland, moor, prairie, savanna, etc., mostly indicate physiognomy.

(6) Pattern of the constituent species.

(7) Various constants (in the mathematical sense) and indices derived directly or indirectly from other criteria have been used, e.g. constants derived from the form of the species-area curve, indices of diversity.

These criteria are not all independent of one another. Quantitative measures and measures of performance have meaning only between stands of broadly similar floristic composition. Growth form and physiognomy, on the other hand, are independent of floristic composition. Two stands of quite different floristic composition may have closely similar physiognomy and growth form spectra. Conversely, it is possible, at least theoretically, for two stands of the same floristic composition to differ in physiognomy and growth form spectrum, as some species differ in their growth form according to the conditions in which they are growing. Pattern, although determinable only in terms of particular species, or groups of species, has meaning and can be compared independently of the species concerned; it is possible for two stands to show similar sets of patterns of the species composing them, although their floristic composition is quite dissimilar. The fundamental criterion in characterizing communities is the essentially subjective one of physiognomy. Only between stands of similar physiognomy are the more detailed and critical criteria relevant. In complex communities of several layers it may even be impossible to place the characterization of different layers on a common basis, e.g. the tree and ground layers of forest. The fundamental importance of physiognomy accounts for the breakdown of attempts to characterize and classify all stands according to a standard technique.

If a considerable number of stands are to be compared the techniques of classification and ordination discussed in the succeeding chapters are appropriate. These are concerned with the overall pattern of relationships among a set of stands rather than with individual comparisons. If, however, a few stands only are being compared, the individual comparisons are important.

Floristic composition*

Floristic composition in terms of species present only, although a valuable basis of classification and ordination, is a crude and insensitive basis of making a single comparison between stands, though useful on a plant geographical scale, e.g. in comparing the floras of different islands. Two coefficients of similarity have been widely used. Jaccard, in a series of papers from 1902 onwards (e.g. see Jaccard, 1912), used the number of species common to two areas expressed as a percentage of the total number of species, i.e. $(a/(a+b+c)) \times 100$ where the two areas contain $a+b$ and $a+c$ species respectively and there are a species in common. Sørensen's (1948) coefficient takes the form of the number of species in common expressed as a percentage of the mean number of species in the two areas, i.e. $(2a/(2a+b+c)) \times 100$. Various other indices of similarity have been proposed (see Dagnelie, 1960; Goodall, 1973) and some are discussed in Chapter 7.

In interpreting such coefficients it must be remembered that the expected similarity for two samples drawn from the same population will be somewhat less than 100%, the expected value depending on the sample size. More importantly in a geographical context, where unequal areas are likely to be compared, the expected value will be still lower if the sample sizes are different. The expected values will depend on the relation between the number of species and the number of individuals in the vegetation under consideration (see below p. 157).

Quantitative measures of abundance and performance

The appropriate methods of sampling in determination of measures of abundance and performance have already been considered in Chapter 2. For measures such as density and biomass for which an estimate of variance for each stand is available, comparison of mean values for a single species is made by a t test, but we are here concerned with assessing the total difference between two stands when all the constituent species are considered. The appropriate test uses the generalized distance of Mahalanobis (p. 138), the multivariate equivalent of the t test (Hughes & Lindley, 1955).

An alternative, less computationally demanding, approach is to use a coefficient of similarity in which the contribution of each species is weighted by a measure of its abundance. Most similarity coefficients can be weighted in this way and measures such as frequency, which cannot be used in the generalized

* A possible future alternative to direct recording of floristic composition in all stands in broad-scale survey may be the use of remote-sensing data, although the results of an attempt by van Hecke *et al.* (1980) to correlate data from aerial photographs with grassland composition were disappointing.

distance, can be used. Strictly, such weighting implies that 'species in common' includes the total weight from the two stands of all species occurring in both. Thus the occurrence of a minimal amount in one stand of a species abundant in the other, will greatly affect the value of the coefficient (Goodall, 1973). It is more satisfactory to regard as common to the two stands only that amount of the species which occurs in both. A widely used coefficient of this type was originally suggested by Czekanowski (1913),

$$\frac{2 \sum_i \min (x_i, y_i)}{\sum_i (x_i + y_i)} *$$

where x_i, y_i are the amounts of species i.

The correlation coefficient between the amounts of different species in two stands has been suggested as a measure of similarity (Motomura, 1952) but has serious disadvantages. It is relatively insensitive at higher values, where the ecological significance is generally greatest (Bray & Curtis, 1957), and its valid use as a measure depends on the data being normally distributed (p. 120). Ghent (1963, 1972) suggested the alternative of using a rank correlation coefficient (p. 116), which is free of these objections.

It may be useful to standardize the data for abundance before making comparisons betweens stands, e.g. by expressing the abundance of each species as a proportion of the total for the stand. The effects of different standardizations in modifying the emphasis placed on different aspects of the composition of vegetation are discussed below (p. 211).

Various workers have attempted, by combining two or more measures into a single item, to make a more comprehensive estimate of the importance of species in a stand than is given by any one of the measures of abundance. For example, Curtis (1947, and see Brown & Curtis, 1952; Curtis & McIntosh, 1951) used an *importance value* obtained by adding together relative frequency (frequency of species as percentage of total frequency values of all species), relative density, and relative 'dominance' (basal area for species as percentage of total basal area). Such derived measures may be handled in comparisons in the same way as simple measures. They are open to the criticism that the mode of combining is quite arbitrary and markedly different situations may give rise to the same combined value. In the present case, if density is constant, the same importance value may be given by a regular distribution of saplings (high

* Often quoted, following Motyka *et al.* (1950) and Bray & Curtis (1957), as $2w/(a + b)$ where a, b are the sums of measures for the two stands and w is the sum of the lesser value for each species.

frequency, low basal area) as by groups of large trees (low frequency, high basal area). However, the use of single measures, especially density and frequency, may likewise fail to detect gross differences in the make-up of the vegetation. Other combined measures are not so arbitrary and represent an approximation to a single measure. For example, Dye and Walker's (1980) importance value for woody plants, $\Sigma(H_iN_iA_i)$, (H_i, mean height of height class i, N_i, number of individuals in height class i, A_i, mean cross-sectional area of stems in height class i) is an estimate of woody biomass, a meaningful single measure. Combined measures should not be used uncritically; their relative advantages and disadvantages should be considered in relation to the vegetation being examined.

Growth and life form

Data on growth or life form usually embrace a record of the number of species falling into each class of some suitable classification, e.g. that of Raunkiaer, or, if certain morphological features only are the main interest, the number of species with and without each feature. These data are based on the species list for the stand, and, like it, provide a relatively insensitive criterion. It is of greater value when applied to regions rather than to stands. Although data are commonly presented as a *spectrum*, giving the percentage of species in each class, the actual number of species of each class in different sets of data must be used in objective comparisons, which can readily be made by analysis of contingency tables.

Table 12 shows data, from Ewer (1932), of the number of species of different life forms according to Raunkiaer's classification, in four regions of the State of Illinois. (A very small number of epiphytes and stem succulents is omitted.) The percentages of species having the different life forms are similar, and the question is posed whether or not there is any real difference between the life form spectra of the four regions. If there is no difference, the numbers in different classes in each region should be in the same proportion as the totals shown in the bottom row of the table. The expected values and the χ^2 contributions resulting from the difference between observed and expected are shown in each cell of the table. The total χ^2 of 51·7596 has 24 degrees of freedom (8×3 from a 9×4 table) and a probability of slightly less than 0·1 %. There is evident discrepancy among the four areas.

The nature of the discrepancy between the areas may be investigated further by examination of the *standardized residuals* (Haberman, 1973; see Everitt, 1977), $e_{ij} = (n_{ij} - E_{ij})/\sqrt{E_{ij}}$, where n_{ij} is the observed and E_{ij} the expected number of observations in the cell in the ith row and jth column of the

Table 12. Numbers of species of different life forms in four regions of Illinois (data of Ewer, 1932)

Region		Megaphanerophyte Mg	Mesophanerophyte Ms	Microphanerophyte Mc	Nanophanerophyte N	Chamaephyte Ch	Hemicryptophyte H	Geophyte G	Helophyte and Hydrophyte HH	Therophyte T	Total
North	Obs.	18	51	82	58	19	686	192	68	193	1367
	Exp.	24·00	70·13	75·11	40·52	13·71	692·85	180·15	62·33	208·20	
	χ^2	1·5000	5·2183	0·6320	7·5407	2·0411	0·0678	0·7795	0·5158	1·1097	
Central	Obs.	16	52	58	30	10	637	165	60	198	1226
	Exp.	21·52	62·89	67·37	36·34	12·30	621·39	161·57	55·91	186·72	
	χ^2	1·4159	1·8857	1·3032	1·1061	0·4301	0·3921	0·0728	0·2992	0·6814	
Mid-South	Obs.	20	55	45	24	6	454	112	31	142	889
	Exp.	15·61	45·61	48·85	26·35	8·92	450·58	117·16	40·54	135·40	
	χ^2	1·2346	1·9332	0·3034	0·2096	0·9559	0·0260	0·2273	2·2450	0·3217	
South	Obs.	23	67	56	18	9	446	109	41	135	904
	Exp.	15·87	46·37	49·67	26·79	9·07	458·18	119·13	41·12	137·68	
	χ^2	3·2033	9·1783	0·8067	2·8841	0·0005	0·3238	0·8614	0·0012	0·0522	
Total		77	225	241	130	44	2223	578	200	668	4386

table. An estimate of the variance of e_{ij} is given by $V_{ij} = (1 - \frac{n_{i.}}{N})(1 - \frac{n_{.j}}{N})$, where

$n_{i.}$ and $n_{.j}$ are the totals of the ith row and jth column respectively and N is the grand total. For each cell an adjusted residual, $d_{ij} = e_{ij}/\sqrt{V_{ij}}$ can be calculated. Thus, for megaphanerophytes in the north region the standardized residual is

$$(18 - 24 \cdot 00)/\sqrt{24 \cdot 00} = -1 \cdot 225,$$

the variance is

$$(1 - 1367/4386)(1 - 77/4386) = 0 \cdot 6762$$

and the adjusted residual is $-1 \cdot 225/\sqrt{0 \cdot 6762} = -1 \cdot 49$.

If the variables of the table are independent the adjusted residuals are approximately normally distributed with zero mean and unit variance, so that any adjusted variable greater than 2 (the approximate 5% point of the normal distribution) can be regarded as showing significant discrepancy. The adjusted variables are shown in Table 13. It is clear that the north region has an excess of chamaephytes and nanophanerophytes and a deficiency of mesophanerophytes and the southern region an excess of megaphanerophytes and mesophanerophytes. (Note that significance cannot be judged from the individual χ^2 contributions.)

In small samples, e.g. particular stands, it is more informative to consider the abundance of plants of different growth (or life) forms rather than the number of species. This may be done by calculating a growth form spectrum in which the contribution of each species to its growth form class is weighted by some measure of abundance. This gives data which are not readily compared and any objective comparison must be of the actual numbers of individuals of different growth forms, or numbers of points or quadrats touched or occupied by plants of different growth forms.

Physiognomy

As already seen physiognomy is the criterion least susceptible to any quantitative approach and must usually rest on verbal description (on descriptive symbols see Christian & Perry, 1953). The only objective approach that has proved of value is the profile diagram introduced by Richards for forest communities (Davis & Richards, 1933–4; see also Richards, 1952). This is prepared by careful measurement of height, girth, crown, etc., of all the individuals in a strip of forest. These individuals are then drawn conventionally, but to scale for the measured features, in their correct positions along a line representing the length of the strip. Such diagrams give a clear impression of many of the features covered by physiognomy.

Table 13. Adjusted residuals for the data of Table 12 (see text).

Region	Life Form								
	Mg	Ms	Mc	N	Ch	H	G	HH	T
North	−1·49	**−2·83**	0·99	**3·36**	**2·08**	−0·45	1·14	0·89	−1·38
Central	−1·41	−1·66	−1·38	−1·26	−0·78	1·05	0·34	0·66	1·06
Mid-South	1·26	1·60	−0·66	−0·52	−1·10	0·26	0·57	−1·72	0·69
South	**2·03**	**3·49**	1·04	−1·93	−0·03	−0·91	−1·12	−0·04	−0·28

Pattern

Although detection and description of pattern, unless it is very pronounced, is dependent on a quantitative approach, there is no entirely satisfactory method of describing pattern within a stand by any simple expression, even for one species. Various measures of departure from randomness based on data from a single size of quadrat have been proposed, but are all more or less unsatisfactory. A more satisfactory description of pattern is obtained from a curve of variance against sample size. (Chapter 3.) Such curves refer to individual species and cannot be integrated for a stand. The technique of multiple pattern analysis (p. 125), however, does allow some summarization of the overall pattern in a stand.

Community constants and indices

There have been a number of attempts to characterize communities in terms of some aspect of the number of taxa involved and the relative contribution of different taxa to the community. The study of *diversity* has produced an extensive literature; useful reviews include those of MacArthur (1965), McIntosh (1967a), Whittaker (1972) and Peet (1974). There is no agreement on a precise definition of diversity, but two elements, of *species richness*, the number of species in the community, and *equitability* (Lloyd & Ghelardi, 1964), the evenness of the contribution of different species to the community, are involved. These can be considered as distinct characteristics, but more commonly the two are combined into a concept of diversity which may usefully be distinguished as *heterogeneity*, which measures not the absolute number of species, but the functional or apparent number (Peet, 1974; see also Hill, 1973c). Thus if five species are present in equal numbers in a population, a random pair of individuals are more likely to be of different species than if the population contained single individuals only of four of the species, and the remaining individuals were all of the fifth species; the first population is said to be more heterogeneous, or diverse, than the second.

Some of the approaches to analysis and measurement of diversity were developed primarily with reference to animal communities. In this case primary data are usually in the form of numbers of individuals of different species in a sample. At least for any one species, number of individual animals of a species is usually an accurate measure of that species contribution to the community. Density of plants is much less satisfactory because individuals may differ widely in size and density should not be used uncritically because a method was developed in relation to counts. Not only is density not necessarily satisfactory, but many plants are in any case not amenable to recording by density. Fortunately, most indices of diversity can be used with other measures of abundance. A further difference between plant and animal data is that the size of sample of animals in which the number of species is determined can normally be described by the total number of individuals. This is not usually possible for plants, but number of individuals can be replaced by area of sample, at least for comparable types of vegetation (Williams, 1950). If the sample is an areal one, however, it must be remembered that pattern may have a distorting effect on the apparent relationship between species number and area.

Two major types of approach to the measurement of diversity can be distinguished. One uses a hypothesis about the relationship between number of species and their relative abundance and if the observed data fit the hypothesis, a parameter of the theoretical relationship provides a measure of diversity, e.g. Williams's (1964) index of diversity. The other seeks an empirical measure of the observed diversity, e.g. Simpson's (1949) index.

If the species present are ranked in order of decreasing abundance, with the first species contributing a proportion k of the total amount, and it is assumed that the next species contributes the same proportion of the remaining amount and so on, the species abundances form a geometric series $p_i = k(1 - k)^{i-1}$, where p_i is the proportional abundance of the ith species (Motomura, 1932; Whittaker, 1965, 1969, 1972).

MacArthur (1957, 1960; Vandermeer & MacArthur, 1966) considered the relationship of abundance if the total is assigned randomly between the species present. This is known as the 'broken-stick' model, from the analogy of a stick broken at random points along its length. If there are S species in a sample of N individuals (or total amount) and the species are arranged in order of increasing abundance, the proportional abundance of the rth species is

$$P_r = \frac{N}{S} \sum_{j=1}^{r} \frac{1}{S-j+1}.$$

The assumptions underlying these two approaches both emphasize the interactions between the species in the community, or lack of such interactions (Whittaker, 1972), and may be expected to be more applicable to relatively homogeneous and integrated samples than to those representing a broader range of vegetation. Whittaker (1972) suggested that the geometric series would be approached by a group of species 'subject to scramble competition and the establishment of strong dominance' and that the broken-stick model represented the opposite limiting case approached by groups 'with contest competition and territorial relations stabilizing populations without development of dominance'. The latter is therefore unlikely to be appropriate to vegetation. It is, however, of interest as a 'base-line' of randomness against which to examine observed species-abundance relations (Hairston, 1964).

Fisher *et al.* (1945; see Williams, 1964) suggested that the logarithmic series is applicable to species abundance relations. This takes the form

$$n, (n/2)x, (n/3)x^2, (n/4)x^3 \ldots \ldots$$

where n is the number of species represented by one individual, and successive terms the number represented by two, three, four etc. individuals. Fisher showed that the ratio n/x is constant for the sample population, whatever the sample size. Williams designated this ratio α and the series may thus be represented

$$\alpha x, \alpha(x^2/2), \alpha(x^3/3), \alpha(x^4/4) \ldots .$$

The series is convergent and the sum to infinity, the total number of species, is

$$S = \alpha \ln\left(1 + \frac{N}{\alpha}\right),$$

or, if N/S, the average number of individuals per species, is large

$$S = \alpha \ln\left(\frac{N}{\alpha}\right).$$

The constant $x = N/(N+\alpha)$, depends on the sample size. From these relationships α and x can be determined (see Williams, 1964). For given numbers of species and individuals, the series is fixed and, provided the data fit the corresponding series, α can be used as an index of diversity, but one that assumes that species richness and equitability are interdependent.

Preston (1948) suggested an alternative model, the lognormal distribution, in which the relationship is not uniquely determined by N and S*. This assumes that the abundances are normally distributed but that abundances are appropriately described on a logarithmic rather than a linear scale. If species are grouped into 'octaves' (groups delimited by successive doubling of the number of individuals, i.e. <1, 1–2, 2–4, 4–8 etc.), the number of species in successive octaves will fall on a normal distribution. The number of species in an octave R octaves distant from a modal octave containing S_0 species is

$$S_r = S_0 e^{-(aR)^2}$$

and the total number of species is

$$S_t = \sum S_r = S_0 \sqrt{\pi/a}$$

where a is a constant, inversely proportional to the standard deviation $(a = \sqrt{0.5}/\sigma)$. The lognormal distribution has thus two independent para-

---

* Bullock (1971) gives a method for fitting the lognormal distribution to data.

meters representing species richness (S_t or S_0) and equitability (a or σ) but assumes a particular pattern of species abundances. In a sample those species with an expected frequency of less than one will be lacking, being cut off by what Preston termed a 'veil line', which shifts to the left as the sample size increases. Thus, if a sample is small relative to the population it represents, there will be a greater number of species with one individual than with two, as with the logarithmic series, but with a large sample this will not necessarily be so.

If data fit either the logarithmic or the lognormal distribution it is possible to calculate the number of species in common between two samples if they are drawn from the same population (Williams, 1947, 1949; Preston, 1962). This is important in comparison of species lists on a geographical scale (above, p. 151). Williams quoted the example of the islands of Guernsey, 24 square miles in area with 804 species, and Alderney, 3 square miles in area with 519 species, which have 480 species in common. On the basis of the logarithmic distribution, the expected number of species in common is 503*, on the basis of the lognormal distribution, 479. Both suggest a close affinity between the two floras, in contrast to the use of Jaccard's coefficient, which has a similarity value of only 56·9%.

The logarithmic and lognormal distributions give rise to *species-area curves* of a defined form. If the number of species is plotted against the logarithm of the area, the logarithmic distribution has a slope at first increasing with area but becoming effectively linear as unity in the expression $S = \alpha \ln(1 + (N/\alpha))$ (or $S = \alpha \ln(1 + (IA/\alpha))$, where A is area and I the number of individuals per unit area) becomes negligible compared with N/α. The lognormal distribution plotted in the same way gives a sigmoid curve, but one the upper part of which is very nearly linear up to the largest areas normally included in observations. It is thus difficult to distinguish between the fit of the two distributions for most data sets. For the smallest areas, discrepancies from either are to be expected. Both distributions assume random distribution of individuals but virtually all vegetation shows pattern; if most species are markedly contagiously distributed, the number of species observed in a sample will be less than if they were randomly distributed, unless the largest scale of pattern of all the species is exceeded.

Jaccard (see Jaccard, 1912) was the first to investigate the relation between species number and area and a number of relationships were suggested

---

* Williams (1964) later suggested, on empirical evidence, that a linear relation between log species number and log area was more appropriate for larger areas. In this case the expected number of species in common is 510.

empirically. There is an extensive literature, mostly concerned with data collected within one vegetation type, which need not be reviewed here (see Goodall, 1952a; Hopkins, 1955; Kilburn, 1966; Dony, 1977). Many sets of data discussed in the literature are unsatisfactory as tests of fit to suggested relationships because the sampling used was not appropriate (Goodall, 1952a). Three methods have been employed: (1) separate samples, each randomly placed; (2) each sample size is obtained by adding a contiguous area to the previous one; (3) the various sizes of sample are made up by adding the data from the appropriate number of randomly placed smaller samples. Only the first is satisfactory, but is the most laborious and has been least used. In the other two the data for different sample sizes are not independent.

The value of the species-area curve in characterizing a community is clearly rather limited, and the collection of valid data to produce it is very time consuming. If the form of relationship between species abundances is known and is the same in communities being compared, the slope of species-area curves provides an estimate of species richness, but these assumptions are rarely, if ever, justified. In practice, for communities of comparable physiognomy, the average species number in samples of a constant size, provides a simple and effective measure of species richness. Attempts to define a 'minimal' or 'characteristic' area for a community from a species area curve are discussed below (p. 167).

Whittaker (1965, 1969, 1972) has suggested the plotting of 'dominance diversity curves' as a method of exploring species abundance relationships. A measure of abundance is plotted against species sequence from the most abundant to the least abundant (Fig. 29). Whittaker considered a logarithmic scale of abundance to be more useful. Plotted in this way, the geometric distribution is linear, the lognormal sigmoid and the MacArthur broken stick also sigmoid but flatter. The shape of the curve thus gives an indication of the type of species-abundance relationship present and the slope of the upper parts is an indication of the degree of dominance, in the sense of one or few species contributing a large proportion to total biomass (cf. p. 4). Their use may be illustrated from Fig. 30, showing curves for an 'old field' succession to forest, which brings out the change from an initial geometric stage to later herbaceous stages approaching lognormal, followed by a steepening with stronger dominance at 20 years as woody species enter.

Two approaches to a numerical characterization of diversity, independent of any assumption about species abundance relationships have been much used. Simpson (1949) (see Williams, 1964) suggested an index $\lambda = \Sigma_i p_i^2$ where p_i is the proportional abundance of the ith out of S species. For a finite sample of N discrete individuals containing n_i individuals of the ith species this

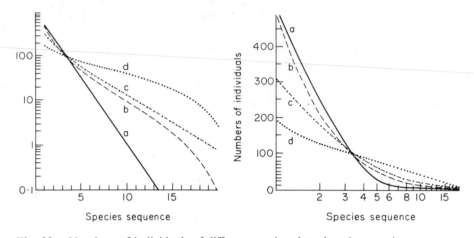

Fig. 29. Numbers of individuals of different species plotted against species sequence in order of abundance for a sample of 1000 individuals in 20 species. (a) Geometric series, (b) lognormal distribution, (c) logarithmic series, (d) MacArthur distribution. Left, number of individuals on logarithmic scale, species sequence linear; right, number of individuals on linear scale, species sequence logarithmic. (From Whittaker, 1972, by courtesy of *Taxon*.)

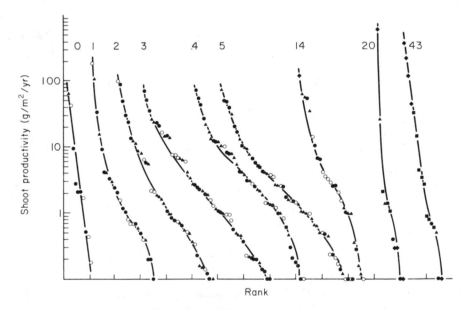

Fig. 30. Shoot productivity, on logarithmic scale, plotted against species sequence in order of productivity for stages in an old field succession. Age of stages in years. (From Whittaker, 1972, by courtesy of *Taxon*.)

is estimated by

$$\frac{\Sigma_i n_i (n_i - 1)}{N(N - 1)}.$$

In either form it can be seen to represent the chance of two successive randomly chosen individuals belonging to the same species (cf. Hurlbert, 1971). It ranges from 1 when all individuals belong to the same species to $1/S$ when every individual belongs to a different species. As Simpson formulated this index it decreases with increasing heterogeneity, and various workers have suggested using instead $1 - \lambda$ (see Peet, 1974). Hill (1973c), however, has argued convincingly for preferring $1/\lambda$.

McIntosh (1967a) derived an index of diversity which, though based on different reasoning, is closely related to Simpson's index (Peet, 1974). The composition of a stand may be described by a point in a hyperspace with as many axes as there are species, the position on any one species axis corresponding to the amount of that species present. The Euclidean distance between the points representing two stands then provides a measure of the dissimilarity between the stands. McIntosh suggested the distance of a stand from the origin as an indicator of diversity, i.e. $\sqrt{\Sigma_i n_i^2}$, where n_i is the number of individuals of the ith species. The maximum value is N, when all N individuals belong to the same species, i.e. when diversity is least. An appropriate measure of diversity is therefore $N - \sqrt{\Sigma_i n_i^2}$, but this is dependent on the size of sample and needs to be standardized. The minimum value of $N - \sqrt{\Sigma_i n_i^2}$ is zero, and maximum value, when there are N species, is $N - \sqrt{N}$. McIntosh therefore suggested as an index of diversity $(N - \sqrt{\Sigma_i n_i^2})/(N - \sqrt{N})$ which has a range from 0 to 1. If the abundances are placed on a proportional basis, $p_i = n_i/N$ (or other proportional basis), when standardization is not necessary, McIntosh's index simplifies to $1 - \sqrt{\Sigma_i p_i^2}$, the one-complement of the square root of Simpson's index.

Indices of diversity based on information theory were introduced into ecology by Margalef (1957) and have been widely used since. Shannon's (see Shannon & Weaver, 1949) measure is $H' = -\Sigma_i p_i \log p_i$ where p_i is the proportional abundance of the ith out of S species. If the data are in the form of counts p_i is estimated as n_i/N for n_i individuals of the ith species out of a total of N individuals.

$$H' = -\Sigma_i \frac{n_i}{N} \log \frac{n_i}{N}$$

$$= \Sigma_i \frac{n_i}{N} (\log N - \log n_i) = \frac{1}{N} (N \log N - \Sigma n_i \log n_i).$$

The base of the logarithm is arbitrary. Logarithms to the base e have commonly been used and some of the earlier work, following the usage in information theory, used logarithms to the base 2.

In the present context Shannon's formula may be regarded in the following way (see Williams, 1976). If we accept n_i/N as the best estimate of p_i, and one individual is picked at random, the probability of that individual belonging to species i is n_i/N, and the probability (strictly the likelihood) of getting n_i individuals of that species in a sample is $(n_i/N)^{n_i}$. To obtain the probability of obtaining the observed set of individuals the corresponding expressions for all species must be multiplied together i.e. $\Pi(n_i/N)^{n_i}$. The value of this expression for the same set of p_j's varies with sample size, but can be standardized by taking the Nth root, i.e. $\sqrt[N]{\Pi(n_i/N)^{n_i}}$. The reciprocal of the probability of getting an observed set of data is an appropriate measure of uncertainty or heterogeneity. This reciprocal could itself be used as a diversity index, and has been (generally represented as exp. H') by some workers, but its logarithm has evident advantages in computation and manipulation, as in partitioning the total information (e.g. Orloci, 1968, 1971, 1976). The logarithm of the reciprocal, again standardized, is, as above,

$$-\Sigma(n_i/N) \log (n_i/N).$$

Shannon's measure is strictly applicable only to an infinite population. For a finite sample of discrete individuals Brillouin's measure (see Brillouin, 1962) is

$$H = \frac{1}{N} \log \left(\frac{N!}{n_1!\, n_2!\, n_3! \ldots n_s!} = \frac{1}{N}(\log N! - \Sigma_i \log n_i!) \right).$$

If all n_i are very large, this approximates to H', as can be seen by using the approximation $1n\ n! = n(1n\ n - 1)$ (Pielou, 1977).

There has been controversy over which is the more appropriate measure to use. If significance tests are to be applied to differences between values obtained, the distinction is important (see Pielou, 1977) but otherwise the broad conclusions drawn from comparison of the two measures for different samples will generally be the same. Peet (1974) quoted an example suggesting that where they differ, the Shannon index accords more with the general concept of diversity. Consider two communities, A, with ten species, each with five individuals, and B with nine species, eight with 110 individuals each and one with 120 individuals. The values of the two measures are:

	Shannon	Brillouin*
Community A	1·00	0·874
Community B	0·954	0·943

* Log factorials, needed in the calculation of the Brillouin index, can be obtained from tables. A table giving all values up to log 1050! is given by Lloyd et al. (1968).

Community A is more diverse, being species richer and having greater equitability. If other measures of species abundance than counts are used, only Shannon's formula is available.

There has also been argument over how appropriate these information measures are as measures of diversity, argument often confused by largely irrelevant analogies with information theory and entropy. Pielou (1966) suggested that heterogeneity can be '... equated with the amount of uncertainty that exists regarding the species of an individual selected at random from a population. The more species there are and the more nearly even their distribution, the greater their diversity'. Since the information measures are essentially measures of uncertainty, they provide appropriate measures of diversity in spite of their apparently irrelevant derivation. Pielou's argument can equally be applied to other measures of abundance, e.g. uncertainty what species a random point will touch, uncertainty what species a random, very small, portion of biomass will belong to.

Different measures of heterogeneity differ in their responses to changes in relative abundance of rare and common species (Hill, 1973c; Peet, 1974). A change in the relative abundance of two common species will have more effect on Simpson's index than a similar change in the relative abundance of two rare species. The converse is true of the information measures. Choice of measure of diversity will thus be influenced by the relative importance attached to rare and common species. With small samples, even if interest centres on rarer species, it must be remembered that the chance occurrence of one or two additional species may give a markedly different value for the information measures.

Simpson's index and the information measures are not unrelated and are only two of a range of possible measures. Good (1953) and Hill (1973c) have both suggested sets of indices. Good proposed indices of the form

$$C_{m,n} = \Sigma_i p_i^m (-\log p_i)^n.$$

Simpson's index is then $C_{2,0}$, Shannon's measure $C_{1,1}$ and the number of species $C_{0,0}$. Hill proposed a series of 'diversity numbers'

$$N_a \text{ (diversity number of order } a\text{)} = (\Sigma_i p_i^a)^{1/(1-a)}.$$

N_0 is the number of species, N_2 the reciprocal of Simpson's index and N_1 is the exponent of Shannon's measure (to the base e). Hill regards a diversity number as 'figuratively a measure of how many species are present if we examine the sample down to a certain depth among its rarities. If we examine superficially (e.g. by using N_2) we shall see only the more abundant species. If we look deeply (e.g. by using N_0) we shall see all the species present'.

Ideally, it would be preferable to separate the species-richness and equitability elements of heterogeneity. As we have seen above the determination of species-richness in any exact sense is not possible. Determination of equitability likewise presents difficulties. An evident approach is to compare an observed diversity measure with the maximum value possible with the observed numbers of species and individuals. A number of such indices of equitability have been proposed, but their value is limited because of their dependence on species number in the community a sample represents. As Peet (1974) pointed out, to use the observed number of species in the sample underestimates the true value and it is, in any case, subject to chance variation (see also Peet, 1975). With vegetation there is the further practical limitation that this approach is only applicable to density data.

Another approach is to use the standard deviation of the observed abundances as an inverse measure of equitability, the lower the standard deviation the greater being the equitability (Fager, 1972). This is applicable to any measure of abundance and, while also affected by the chance of occurrence of rare species in the data, is a practical measure which is simple to determine, and may appropriately be used for samples of the same size from similar vegetation types.

Hill (1973c) suggested an approach based on the diversity numbers discussed above. For complete evenness, where all the n species have the same abundance, the diversity numbers of all orders are equal to n. The less even the distribution of abundances, the more the diversity numbers of different orders diverge and Hill suggests the ratio of one diversity number to another as an index of equitability. The most appropriate ratio to use has apparently not been considered. Hill pointed out that Sheldon (1969) had previously suggested the ratio of N_1 to N_0 and Peet (1974) gave an example of the use of the ratio of N_1 to N_2 with artificial data.

A limitation of the indices so far discussed is that they take no account of spatial variation in the community being considered, i.e. of pattern. While this may not be important in considering assemblages of mobile animals, it is a very serious limitation in vegetation; the local uncertainty which species a randomly chosen individual will belong to or whether two randomly chosen individuals will belong to the same species may be very different from that estimated for the community as a whole. Peterson (1976) has suggested a potentially valuable 'index of local diversity' applicable to vegetation made up of distinct individuals, based on the identity of nearest neighbours. For a random sample of individuals the number, $X(n)$ of the n nearest neighbours belonging to the same species as the 'focal' individual is determined. The ratio of the average number to n is then an estimate of the likelihood of a nearest neighbour being of the same species. The reciprocal of this ratio, $1/\{\bar{X}(n)/n\}$ is then an index of local diversity. If pattern is present, the value of the index will differ for different values of n, which must be determined appropriately for the vegetation being considered.

From the same data, an index of segregation (cf. p. 122) can be derived. For each species i the average probability of finding it as a neighbour of itself may be compared with the probability, p_i, of finding it in the population, so that a measure of its segregation relative to the remaining species is $\bar{X}(n)_i/np_i$, which will be unity in the absence of segregation, >1 for positive and <1 for negative segregation. The probability p_i will normally have to be estimated from the observed number in the collection. Strictly, the expected probability of finding a second individual in the absence of segregation is $(L_i-1)/(M-1)$, where L_i is the total number of individuals of species i and M the total size of the collection.

The indices of segregation for individual species may be combined to form a community segregation index. Peterson suggests weighting the index for each species by its relative abundance to give an index

$$\Sigma_i p_i\{\bar{X}(n)_i/np_i\} = \Sigma_i \bar{X}(n)_i/n.$$

Whittaker (1960) distinguished three types of diversity. *Alpha* diversity refers to a particular stand or community sample, *beta* diversity to the extent of change in composition along an environmental gradient and *gamma* diversity to the diversity of a number of stands which have been combined. Both alpha and gamma diversities can be measured by the indices which have been discussed above, but whereas alpha diversity may be regarded as a real characteristic of the stand described, gamma diversity will depend on the range of stands which have been combined and is, to a considerable extent, a property of a particular data set. Since there can be no measure of the 'length' of an environmental gradient which would allow change to be expressed per 'unit length' of gradient, beta diversity is entirely a property of the data set.

The concepts of *minimal area* and *representative area* have long figured in the literature and are closely related to the form of the species-area curve. Though varying terms and definitions have been applied to these concepts, they have a common basis in the idea that the true characteristics of a plant community only appear when a certain minimum area of it is examined. In broad terms this is a truism; it is evident that the larger the sample examined the greater will be the information obtained, whether about the species present, their quantitative representation or their patterns. Likewise a law of diminishing returns will apply in that successive equal increases in size of sample will give successively smaller amounts of additional information. The validity of determining, other than empirically, any precise area to be used in describing vegetation depends on whether or not there is, at some point, a sudden decrease in the amount of additional information obtained.

As in consideration of species–area curves, it is not necessary to review the rather extensive literature; reference may again be made to Goodall's (1952a) comprehensive review. There have been three approaches to the determination of characteristic areas, based respectively on species composition, species frequency, and homogeneity of composition. The species–area curve, as we have seen, at first rises sharply and then flattens off, eventually becoming nearly horizontal. The area above which it becomes nearly horizontal has been regarded by various ecologists as an indication of the minimal area. Cain (1934) claimed that there is a 'break' in the curve at which its slope decreases sharply, and many curves from field data do show this 'break'. Cain (1938) himself later realized that the position of the break depends on the relative scales of the ordinate and abscissa. Realization of this difficulty led to attempts to determine a characteristic area from the species–area curve in some other way. Cain (1938) suggested the point where the slope was equal to the average slope for the whole curve, or to some fraction of it (Cain, 1943). The value obtained, however, depends on the largest area examined, as does Archibald's (1949b) use of the area containing 50 % of the total number of species. This objection does not apply to Vestal's (1949) use of a pair of points, such that the ratio of the corresponding areas is 1 : 50 and of the species numbers 1 : 2, to define the minimal area as five times that corresponding to the lower of the pair of points. These points are readily determined, particularly if a graph is drawn on a log basis, but it is difficult to see their particular significance in relation to the vegetation. Even if this concept of minimal area is accepted, it is evident that its value for a particular stand can only be determined very approximately. What has been said above on the species–area curve and on attempts to determine a minimal area from it does not detract from the value in some circumstances, e.g. in species-rich tropical forest, of plotting such a curve as a guide as to whether most of the species present have been recorded or not.

Archibald (1949a) suggested the minimum quadrat size which shows 95 % frequency for at least one species as a minimal area (with particular reference to comparison of 'dominants'). The principal elaboration of a minimal area concept based on frequency is that of the Uppsala school, e.g. Du Rietz (1921). It is claimed that, if frequency is estimated with increasing size of quadrat, and if the number of species having a frequency of more than 90 % is plotted against quadrat size, the curve becomes and remains horizontal, at least for a considerable further increase in area. Those species with frequency greater than 90 % are referred to as constants and the minimal area is defined as that at which the full number of constants is attained or, at least, the number of constants remains the same for a considerable further increase in area. The crux of this approach is the doubtful validity of the assumed relationship between the number of constants, as defined, and area. There has been marked disagreement over this (see Pearsall, 1924; Goodall, 1952a); it is relevant to point out that the data put forward in support of the supposed relationship were obtained from quadrats selectively placed and not from random sampling.

Both the scale and intensity of pattern of the different species present will affect the size of minimal area found, if, indeed, any definite area can be determined. A species will appear at a relatively small size of quadrat if the only pattern it exhibits is small scale. If large-scale pattern is present, the size at which it appears will depend on its intensity; a species with dense clumps separated by spaces in which it is absent (high intensity of pattern) will tend to appear only in large quadrats. Conversely, a species with large-scale pattern of a mosaic of patches with higher and lower density (low intensity of pattern) will appear at a smaller quadrat size. Minimal area is thus not related to pattern in any simple manner; the same minimal area may be found for stands of quite different structure.

Goodall (1954b) has put forward a different approach to minimal area, based on homogeneity of values of a quantitative measure of the various species in replicate samples at different spacings. He suggested as the minimal area 'the smallest sample area for which the expected differences in composition between replicates are independent of their distance apart', i.e. are due to random variation only. He presented data for salt marsh in support of the view that no such area can be found or, at least, that it is greater than the largest area examined ($10 \, m^2$). Later he found some evidence for a minimal area of a quarter of a mile square in Uganda rain forest but no evidence for minimal area in mallee nor in more extended samples from salt-marsh (Goodall, 1961). This approach is, in effect, a direct examination of scale of patterning and attempts to determine the maximum scale of pattern present for each species and hence that for all species in the stand (Chapter 3). The minimal area, as defined, refers to a part only of the information obtained and it scarcely seems desirable to single it out for special consideration.

We have been considering minimal area as one characteristic of the community. Before leaving the subject it may be well to emphasize that, in spite of some confusion in the literature, it bears no necessary relationship, however it is determined, to the most suitable size of sample area for determination of other characteristics of the community. The most suitable quadrat size for determination of frequency, or density, or any other quantitative measure of species rests upon the considerations of edge effect, ease of recording, minimizing of variance of mean, and shape of distribution curve obtained, matters already dealt with in Chapter 2.

CHAPTER 7

Plant Communities—II. Classification

Having considered in the last chapter techniques of describing and comparing stands, we may turn to techniques of summarizing and ordering the available information on a set of stands. To do so may be of interest as an end in itself, but more importantly provides a basis of comparison with the environmental and other factors affecting the composition of vegetation. This in turn leads to generation of hypotheses about the causation of the composition of stands as we find them.

Two approaches to this problem are available. *Classification* involves arranging stands into classes the members of each of which have in common one or more characteristics setting them apart from the members of other classes. Classification is a very familiar concept and has been applied throughout the history of vegetation study, though until relatively recently it was largely subjective. *Ordination* attempts to place each stand in relation to one or more axes in such a way that a statement of its position relative to the axes conveys the maximum information about its composition. Ordination, although it had roots in earlier work on vegetation, was dependent on numerical techniques for its effective development.

Though classification of vegetation has, usefully in some contexts, been extended to include widely and clearly diverse types within one scheme, a level at which it must be based on physiognomy, the major problems are at levels where the species present provide the appropriate criteria. Traditional systems of classification operate either by grouping stands together on subjective assessment of similarity, or by dividing the whole set of stands into two or more groups on the basis of the presence of one or few species subjectively selected as likely to give a useful classification, e.g. classification on the basis of dominant species. The contribution of numerical procedure is, in the one case, to provide an objective measure of similarity and, in the other, to allow the data themselves to indicate which species will provide the most satisfactory division criteria. Though one particular numerical procedure can give only one result for a particular set of data, the decision which procedure should be used remains a subjective one. Any simplification or summarization of a set of stand data involves rejecting some of the information in the original data;

171

what information is retained will depend on the procedure used. While a classification that retains little information is unlikely to be generally useful (though it may be highly efficient for a single, narrowly defined purpose), precisely which information it is desirable to retain is a matter of opinion and may, in any case, vary with the use that is to be made of the classification. Clearly, there can be no accepted 'best' procedure of numerical classification, though it may be hoped that a consensus will eventually be reached on suitable techniques for general purpose use. There are evident advantages in communication between different workers if a limited number of procedures are generally used, even if the classification presented is not quite optimal in a particular case.

Traditional procedures of classification of vegetation are aimed at the production of a comprehensive system in which, in principle, any stand can be placed. Numerical procedures necessarily operate on a finite number of stands. This has undoubtedly influenced development of techniques, most of which have been devised on the implicit assumption that the problem is to divide a 'universe' of stands efficiently into groups, rather than to divide a sample of stands into groups themselves representing samples of wider groups. For some workers the objective is to use a classification of a set of stands as an aid to the elucidation of the ecological relationships of those particular stands (Lambert & Dale, 1964). From this viewpoint it is irrelevant if analysis of a further set of stands of comparable vegetation produces a different classification. Others, however, use numerical procedures as a basis of a comprehensive classificatory system, comparable to traditional systems. In practice, some procedures readily permit the incorporation of further stands on the criteria used in deriving the classification, and the results of doing so may empirically prove satisfactory, but strictly the validity of using in this way a classification based on an analysis of a finite data set treated as a universe is questionable*.

If a set of stands is to be classified an initial choice must be made between hierarchical and non-hierarchical (reticulate) systems. Hierarchical systems aim to subdivide the population successively by the most efficient steps; non-hierarchical systems aim to produce the most efficient groupings regardless of the route by which they are derived. Maximal efficiency of grouping implies,

---

* Because the procedures operate on a finite set of stands, not regarded as a sample, some workers prefer the term 'cluster analysis' to 'classification', particularly when an agglomerative strategy (below) is used. Others, e.g. Lance and Williams (1967a), restrict 'cluster analysis' to non-hierarchical procedures. Because numerical procedures will divide into groups a set of stands forming a continuum of variation, the term 'dissection' has also been used.

for most purposes, maximum homogeneity of groups within any constraint of size or number of groups and this is clearly desirable, but no method is available that maximizes both group homogeneity and hierarchical efficiency (Williams *et al.*, 1966; Lance & Williams, 1966a). Sørensen (1948), in one of the earliest attempts to use a numerical procedure of classification, used a non-hierarchical system (though with some hierarchical structure imposed on it by further non-hierarchical grouping of the initial groups into '2nd order groups') (see also del Moral, 1975; Yarranton *et al.*, 1972). The great majority of systems proposed for vegetation have, however, been hierarchical. Not only are hierarchical systems in general less cumbersome but they have the overwhelming advantage of being more readily interpreted ecologically. Particularly at the higher levels of a hierarchy it is generally possible, by examining the means and ranges, or presence and absence, of environmental features, to erect hypotheses about the factors controlling the composition of the vegetation being analysed (Fig. 31).

Accepting that a hierarchical system is preferable, two further choices of strategy are required (Williams *et al.*, 1966). The population may be progressively subdivided into groups of diminishing size (*divisive* strategy); alternatively a hierarchy may be built up by fusing individuals progressively into groups of increasing size until the entire population is fused into a single group (*agglomerative* strategy). The relative advantages of divisive and agglomerative strategies are considered below (p. 190), but it is useful to note at this point that a divisive strategy allows the hierarchy to be cut off at whatever level is appropriate to the problems under consideration.

The second choice is between *monothetic* and *polythetic* strategies. In a monothetic strategy each node in the hierarchy is based on a single attribute, i.e. division into subgroups is made on the presence or absence of a single attribute (species), fusion on the common presence of a single attribute in the groups being fused. Thus every group, at all stages, is definable in terms of the presence or absence of specified attributes. In a polythetic strategy the dichotomies of the hierarchy are each based jointly on a number of attributes, so that groups are defined in terms of the overall similarity of their members; only coincidently may it be possible to define them in terms of specified attributes. As Williams *et al.* (1966) pointed out, agglomerative-monothetic methods cannot exist except in a trivial sense, as fusion of groups cannot continue beyond the stage at which no two groups, taken together, have an attribute present in all individuals. The remaining three possible systems are potentially useful.

The first widely used procedure was divisive-monothetic, the 'association analysis' of Williams and Lambert (1959, 1960). Although this has been

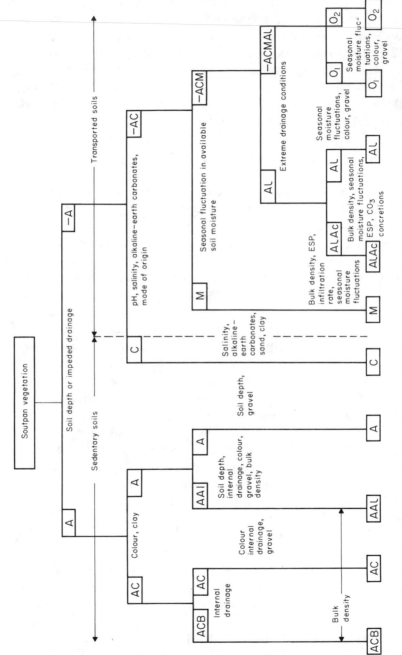

Fig. 31. Major divisions in a numerical classification of 1120 stands of bushveld vegetation, with correlated environmental differences indicated. (From Grunow, 1967, by courtesy of *Journal of Ecology*.)

superseded by other more satisfactory procedures, it is worth discussing in some detail not only because it has been widely, and often uncritically, used, but also because it provides an introduction to the general problems of divisive procedures in a straight-forward context.

Association analysis was a development from the pioneering work of Goodall (1953a), who suggested the use of interspecific association as a basis of classifying a set of stands into groups exhibiting no internal heterogeneity in composition. Association between species was examined in terms of presence and absence, by calculating χ^2 for all possible 2×2 contingency tables between pairs of species (Chapter 4)*. Only positive associations were considered. He suggested four possible procedures. These were to take as a potentially homogeneous group (1) all those stands containing one of a pair of species showing associations, (2) all those not containing one of the pair of species, (3) all those containing both the species, and (4) all those containing neither of the species. After trial of all four methods Goodall concluded that the first was the most successful in reducing the amount of association within the group formed and also the least laborious. At each stage in division there is a choice of species showing association; division was made on the most frequent species showing positive associations. The course of Goodall's analysis is shown in Fig. 32. After the first division, on *Eucalyptus dumosa*, those stands containing it were tested for association and again divided on the basis of the most frequent species showing positive association, *Triodia irritans*. This procedure was repeated until no positive association remained (Group A). Those stands rejected at each stage were pooled and the procedure repeated on the pooled residue. Goodall's final step was to test whether any pair of final groups could be combined without reintroducing positive association. In the test data groups A and E could be so combined.

Goodall's procedure has the considerable disadvantage that as a result of the pooling of negatively defined residues the final analysis is not hierarchical, with the limitations on ecological interpretation that this implies. Williams and Lambert (1959, 1960; see also Williams & Lance, 1958) examined the approach more critically. They pointed out that Goodall's definition of

---

* There is no fundamental distinction between the heterogeneity shown by a mosaic of relatively small-scale patches within what is commonly regarded as one community and the larger-scale mosaic formed by patches of different communities. The interpretation of a particular case depends on further information about the vegetation, e.g. whether or not there is evidence of dynamic relationship between the patches of different composition, and on the size of the sample areas being analysed. If the sample areas are large, it is reasonable to suppose that each one will include all the phases that have been regarded earlier as coming under the heading of pattern.

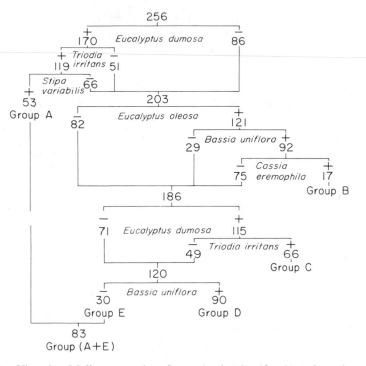

Fig. 32. Victorian Mallee vegetation. Stages in the classification of quadrats into groups by taking as a potentially homogeneous group all quadrats containing one of two species showing positive association. At each stage the number of quadrats in which the species in question is present (+) and absent (−) is indicated. (From Goodall, 1953a, by courtesy of *Australian Journal of Botany*.)

'homogeneous groupings' involves two independent concepts; significant associations may be absent in a grouping either by being indeterminate (one or both species being present or absent in all samples) or by association falling below a specified level. These are not of equal importance, an association below the specified level having been tested and rejected, whereas an indeterminate association cannot be tested. They therefore preferred, if the choice has to be made, to reduce the level of association rather than render it indeterminate. Goodall used only positive associations; Williams and Lambert pointed out that, at least with markedly heterogeneous data, major divisions are likely to be reinforced by strong negative associations and therefore included both positive and negative associations.

The objective is so to sub-divide the population of stands that all associations between species disappear. Williams and Lambert pointed out that in general there will be a number of alternative subdivisions which will achieve this. They therefore introduced the concept of efficient subdivision, i.e. subdivision at any stage based on that species which produces the smallest amount of association in the two subgroups. At that time it was not practicable for any but the simplest sets of data to test the associations in the sets of subgroups that result from subdivision on all possible species. It was therefore necessary to choose some parameter A as a measure of association. Williams and Lambert showed, on theoretical grounds, that division on the species with the maximum ΣA will tend to reduce the residual ΣA in the resulting subgroups to a minimum, i.e. to give efficient subdivision (see also Lance & Williams, 1965).

Various parameters might be used as the index A. Having considered χ^2 (or χ^2/N since N is constant at any one stage in the analysis), $\sqrt{(\chi^2/N)}$ summing all values in both cases, and χ^2 with Yates's correction, summing only values exceeding that corresponding to a selected probability level, Williams and Lambert selected $\sqrt{(\chi^2/N)}$. Lance and Williams (1965) later showed from theoretical considerations that $\Sigma \chi^2$ carries more information than $\Sigma \sqrt{(\chi^2/N)}$, but it tends to fragment the analysis by initially splitting off outliers from the population ('chaining'), making it less useful for ecological interpretation. The less efficient $\Sigma \sqrt{(\chi^2/N)}$ tends in practice to give a more even split and Lance and Williams therefore recommended it for general use. This effect is due to the greater weight given by $\Sigma \chi^2$ than by $\Sigma \sqrt{(\chi^2/N)}$ to the occurrence of rare species and the absence of common species relative to the remainder of the information in the data (Gower, 1976b). The empirical coefficients $\Sigma |ad - bc|$ and $\Sigma (ad - bc)^2$ have also been suggested (Lance & Williams, 1965); the latter coefficient was used by Crawford et al. (1970) in a comparative study. Podani (1979a) has suggested using the likelihood ratio criterion (p. 112) instead of χ^2. It is not always clear in published uses of association analysis whether χ^2 with or without Yates's correction has been summed. Noy-Meir et al. (1970) preferred the corrected $\Sigma \chi^2$ on the ground that Yates's correction counteracts the tendency for uncorrected $\Sigma \chi^2$ to be dominated by associations between rare species and hence to produce a fragmented classification.

Williams and Lambert acknowledged the disadvantages inherent in the pooling of negative residues used by Goodall, adding that hierarchical division is likely to produce substantially the same groups by a shorter route. A practical test confirmed this*. Hierarchical division is clearly to be preferred on all grounds. In hierarchical division it is clear that the relative importance of

* Wellbourn and Lange's (1967) demonstration of markedly dissimilar groupings obtained from the same data by Goodall's and by Williams and Lambert's procedures is irrelevant to the question of pooling versus hierarchical division as the type of division is confounded with the choice of division species.

different subdivisions of the same order may vary. Some measure of the heterogeneity of any group under examination is therefore desirable as a means of assessing the fall in level of heterogeneity produced by a subdivision. After consideration of other possible measures, Williams and Lambert concluded that the highest individual χ^2 within a group is the most suitable measure. Thus in Fig. 33 the initial division reduces the highest single χ^2 from c. 54 in the whole population to c. 14 in the 'J' (*Agrostis* present) class and to c. 26 in the 'j' (*Agrostis* absent) class. In assessing such dendrograms it must be remembered that the maximum individual χ^2 possible cannot be greater than N. Thus if a large heterogeneous group is divided into two groups of very uneven size, the smaller one is likely to have a very much lower largest single χ^2, even if two

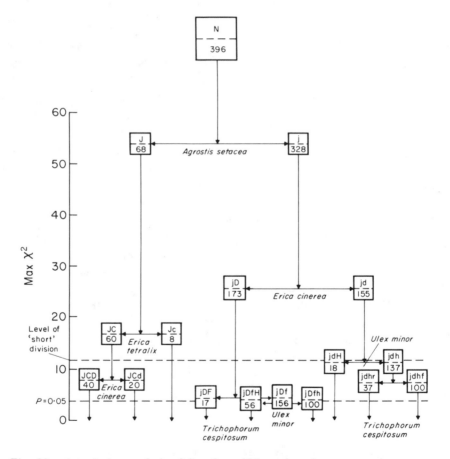

Fig. 33. Association-analysis of data from 396 quadrats in a community containing ten species.(From Williams & Lambert, 1960, by courtesy of *Journal of Ecology*.)

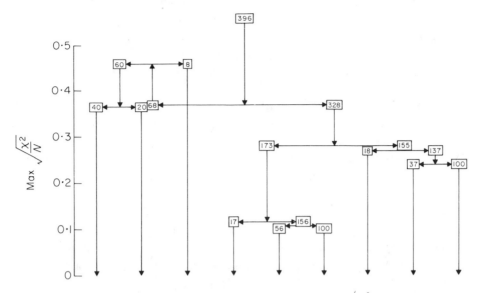

Fig. 34.   The same analysis as in Fig. 33, but with maximum $\sqrt{(\chi^2/N)}$ as the measure of heterogeneity.

species show very strong association within it. It might be argued that the largest single value of χ^2/N or of $\sqrt{(\chi^2/N)}$ (Fig. 34) would give a visually more satisfactory dendrogram. In practical interpretation, however, little detailed attention is generally paid to heterogeneity levels. Moreover, reversals of heterogeneity level are likely to be more frequent*, as in Fig. 34. Other measures of heterogeneity have been used for the vertical scale of the dendrogram, e.g. maximum species $\Sigma\chi^2$ (Madgwick & Desrochers, 1972), number of associations above a specified χ^2 level (Sarmiento & Monasterio, 1969), but the former is similarly related to group size, and the latter to the number of determinate χ^2 values, and they do not have any compensatory advantages.

When a hierarchical system of classification is used, a decision is needed on the lowest level at which the interpretation is to be attempted. If the hierarchy is complete, i.e. is continued down to single stands, there will necessarily be groupings at the lowest levels which are trivial. With an agglomerative strategy, when the hierarchy is necessarily complete, decision on the lowest level of interest can be made subjectively though some guiding rule is desirable. With a

---

* It was for this reason that Williams and Lambert (1960) rejected the largest species $\Sigma\sqrt{(\chi^2/N)}$ as the measure of heterogeneity.

divisive strategy a stopping rule is needed if superfluous computation is to be avoided.

In their initial work, Williams and Lambert terminated subdivision when no individual χ^2 exceeded 3·84, corresponding to $P = 0·05$ with one degree of freedom. This criterion does *not* imply a 5% probability of rejection of the hypothesis of homogeneity of the group, but a considerably higher one, because what is at issue is the highest single χ^2 among a matrix of χ^2*. Thus Madgwick and Desrochers (1972), quoting from Jensen *et al.*'s (1968) tables, pointed out that simultaneous tests of only thirty χ^2 values with one degree of freedom require a test value of 9·885 for significance at the 5% level. In the absence at that time of any simple, valid test for a matrix of correlations, Williams and Lambert chose this criterion as a straightforward, though stringent one. Since increase in the number of stands drawn from the same range of variation in composition will increase the number of associations reaching the $\chi^2 = 3·84$ level, bringing further subdivision into consideration, Williams and Lambert suggested the arbitrary termination value for 'short division' of $\chi^2 = N.2^{-5}$, which experience indicated would uncover most divisions likely to be interpretable. Other workers have found that the lowest levels of usefulness are even higher e.g. Grunow (1967), $\chi^2 = 50$ ($N.2^{-5} = 35·0$); Greig-Smith *et al.* (1967), $\chi^2 = 10$ ($N.2^{-5} = 3·4$); Kershaw (1968), $\chi^2 = 40$ ($N.2^{-5} = 13·2$), but too much importance should not be attached to this variation in interpretability; ecological interpretability of a set of data certainly depends on the adequacy of information on environmental factors and may be affected by the average range of tolerance of the species concerned. There is a further difficulty in the use of the maximum individual χ^2 that because the number of species under consideration will vary in different parts of the hierarchy the use of a given value of maximum individual χ^2 will not be comparable over the whole hierarchy (Cormack, 1971). Macnaughton-Smith (1965) suggested a limit depending on the species-richness at each stage, terminating division when the largest single χ^2 is not greater than the 5% χ^2 with degrees of freedom one less than the number of species involved in determinate χ^2. Possible termination values other than the largest single χ^2 include the maximum species $\Sigma\chi^2$ and the total $\Sigma\chi^2$. Noy-Meir *et al.* (1970) considered these, but rejected both on the ground that they aggravate the lack of balance in efficiency between species-rich and species-poor parts of the data characteristic of association analysis.

Lambert and Williams (1966) discussed stopping rules more generally. They drew an important distinction between the extent to which the

* This point has been overlooked by some users of association-analysis, although it was clearly made by Williams and Lambert (1959) (see also Macnaughton-Smith, 1965).

heterogeneity of a group can be reduced by division, and the degree of heterogeneity actually exhibited by the group. A 'Type 1' rule disregards intrinsic heterogeneity and uses as a criterion the degree of reduction possible by efficient subdivision; a reduction of less than a specified amount is regarded as unprofitable. A 'Type 2' rule, on the other hand, provides an upper limit of heterogeneity within a group which the user is prepared to tolerate; if this is exceeded the group is subdivided as efficiently as possible without regard to the extent of the consequent reduction in heterogeneity.

Lambert and Williams further distinguished between non-probabilistic (in the sense that normal significance tests cannot be applied) and probabilistic rules. When the data are regarded as finite and interest centres exclusively on interpretation of these data, the situation is non-probabilistic, and whichever type of rule is used, the user can only declare a level below which he is not interested. If the set of stands being analysed is regarded as representative of the area from which they are drawn, they represent a sample of a wider population, and the situation is probabilistic. 'For a Type 1 rule, the null hypothesis is that the two products of a division differ by no more than would be expected of two independent random samples from the same population; for a Type 2 rule, the null hypothesis is that the degree of heterogeneity exhibited by the sample could have arisen as a result of sampling from a random population'. It has since become clear that most suggested probabilistic rules are invalid (cf. Bottomley, 1971; Williams, 1976).

The choice between Type 1 and Type 2 rules is more important and has apparently often not been considered by users. When interest centres on correlation between classes and the environmental or other controlling factors determining their composition, a Type 1 rule would seem to be more appropriate; only if there is a marked drop in heterogeneity on division is ecological interpretation likely to be possible. If the classes themselves are of more interest as, for instance, in a management context, then a Type 2 rule is more relevant; it is important that no class could be too heterogeneous.

A cruder procedure for terminating division, which commonly works well in practice, is to declare in advance how many final groups are required (Williams, 1976). A decision is needed at each stage which group should next be divided. This may be made by either a Type 1 or a Type 2 rule, but a Type 1 rule, requiring trial division of each group, involves more computation.

In the most usual form of association analysis, using the maximum single χ^2 as the criterion for termination, the stopping rule is Type 2 and, as Lambert and Williams showed, is non-probabilistic. The common failure to realize that this is so and that, in any case, the probability to be attached to the maximum single χ^2 in a set is much greater than indicated from tables for one degree of freedom

(p. 180 above) has led to misunderstanding of association analysis. Division has commonly been carried too far down a hierarchy with the result that the lower dichotomies are, not surprisingly, ecologically uninterpretable. Although Lange *et al.* (1965) do not state the number of 'species' used in their data, their claim to have obtained 'significant' divisions in simulated random data can almost certainly be explained in this way.

Association analysis not only has the limitations of all monothetic methods (see below, p. 187) but has a further disadvantage that it involves standardiz-ation of each species to unit variance, so that rare and common species carry the same weight (Gower, 1967b). Not only may this be undesirable in relation to the objectives of a classification, but the data are restandardized at each division. Thus the course of the analysis is unduly influenced by the chance occurrence of a species rare in the particular subset of stands being considered or the chance absence of a species common in the subset.

If $\Sigma \chi^2$ is used as the division criterion, association analysis is a particular case (for species-standardized data) of division on the species that gives maximum between-group sum of squares, equivalent to maximum between-group Euclidean distance, $\sqrt{\Sigma(x_1 - x_2)^2}$ (see below, p. 194). Direct use of division based on maximum between-group sum of squares not only leaves open the option of standardization, but if species standardization is desirable, allows the same standardization to be maintained throughout the analysis.

The usual measure of heterogeneity, maximum single χ^2, is not additive over the hierarchy, i.e. the sum of the heterogeneity values for the two subgroups may be greater than that for the parent group. This prevents the use of a Type 1 stopping rule. One subgroup may even show heterogeneity greater than that of the parent group, so that change in heterogeneity is not even monotonic, complicating interpretation.

An alternative approach to monothetic divisive classification is to test at each stage the subgroups produced by division on each species in turn. That species which produces the greatest decrease in heterogeneity in the resulting subgroups is then accepted as the division criterion. When association analysis was put forward, the speed and capacity of available computers made such an approach impractical. The advantage of association analysis was that the computation of one set of association values between all pairs of species indicated the species to be used as the division criterion without any test of the results of using the remaining species. With increasing availability of more advanced computers, it became possible to make use of the computationally more demanding approaches.

The strategy of testing each species in turn is straightforward, but decision is required on the criterion of heterogeneity to be used. One possibility, referred to above, is to use the within-group sum of squares and take as the division-criterion that species which minimizes the within-group sum of squares, i.e. maximizes the between-group sum of squares. This was used by Crawford *et al.* (1970) in a comparative study of different methods of analysis, and examined more critically by Lambert and her co-workers (see Lambert *et al.*, 1973).

The use of an information statistic attracted more attention (Macnaughton-Smith, 1965; Lance & Williams, 1968a), following the earlier use with an agglomerative strategy (Williams *et al.*, 1966; Lambert & Williams, 1966). Information statistics were developed in relation to the very different field of communication theory and the term is to some extent misleading in the present context*. Here they may be regarded as measures of the unevenness of species contributions to members of a set of samples, an intuitively statisfactory criterion of heterogeneity.

Consider a population of n samples containing p species and let a_j be the number of samples containing the jth species. Two measures of the total information in the population are available.

$$I_1 = \Sigma_j \{\log n! - \log a_j! - \log(n - a_j)!\}$$
$$= p \log n! - \Sigma_j \{\log a_j! + \log(n - a_j)!\} \quad \text{(Brillouin measure)}.$$
$$I_2 = \Sigma_j \{n \log n - a_j \log a_j - (n - a_j) \log(n - a_j)\}$$
$$= pn \log n - \Sigma_j \{a_j \log a_j + (n - a_j) \log(n - a_j)\} \quad \text{(Shannon measure)}.$$

Shannon's measure may be regarded in the following way (see Williams, 1976). In a randomly chosen sample the best estimate available of the probability that it contains the jth species is a_j/n and the probability that it does not contain the jth species is $(n - a_j)/n$. The probability that a_j out of n samples contain the species and $(n - a_j)$ do not is then $\left(\dfrac{a_j}{n}\right)^{a_j} \left(\dfrac{n - a_j}{n}\right)^{n - a_j}$. This is for one species only. The probability of obtaining the observed set of frequencies of all species is

$$\Pi_j \left(\frac{a_j}{n}\right)^{a_j} \left(\frac{n - a_j}{n}\right)^{n - a_j}.$$

The reciprocal of this probability is a measure of heterogeneity and taking logarithms the information content is

$$I_2 = -\Sigma \left\{ a_j \log\left(\frac{a_j}{n}\right) + (n - a_j) \log\left(\frac{n - a_j}{n}\right) \right\}$$
$$= -\Sigma \{a_j \log a_j - a_j \log n + (n - a_j) \log(n - a_j) - (n - a_j) \log n\}$$
$$= -\Sigma \{a_j \log a_j + (n - a_j) \log(n - a_j) - n \log n\}$$
$$= \Sigma \{n \log n - a_j \log a_j - (n - a_j) \log(n - a_j)\}.$$

* For general background to information theory and statistics, see Kullback (1959).

Strictly, I_1 is appropriate to a finite population and I_2 to a sample from an infinite population (Pielou, 1977), but the difference is small except in small samples. In practice most users have accepted the Shannon measure in both divisive and agglomerative information analysis.

Both measures are obviously zero for a set of identical samples. They have a maximum value when each species is present in 50% of the samples. The maximum value is clearly dependent on n, so that, as with association analysis, if a heterogeneous group is divided into two subgroups of very uneven size, the smaller subgroup will tend to have a lower value of I than the larger subgroup, even if their heterogeneity is comparable. If a group A is divided into two subgroups B and C, the decrease in information as a result of the division is

$$\Delta I = I_A - I_B - I_C.$$

By definition the ΔI's are additive over the hierarchy. The value of ΔI provides a Type 1 stopping rule and the information content of the subgroups a Type 2 rule. Lambert & Williams (1966) pointed out that this is not the only possible Type 2 rule and considered the use of I/n (entropy) or I/pn (mean entropy per species). The latter varies between known limits and is thus independent of group size and species number, but neither measure is necessarily monotonic.

Divisive information analysis as described above, like association analysis, operates only on presence and absence data. Dale *et al.* (1971) have suggested alternative ways of extending agglomerative information analysis to quantitative data, one of which would be applicable to divisive strategy, but has apparently not been used in this way. It involves defining a_j as the amount of species j in the set of samples and n as $\Sigma_j a_j$. The information statistic is then $I = n \log n - \Sigma_j a_j \log a_j$.

Use of minimum within-group sum of squares as the criterion of division is immediately applicable to quantitative data (cf. Lambert *et al.*, 1973). Within-group sum of squares is affected by group size and species number. It would be possible to use mean sum of squares per species, which would be independent of group size and species number, (comparable to mean entropy per species in information analysis) but, unlike sum of squares, it would not be additive over the hierarchy.

An alternative to division on that species giving maximum decrease in heterogeneity in the resulting subgroups, is to divide on that species which produces minimum similarity, or maximum dissimilarity, between the subgroups (see below p. 191). Any of the very wide range of measures of similarity or dissimilarity that have been suggested could be used in this way. A non-probabilisitic stopping rule (essentially a type 1 rule) is readily provided by the user declaring the maximum similarity or minimum dissimilarity in which he is interested.

The techniques so far considered aim essentially to reduce the heterogeneity in the resulting sub-groups at each division. Crawford and Wishart (1967, 1968) introduced a monothetic divisive procedure, group analysis, with a somewhat differently expressed aim, to identify major groups of species which occur together, and the stands which contain them. They consider two characteristics of a species in a set of stands, P_x, the probability of its occurrence i.e. the proportion of stands in which it occurs, and the *mean sample density*, V_x, the average number of species present in stands containing the species. On the grounds that 'it is the species which occur frequently with high mean sample density that determine an ecological group and not those which are frequent but isolated, or infrequent yet occurring in floristically rich areas', they define

$$W_x = (P_x . V_x)/\overline{V},$$

where $\overline{V}$ is the mean sample density of all stands, as the *group element potential* (GEP) of the species, a measure of its significance as a contributor to a potential group. If the GEP of the species present in a stand, J, are summed, this value (S'_J) may be taken as a measure of the group attributes of that stand. The maximum value of S'_J, S'_{Jmax} occurs when all species are present and Crawford and Wishart define as the *set element potential* (SEP)

$$S_J = S'_J/S'_{Jmax}.$$

Each stand has a *non-set element potential*, $\overline{S}_J$, the complement of S_J, representing its negative group attributes, with

$$S_J + \overline{S}_J = 1.$$

If the set of N stands is divided on the presence or absence of a species X, present in f_x stands, the interaction between division on that species and the sums of the SEP and the non-SEP for the two groups resulting can be examined in 2×2 table

	Species X	
	−	+
ΣS	A	B
$\Sigma \overline{S}$	C	D
	$N - f_x$	f_x

Crawford and Wishart propose to measure the interaction by the sum of the squared cell deviations from expectation, and divide on the species showing the maximum interactions. Each subgroup is then reanalysed in the same way (Fig. 35). Termination of division is dependent on ΣS. The *group coefficient* $c = N^{-1} \Sigma S$, provides a measure of the 'significance' of a set of stands, division being terminated when c rises above an arbitrary level. This is essentially a non-probabilistic type 2 stopping rule. To avoid continuous division of sets that fail to reach 'significance', division is also terminated if ΣS falls below an arbitrary level (Crawford & Wishart used $\Sigma S = 10$).

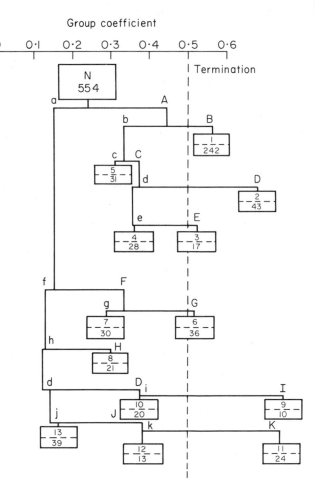

Fig. 35. Group analysis based on 554 stands and 98 species. The lower numbers enclosed in boxes denote the number of stands in final groups. A etc., presence of a species; a etc., absence of a species. (See text). (From Crawford & Wishart, 1967, by courtesy of *Journal of Ecology*.)

Group analysis is not as dissimilar from association analysis as may appear at first sight. Cormack (1971) has pointed out that the group element potential of a species represents the sum of all cross products in the original data matrix that involve the species in question (though then scaled by division by the sum of the matrix). In association analysis the $\Sigma \sqrt{(x^2/N)}$ for a species is also the sum of cross products, but after centring the individual items in the matrix and standardizing them by standard deviates (and then scaled by division by the number of stands), with the effect of removing differences in abundance of species, which are retained in group analysis.

Group analysis, unlike association analysis, distinguishes two types of terminal groups, those with high values of c, regarded as 'significant' groups and those with low values of c, which have failed to attain 'significance'. It also, from the nature of the division procedure, tends to show 'reversal' in the value of c at most divisions. This makes the hierarchy as a whole less readily interpretable. Group analysis was, however, devised with the primary object of identifying groups of species of similar ecological requirements and the sites of which they are most characteristic, preparatory to physiological investigations. For this purpose the pin-pointing of the most interesting groups is valuable and the routes by which they are derived are largely irrelevant.

Any divisive-monothetic procedure has, at least in theory, the disadvantage that a stand may be misclassified if, by chance, it lacks the particular species serving as the division criterion, although its overall composition is more like that of the group of stands containing the species or, vice versa, a stand may be misclassified by containing the species. How important the normally fairly slight degree of misclassification present in a divisive-monothetic classification is will depend on the use to be made of it; where the objective is to formulate hypotheses about the main controlling factors, some degree of misclassification may be unimportant, but in more detailed studies it may seriously hamper interpretation.

Some attempts have been made to refine the results of divisive-monothetic classifications by examining the final groups produced e.g. Greig-Smith et al. (1967) used an ordination to modify an initial association analysis. Such modification of an initial classification is, however, unsatisfactory in that it involves a different criterion from that used in the initial classification and is, in any case, computationally clumsy. Theoretically less unsatisfactory is Crawford and Wishart's (1968) testing of stands having low set element potential for the groups to which they have been assigned by group analysis for their potential for other groups, with relocation until stability is achieved. Lambert et al. (1973) suggested improving the results of a monothetic split by relocating individuals until further relocation produces no improvement on whatever criterion of division efficiency is being used.

Modification of the initial classification is to some degree polythetic and it is more logical, if monothetic classification is unsatisfactory, to consider a polythetic strategy. A set of n stands can be divided into two groups in $2^{n-1} - 1$ different ways. Edwards and Cavalli-Sforza (1965) made a practical trial, using maximum between-group sum of squares as the criterion, and concluded that examination of all possible splits was not practical for more than 16 individuals. Gower (1967b) quoted 54,000 years as the time needed for 41 individuals. Thus even with considerably increasing computing speed, unrestricted consideration of all possible splits will be unpractical for the foreseeable future and some form of 'directed search' is required (Williams & Dale, 1965).

Macnaughton-Smith et al. (1964) selected first that individual having maximum dissimilarity with the remainder. Each of the pairs consisting of this 'best' individual and one other was then tested and that pair having maximum dissimilarity with the remainder selected. The triad consisting of the 'best' pair and one other that had maximum dissimilarity with the remainder was then selected, and so on until any further transfer to the increasing group decreased between group dissimilarity. Gower (1967b) proposed aggregating onto the pair of individuals which are most dissimilar, adding individuals in turn to that group from which they are least dissimilar. (Recalculation of dissimilarity at each stage is involved, because the centroid of a group changes after addition of a further individual.) Lambert and her co-workers examined these and other procedures of selecting the starting point for assignment into two groups and concluded that examination of all possible pairs of individuals to initiate two groups (where Gower took one predetermined pair) is the most efficient, but computationally, most demanding of the procedures tested (J. M. Lambert, personal communication; Lambert, 1972).

A different approach to polythetic-divisive classification is to extract an axis of variation, ideally that which accounts for as much as possible of the total variation, and, with the stands arranged linearly along this axis, consider the ordered $n-1$ possible splits. This is equivalent to obtaining the first axis of an ordination (Chapter 8) of the data.*. Lambert and her co-workers examined a range of variants of this technique, involving different procedures of relocation of individual stands after the initial split (J. M. Lambert, personal communication). They reported on a technique, AXOR, which was particu-

* The potential use of an ordination axis in this way was indicated by Goodall (1954c), who presented a diagram showing a bimodal distribution of scores on the first axis of a principal component analysis.

larly successful, though computationally demanding (Lambert *et al.*, 1973). This involves principal component analysis of the covariance matrix of species (or the equivalent principal coordinate analysis of stands, depending on the shape of the initial data matrix). The $n - 1$ splits on the first axis are examined in turn, and the 'best' split accepted (on the criterion of minimum within-group sum of squares or minimum within-group value of Shannon's information statistic, modified if necessary for quantitative data). Improvements on the split are then made by relocation of all individuals, taken in order from the stand with the highest score on the second axis to that with the lowest, and so on with subsequent axes until no further improvement results.

Noy-Meir (1970, 1973b) considered the use of both centred and non-centred principal components, and centred and non-centred varimax components from the same data. He preferred to base division on minimum sum of within-group variances rather than within-group sum of squares on the grounds that if two clusters differ markedly in size, the dependence of sum of squares on number will lead to suboptimal division. He divided on each axis in turn and did not attempt any relocation of stands.

In 'indicator species analysis' (Hill *et al.*, 1975) the first axis produced by reciprocal averaging ordination (Hill, 1973b) is divided at the centroid. The five species whose occurrences are most nearly confined to stands on one or other side of the division are identified and the division point is adjusted so as to maximize the indicator efficiency of these indicator species in subsequent placing of additional stands into the classification. Each subgroup is then subjected to reciprocal averaging ordination as the basis of further division and the procedure repeated to give as many final groups as is appropriate.

In the wider context of numerical classification in general, agglomerative-polythetic procedures have been used much more extensively than divisive procedures, and in some fields it appears to have been accepted that only an agglomerative strategy is worth considering. Divisive procedures have, however, proved useful tools in analysing vegetation and the relative merits of the two approaches in this context require consideration.

Divisive procedures are, in principle, based on information drawn from the whole of the data, though if division is monothetic, much of the information is disregarded in making a division. Divisive-monothetic procedures have the disadvantage, referred to above (p. 187), that they are liable to misclassify some stands, but, if the classification being produced is regarded as a potentially general one into which further stands are to be placed, they provide an immediate means of doing so by dichotomous keying. Divisive-polythetic procedures carry a lesser risk of misclassification; they do not necessarily preclude the derivation of 'key' species for placing further stands (R. R. Sokal quoted in Greig-Smith, 1971a), though this is only explicitly allowed for in indicator species analysis. Though not relevant to the assessment of the strengths and defects of divisive-monothetic procedures, the parallelism with concepts of the indicator value of species probably contributed in practice to their readier acceptance in plant ecology than in other fields.

Though not in principle proof against misclassification as a result of chance presence or absence of species, agglomerative-polythetic procedures, involving fusion between stands or groups on the basis of overall similarity, are in general less likely to lead to such misclassification. Against this advantage must be set the disadvantage that, in most agglomerative procedures, a fusion is based only on the information carried by the stands or groups being fused. Vegetational data matrices commonly have a large proportion of empty cells, so that a large proportion of the data on a particular stand is negative, i.e. the absence of species. Absence of a species may be informative, indicating that the characteristics of that stand are outside the range of tolerance of the species, or uninformative, resulting from chance absence from a stand providing a potential habitat for it. Thus, in comparison of stands or, to a lesser extent as the groups formed become larger, of groups, fusion is not only based on a small proportion of the total data, but the data used contain an unknown and possibly high proportion of irrelevant information*. The position contrasts with that in the analysis of taxonomic data where attributes are often multistate rather than positive or negative, e.g. leaves simple or compound, and even when negative, e.g. absence of stipules, are rarely the result of chance. Conclusions about the efficacy of agglomerative procedures based on experience with taxonomic data should be accepted uncritically for vegetational data.

Agglomerative procedures necessarily involve construction of the complete hierarchy, although only the upper part will normally be of interest; with large bodies of data much of the total computing time will be used on deriving groupings and linkages of no interest in themselves. Divisive procedures, by contrast, can be stopped as soon as the required level is reached and only groupings and linkages of interest need be derived.

Agglomerative procedures involve the successive fusion of those separate individuals or groups which are more alike on some criterion at each stage.

---

* Species with a large chance element in their occurrence are unlikely to appear as division criteria in divisive-monothetic procedures, because they will tend not to have strong correlation with other species.

Two decisions are necessary in selecting the procedure to be used, the criterion of similarity and the strategy of fusion.

Coefficients of similarity have been used in isolation for individual comparisons between communities at least since Jaccard introduced his simple coefficient of community in 1902 (p. 151), but relatively little consideration was given to their properties as long as they were confined to this function. When the set of similarity (or dissimilarity*) coefficients between members of a set of stands is used as the basis of deriving a classification, or an ordination, some coefficients are clearly likely to be more suitable than others and some may be quite inappropriate. Further, choice of coefficient for classification cannot be considered in isolation from choice of strategy of fusion.

Williams and Dale (1965) urged as a minimum requirement that the distance measure used should be a metric (see also Orloci, 1972a). A metric has the following properties.

(1) It cannot be negative.

(2) It is symmetric i.e. the similarity of A to B is the same as that of B to A (or the distance in the space defined by the measure is the same whether measured from A to B or B to A).

(3) It possesses the triangle inequality, i.e. for any three points in the space any one side of the triangle they form must not be greater than the sum of the other two sides.

These properties are intuitively desirable. In relation to a spatial model of stand relationships, which dissimilarity measures imply, a negative distance between a pair of stands has no straightforward meaning. Although the possible value of asymmetric coefficients has been suggested by Goodall (1953b) in relation to concepts of fidelity (cf. Williams & Dale, 1965), in most

---

* Measures of similarity with a maximum possible value can be converted to a measure of dissimilarity by subtraction from the maximum. Many similarity measures, e.g. correlation coefficient, are of the Euclidean or cross-product form

$$s_{ij} = \sum_k y_{ik} y_{jk}$$

where y_{ik}, y_{jk} are the amounts of species k in stands i and j, with or without initial transformation. For such measures the corresponding dissimilarity, or distance, is

$$d_{ij} = \sqrt{(s_{ii} + s_{jj} - 2s_{ij})} \text{ (Gower, 1966).}$$

For non-Euclidean measures of the form

$$s_{ij} = 2\sum_k \min(y_{ik}, y_{jk}),$$

e.g. matching coefficient, an analogous relation applies $d_{ij} = s_{ii} + s_{jj} - 2s_{ij}$ (Noy-Meir & Whittaker, 1977).

contexts symmetry is obviously desirable. If the triangle inequality is not satisfied, two stands compared directly may appear more dissimilar from one another than if the comparison is made in two stages via a third stand. This produces an uncertainty about the apparent relationships between members of a set of stands because their relationships may be altered by the addition of further stands. Some widely used coefficients fail on this requirement, e.g. Sørensen's coefficient, the one-complement of which, giving interstand dissimilarity, is for presence/absence data in the usual symbolism for a 2×2 contingency table (p. 107) $(b + c)/(2a + b + c)$. Consider the following case:

	Stand		
	1	2	3
Species X	+	+	−
Y	+	−	−
Z	−	+	+

The interstand distances are

Stands	1,2	$\frac{1}{2}$
	1,3	1
	2,3	$\frac{1}{3}$

and $1 > \frac{1}{2} + 1\frac{1}{3}$.

A large number of coefficients have been suggested and useful comparative accounts have been given by Dagnelie (1960), Goodall (1973) and Sneath and Sokal (1973) (see also Cormack, 1971), but relatively few have been much used or received serious consideration. Table 14 lists some examples.

Lance and Williams (1967a) distinguished three types of measures of groups and the similarities between them. Suppose two groups (i) and (j) fuse to form a group (k). Then (i) measures define a property of a group and are directly usable only in non-hierarchical cluster-analysis; (ij) measures define a resemblance between or difference between two groups; (ij, k) measures define a difference between the original two groups, considered jointly, and that formed by their fusion. Information gain is clearly an (ij, k) measure, since it defines the difference between the information content of the group (k) and sum of the information contents of groups (i) and (j) (p. 198). The measures in Table 14 are clearly (ij) measures. The advantage of an (ij, k) measure is that its calculation necessarily involves the calculation also of an (i) measure for each group formed, and thus a directly related measure of the hierarchical level of each group. With most (ij) measures, which are not additive over the hierarchy, no such direct measure of hierarchical level is available and it has been conventional, as Lance and Williams pointed out, to use the (ij) measure itself. While this has in practice proved reasonably satisfactory, it is not strictly justifiable.

The between-group sum of squares ('incremental sum of squares') is an (ij) measure of dissimilarity, but with centroid sorting (see below), the only fusion strategy with which it is meaningful, it is additive over the whole hierarchy, providing an appropriate (i) measure for each group, the corresponding within-group sum of squares.

Measures differ in whether they include or exclude double matches (the common presence or absence of species) and in the effect on them of the species-richness of the stands being compared. The question of whether double zero matches (the common absence of species) should be included is well recognized, but cannot be answered without reference to the type of data being analysed. When the stands include clearly distinct vegetation types, the common absence of species is likely to reflect such major distinctions; thus, if interest centres on major divisions among a wide range of stands, inclusion of double zero matches is likely to be useful. On the other hand, between closely similar stands, particularly if stands are relatively small, absence of particular species is more likely to be a chance effect and common absence of a species may be misleading, resulting from chance absence from both the stands or, more seriously, from it being absent from one stand by chance and from the other because it is incapable of growing there. Thus, when interest centres on the detailed classification of broadly similar stands, double zero matches are probably better excluded from the measure.

That the inclusion of double positive matches also raises problems is rarely explicitly acknowledged. Direct measures of stand dissimilarity (as opposed to those obtained as the one-complement of similarity) are based on differences between the stands, ignoring common presence. Consider comparisons 1, 2 and 3 in Table 15. Comparison 2 differs from 1 in the addition of six further species present in both stands; comparison 3 differs from 1 in the addition of six species absent in both stands. The directly dissimilarity measures give the same value for all three comparisons; the correlation coefficient and the matching coefficient give the same values for 2 and 3, indicating the stands of 2 and 3 as being more alike than those of 1; the Jaccard and Sørensen coefficients do not distinguish between 1 and 3, but indicate 2 as being more alike. In terms of interpretation most ecologists would probably regard 2 as being more alike, but whether they would distinguish between 1 and 3 would depend on the context, as indicated above. The correlation coefficient, unlike the matching

Table 14. Measures of similarity and dissimilarity (distance) between stands

			Metric
Derivatives of Manhattan metric			
Manhattan metric	D	$\Sigma\lvert x_{1p}-x_{2p}\rvert$	+
Czekanowski coefficient (Bray & Curtis, 1957)	S	$\dfrac{2\Sigma\min(x_{1p},x_{2p})}{\Sigma(x_{1p}+x_{2p})}$	
	D	$\dfrac{\Sigma\lvert x_{1p}-x_{2p}\rvert}{\Sigma(x_{1p}+x_{2p})}$	−
		———	−
Canberra metric (Lance & Williams, 1967b)	D	$\displaystyle\sum\frac{\lvert x_{1p}-x_{2p}\rvert}{(x_{1p}+x_{2p})}$	+
Gower metric (Gower, 1971)	D	$\displaystyle\sum\frac{\lvert x_{1p}-x_{2p}\rvert}{\text{range }p}$	+
Standardized by stand total (Orloci, 1974)	D	$\displaystyle\sum\left\lvert\frac{x_{1p}}{\Sigma x_{1p}}-\frac{x_{2p}}{\Sigma x_{2p}}\right\rvert$	+
Derivatives of Euclidean distance			
Euclidean distance	D	$\sqrt{\Sigma(x_{1p}-x_{2p})^2}$	+
Sum of squares	D	$\frac{1}{2}\Sigma(x_{1p}-x_{2p})^2$	−
Standardized by species standard deviate	D	$\displaystyle\sqrt{\sum_{p}\frac{(x_{1p}-x_{2p})^2}{\Sigma_q(x_{qp}-\bar{x})^2}}$	+
Standardized by species range	D	$\displaystyle\sqrt{\sum\left(\frac{x_{1p}-x_{2p}}{\text{range }p}\right)^2}$	+
Standardized by stand vector (Standard distance of Orloci, 1967a)	D	$\displaystyle\sqrt{\sum\left(\frac{x_{1p}}{\sqrt{\Sigma x^2_{1p}}}-\frac{x_{2p}}{\sqrt{\Sigma x^2_{2p}}}\right)^2}$	+
Correlation coefficient	S	$r=\dfrac{\Sigma(x_{1p}-\bar{x}_1)(x_{2p}-\bar{x}_2)}{\sqrt{\Sigma(x_{1p}-\bar{x}_1)^2\,\Sigma(x_{2p}-\bar{x}_2)^2}}$	
	D	$=(1-r)$	−
Matching coefficient			
		———	+

x_{qp}, amount of species p in stand q; S, similarity measures; D, dissimilarity measures.

Sensitive to equal quantities of a species	Sensitive to common absence of a species	Equivalent for presence/absence
−	−	$b+c$
+	−	$\dfrac{2a}{2a+b+c}$ (Sørensen coefficient) $\dfrac{b+c}{2a+b+c}$
+	−	S $\dfrac{a}{a+b+c}$ (Jaccard coefficient) D $\dfrac{b+c}{a+b+c}$
−	−	$b+c$
−	−	$b+c$
(+)	−	$\dfrac{b}{a+b}+\dfrac{c}{a+c}$
−	−	$\sqrt{(b+c)}$
−	−	$\tfrac{1}{2}(b+c)$
−	−	−
−	−	$\sqrt{(b+c)}$
(+)	−	$\sqrt{\left\{2\left(1-\dfrac{a}{\sqrt{(a+b)(a+c)}}\right)\right\}}$
		$r=\dfrac{(ad-bc)}{\sqrt{(a+b)(c+d)(a+c)(b+d)}}$ $(1-r)$
+	+	S $\dfrac{(a+d)}{(a+b+c+d)}$
+	+	D $\dfrac{(b+c)}{(a+b+c+d)}$

coefficient, is asymmetric in relation to double positive and double zero entries, being more sensitive to double positives in comparisons between species-poor stands and to double zeros in species-rich stands (comparison 4–6). Only when both stands have 50% of the total number of species, as in comparison 1, is the correlation coefficient symmetric.

A further problem in choice of measure of similarity or dissimilarity arises if the stands vary widely in species richness. Field (1969) pointed out that the information statistic tends to give greater similarity for comparison of species-poor stands than for comparison of species-rich stands. This is true of all direct measures of dissimilarity. Comparisons 7 and 8 in Table 15 have the same proportions of species present or absent in both stands and species occurring in only one stand, but the stands of 8 have twice as many species involved as those of 7. Between-group sum of squares, Euclidean distance and information gain all show a greater interstand distance for 8, but the remaining measures do not distinguish between 7 and 8. Whether it is desirable for the measure used to be sensitive to such differences in species-richness will depend on the nature of the data and the objectives of the analysis. Generally, if diverse habitats are included, sensitivity to species-richness is liable to result in grouping of stands at a relatively low level only because they are species-poor (and may have no species in common), and a measure not sensitive to differences in species-richness may be preferred. With less diverse sets of data, or if interest centres on the detailed relationships within a diverse set of data, differences in species-richness, especially the occurrence of a small number of species in a stand, may be ecologically meaningful, and a measure sensitive to such differences may be desirable.

Similarity and dissimilarity measures have so far been discussed in terms of presence and absence of species in stands. If stands are described in terms of quantities of each species present, a variety of measures are again available (Table 14). A number of direct measures of dissimilarity have been derived from the Manhattan metric (absolute distance), $\Sigma |x_1 - x_2|$, and from Euclidean distance, $\sqrt{\Sigma (x_1 - x_2)^2}$ (Table 14). Both these measures are very sensitive to relative abundance between stands and may result in stands with no species in common appearing more alike than stands with the same species present in both, e.g.

	Stand		
	1	2	3
Species X	–	1	–
Y	3	–	1
Z	5	–	2

Table 15. Interstand distance for pairs of stands by different measures (Presence/absence data)

	1	2	3	4	5	6	7	8
	+ −	+ +	− −	+ +	+ +	+ +	+ +	+ +
	+ −	+ +	− −	+ +	+ +	+ +	+ +	+ +
	+ −	+ +	− −	+ −	+ +	+ −	+ −	+ +
	− +	+ +	− −	− −	+ −	− −	− −	+ −
	+ +	+ +	− −	− −	− −	− −	− −	+ −
	+ +	+ +	− −	+ −	+ −	+ −	+ −	+ −
		− −	+ +	− +	− +	− +	− +	− +
		− −	+ +	− −	− −	+ −	+ −	− −
		− +	+ +	− −	− −	− −	− −	− −
		+ +	− −	− −	− −	− −	− −	− +
			− −	− −	− −	− −		+ −
			− −	− −	− −	− −		+ +
								+ +
								− +
								+ −
								− −
								− −
Correlation coefficient $(1-r)$	2	1·33	1·33	0·85	0·75	0·83	1·26	1·26
Sum of squares	3	3	3	3	3	3	2·5	5
Euclidean distance	2·45	2·45	2·45	2·45	2·45	2·45	2·24	3·16
Information gain	12 log 2	12 log 2	12 log 2	12 log 2	12 log 2	12 log 2	10 log 2	20 log 2
Matching coefficient	1	0·5	0·5	0·37	0·35	0·35	0·62	0·62
Jaccard coefficient	1	0·5	1	0·75	0·67	0·75	0·71	0·71
Sørensen coefficient	1	0·33	1	0·6	0·5	0·6	0·56	0·56

		Manhattan metric	Euclidean distance
Stands	1,2	9	5·92
	1,3	5	3·61
	2,3	4	2·45

Although it may, or may not, be desirable that differences in standing crop should influence an analysis, it is clearly unsatisfactory that such differences should be able to override floristic similarity completely. Euclidean distance is also very sensitive to extreme values. Some standardization is therefore needed; choice of standardization is discussed below (p. 211).

The desirability of including double zero and double positive matches in a measure of similarity or dissimilarity has been discussed in relation to presence/absence data. The occurrence of equal quantities of a species in the two stands being compared is the equivalent for quantitative data of a double positive match. Of the direct similarity measures discussed, only the Bray and Curtis measure and those involving stand standardization are sensitive to equal quantities of the same species.

After the initial stages of fusion, dissimilarity measures between groups of stands, or between a stand and a group, are required. With some fusion strategies the dissimilarity between groups is defined in terms of dissimilarity between individual stands, e.g. nearest-neighbour sorting; with others a group is regarded as being replaced by a single 'average stand', e.g. centroid sorting (see below). In the latter case presence/absence data for single stands necessarily produce quantitative data for groups and the appropriate quantitative measure is used. (With two measures, between-group sum of squares and information gain, an average stand is not explicitly defined, although the sorting is effectively centroid.)

With both presence/absence and quantitative data, the sum of squares between individual stands is half the corresponding squared Euclidean distance, but there is no fixed relation between the sum of squares between two groups and the Euclidean distance between their centroids and between-group sum of squares must be considered a distinct measure ('optimal agglomeration' of Orloci, 1967a).

Information gain between single stands is a multiple of squared Euclidean distance and its use with strategies depending throughout on comparison of individual stands is pointless. Between-group information gain is obtained in the same way as information fall in divisive analysis (p. 184), as the difference between the information content of the combined group and the sum of the information contents of the two subgroups. Dale (1971; see also Dale *et al.*

1971; Dale & Anderson, 1972; Orloci, 1978) discussed the extension of information gain to quantitative data. For data of frequency form, a stand can be regarded as a pre-grouping of the sub-units used in determining frequency. In contrast to presence/absence data, where the information content of a single stand is necessarily zero, for frequency data, a single stand has a defined information content. For example, a stand sampled by 25 quadrats with three species having frequencies of 80%, 40% and 8% respectively will have an information content calculated as follows.

No. of sub-units		
Species present	Species absent	Contribution to information content
20	5	$25 \log 25 - 20 \log 20 - 5 \log 5 = 5\cdot43$
10	15	$25 \log 25 - 10 \log 10 - 15 \log 15 = 7\cdot31$
2	23	$25 \log 25 - 2 \log 2 - 23 \log 23 = 3\cdot03$
		$15\cdot77$

In this approach each species contributes a distinctive element to the total information for the stand. With density data a different approach is necessary. The stand may be regarded as made of as many sub-units as there are individuals, with each individual being assigned to one of as many 'states' as there are species, i.e. instead of there being only two possible 'states' for an observation (present or absent) there are several possible 'states'. The information content of the stand is then

$$n \log n - \sum a_j \log a_j,$$

where a_j is the number of individuals of the jth species and $n = \Sigma\, a_j$. Here the species do not make independent contributions to the total information content, which depends not only on the total numbers of each species but also on the relative numbers of different species. This form can be extended to continuous variables such as yield measures, with a_j as the measured amount of the jth species (The 'individuals' under consideration then represent the size of the smallest amount discriminated by the method of measurement.)

Information measures are sensitive to group size. Dale and Anderson (1972) suggested dividing the information measure by the number of individuals involved, to put it on a 'per individual in group' basis. The increase in information on fusing groups A and B (containing a and b stands respectively) into a single group $(A + B)$ then becomes

$$I = \frac{2I_{(A+B)}}{(a+b)} - \frac{I_A}{a} - \frac{I_A}{b}$$

This is a convenient approximate correction, but they point out that a strictly size-free measure would be based on probabilities rather than observed values. The measure

$$\Sigma\,(n\log n - a_j\log a_j + (n - a_j)\log(n - a_j)$$

is then replaced by

$$-\Sigma\left(\frac{a_j}{n}\log\frac{a_j}{n} + \left(\frac{n-a_j}{n}\right)\log\left(\frac{n-a_j}{n}\right)\right)$$

and $\quad n\log n - \Sigma\, a_j\log a_j$

by $\quad -\Sigma\dfrac{a_j}{n}\log\dfrac{a_j}{n}.$

Table 16 shows the values of a selection of measures for comparisons based on quantitative data. Comparison 2 differs from comparison 1 by the addition

Table 16. Interstand distance for pairs of stands by different measures (quantitative data)

	1	2	3	4
	-3	-3	-3	$--$
	1 2	1 2	1 2	1 2
	2 1	2 1	2 1	2 4
	2 1	2 1	2 1	2 4
	3 $-$	3 4	3 4	3 6
	4 $-$	4 $-$	4 $-$	4 8
	-1	-1	-1	
	$--$	$--$	3 3	
	$--$	$--$	2 2	
Manhattan metric	14	14	14	12
Bray and Curtis measure	0·70	0·70	0·47	0·33
Canberra metric	5	5	5	1·67
Manhattan metric standardized by stand total	1·33	1·33	0·99	0
Euclidean distance	6·16	6·16	6·16	5·83
Sum of squares	19	19	19	17
Euclidean distance standardized by stand vector	1·22	1·22	1·00	0
Correlation coefficient $(1 - r)$	1·80	1·37	1·45	0
Information gain*	3·36	3·36	3·50	0

* $I = n\log n - \Sigma_j a_j\log a_j$, where $n = \Sigma_j a_j$ (Dale *et al.*, 1971).

of two double negative matches, and comparison 3 by the addition of two species with the same amount in each stand (comparable to a double positive match with qualitative data). Comparison 4 is of two stands containing the same species in the same proportions, but differing in total amount.

Understanding of the different strategies of fusion owes much to Lance and Williams (1966b, 1967a) on whose discussion the following account is based (see also Cormack, 1971).

Starting with the matrix of interstand distances, the first operation is clearly to fuse the two nearest (most similar) stands, or if there are several pairs of stands having the same smallest interstand distance, to fuse the members of each pair. Two stands thus fused initially form a group and it is necessary to define the distance of other stands from this group (and, more generally, the distance between two groups). Lance and Williams considered five possible definitions (Fig. 36).

(a) *Nearest-neighbours.* The distance between two groups is defined as the shortest distance between a pair of stands, one in each group.

(b) *Furthest-neighbour.* The distance between two groups is defined as the greatest distance between a pair of stands, one in each group.

(c) *Centroid.* A group is replaced on formation by the co-ordinates of its centroid i.e. by the same number of stands each having the average composition of the group. The distance between two groups is then the distance between their centroids.

(d) *Median.* A disadvantage of the centroid strategy is that if the two groups fused are very disparate in size, the centroid of the new group will be very close to that of the larger of the two and the characteristic properties of the smaller group will be virtually lost. To overcome this the new group may be placed midway between the positions of the two groups forming it; a group is then replaced by a single 'stand' at the appropriate position and every fusion is, in effect, a fusion between two individual 'stands'.

(e) *Group-average.* The distance between two groups is defined as the average distance between all possible pairs of stands, one from each group. This has a similar disadvantage to that of centroid when groups of markedly dissimilar size are fused. A modification, *weighted average* (Cormack, 1971) ('unweighted group average' of Pritchard & Anderson, 1971), is available in which the group average distances of a third group to each of the two fusing groups are weighted equally. The relationship of weighted average to group average is comparable to that of median to centroid.

Orloci (1967a) suggested fusing at each stage those two groups the fusion of which results in minimum increase in the total within-group sum of squares, a strategy which he termed *optimal agglomeration* (*minimum variance* strategy of

Cormack, 1971). This is equivalent to maximizing the remaining between-groups sum of squares and is the agglomerative analogue of divisive procedures based on sum of squares. Orloci (1972b) later suggested that at each stage fusion should take place wherever there is a local minimum distance (a value minimal for both the row and column in which it occurs) and claimed that this resulted in a more balanced hierarchy.

(a)
Nearest neighbour

Furthest neighbour

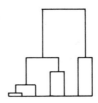

Centroid

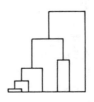

Median

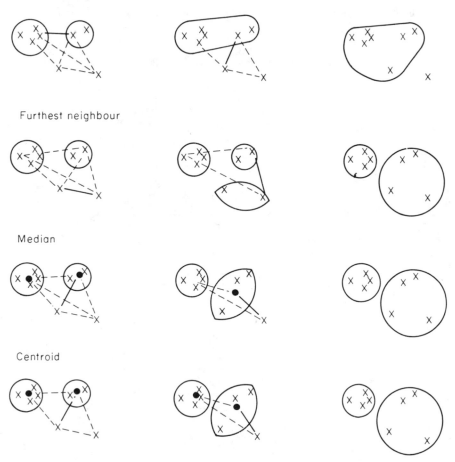

Fig. 36. Fusion strategies. (a) Linear arrangement of stands, showing successive fusions and hierarchy for different strategies. (b) Two-dimensional arrangement of stands, showing later fusions for different strategies. (---) Distances to be considered; (——) distance determining fusion.

Lance and Williams (1966b, 1967a) considered three important properties of these fusion strategies. In a *compatible* strategy measures calculated later in the analysis are of exactly the same kind as the initial interstand measure. In an *incompatible* strategy this is not so and the resultant difficulties of interpretation make incompatible strategies undesirable.

Consider two groups (i) and (j) of n_i and n_j stands respectively and with intergroup distance d_{ij} the smallest distance remaining to be considered, so that they will be fused to form a new group (k) of $n_k = n_i + n_j$ stands. Consider a third group (h). If the distance, d_{kh} of (h) from the new group (k) can be calculated from d_{hi}, d_{hj}, d_{ij}, n_i, and n_j, the strategy is said to be *combinatorial*. Since the original data need not be stored after calculation of the initial interstand distances, combinatorial strategies have evident computational advantages. Lance and Williams postulated the relationship

$$d_{hk} = \alpha_i d_{hi} + \alpha_j d_{hj} + \beta d_{ij} + \gamma |d_{hi} - d_{hj}|.$$

The combinatorial properties of a strategy depend on the measures used (Table 17). Table 18 shows the values of the coefficients for different strategies*. It is a serious disadvantage in interpretation of the hierarchy if the interstand and intergroup distances associated with successive fusions are not monotonic. Nearest-neighbour and furthest-neighbour strategies are monotonic by de-

Table 17. Properties of different agglomerative strategies with different measures

	Nearest-neighbour	Furthest-neighbour	Centroid
Between-group sum of squares	($\equiv D^2$)	($\equiv D^2$)	($-$)
Squared Euclidean distance (D^2)			Compatible Combinatorial Space-conserving Not monotonic
Correlation co-efficient	Compatible Combinatorial Space-contracting Monotonic	Compatible Combinatorial Space-dilating Monotonic	Compatible Combinatorial† Space-conserving Not monotonic
Sørensen co-efficient			Compatible Non-combinatorial Space-conserving Not monotonic
Information statistic	($\equiv D^2$)	($\equiv D^2$)	Compatible Non-combinatorial (but derived from definition of groups as they are formed) Behaves as space-dilating Monotonic

† Indirectly, by storing covariances and variances, and calculating correlation coefficient when required.

* Podani (1979b) gives a similar combinatorial formula applicable to some within-group homogeneity measures rather than to between-group distance measures.

finition. Lance and Williams pointed out that strategies for which $\gamma = 0$ are monotonic provided $\alpha_i + \alpha_j + \beta \geqslant 1$.

Lastly, Lance & Williams considered whether a strategy is *space-conserving* or *space-distorting*. The initial interstand distances may be regarded as defining a space with known properties, but as groups form, the intergroup distances do not necessarily define a space with the original properties. If they do so, the strategy is space-conserving. Some strategies result in apparent distortion of the space as grouping proceeds, e.g. with nearest-neighbour sorting, as a group forms it appears to become nearer to some or all of the remaining stands. The strategy is *space-contracting* and there is a tendency for individual stands to

Median	Group-average	Minimum variance (optimal agglomeration)	Flexible
(−)	(≡ D^2)	Compatible Combinatorial Space-conserving Monotonic‡	(−)
Compatible Combinatorial Space-conserving Not monotonic		(−)	Compatible Combinatorial Variably space-distorting Monotonic
(Incompatible)	Compatible Combinatorial ± Space-conserving Monotonic	(−)	(Incompatible)
Compatible Combinatorial Space-conserving Not monotonic		(−)	Effectively compatible Combinatorial Variably space-distorting Monotonic
(Incompatible)	(≡ D^2)	(−)	(Incompatible)

‡ If variance, rather than within-group sum of squares, is used as the measure of heterogeneity, reversals may appear in the hierarchy (e.g. Orloci, 1967a; Austin & Greig-Smith, 1968).

Table 18. Coefficients for different combinatorial strategies (see text)

	α_i	α_j	β	γ
Nearest-neighbour	$\frac{1}{2}$	$\frac{1}{2}$	0	$-\frac{1}{2}$
Furthest-neighbour	$\frac{1}{2}$	$\frac{1}{2}$	0	$\frac{1}{2}$
Centroid	$\dfrac{n_i}{n_k}$	$\dfrac{n_j}{n_k}$	$-\alpha_i\alpha_j$	0
Median	$\frac{1}{2}$	$\frac{1}{2}$	$-\frac{1}{4}$	0
Group-average	$\dfrac{n_i}{n_k}$	$\dfrac{n_j}{n_k}$	0	0
Weighted average	$\frac{1}{2}$	$\frac{1}{2}$	0	0
Minimum variance (optimal agglomeration)	$\dfrac{n_i+n_h}{n_k+n_h}$	$\dfrac{n_j+n_h}{n_k+n_h}$	$-\dfrac{n_h}{n_k+n_h}$	0
Flexible sorting	$\frac{1}{2}(1-\beta)$	$\frac{1}{2}(1-\beta)$	$\beta(<1)$	

join an existing group rather than act as the nucleus of a new group, resulting in a 'chained' hierarchy of much less utility in ecological interpretation than a more evenly divided one (Fig. 36). By contrast, with furthest-neighbour sorting a group as it forms appears to move further away from some or all of the remaining stands. The strategy is *space-dilating*, with a tendency for remaining stands to form the nucleus of new groups rather than join existing ones, giving a more readily interpretable hierarchy (Fig. 36).

The remaining strategies considered above are more or less space-conserving. Furthest-neighbour sorting is very strongly space-dilating. The potential value of a varying degree of space distortion, particularly of a strategy less strongly space-dilating than furthest-neighbour sorting, led Lance and Williams to propose a *flexible* strategy with the constraints $\alpha_i + \alpha_j + \beta = 1$, $\alpha_i = \alpha_j$, $\beta < 1$, $\gamma = 0$. This is combinatorial by definition and compatible for Euclidean distance (and effectively so for Sørensen's coefficient) but not for the correlation coefficient. Its space-distorting properties vary with the value of β (Fig. 37). As β approaches unity the system becomes increasingly space-contracting and, apart from initial ambiguities (where two or more pairs of stands have the same, smallest interstand distance) can be made to chain completely by taking β sufficiently close to unity (cf. $\beta = +0.98$ in Fig. 37). As β falls to zero and then becomes negative, the system ceases to be space-contracting and becomes increasingly intensely-grouped*.

* Weighted average sorting, mentioned above, is a particular case of flexible sorting, for which $\beta = 0$.

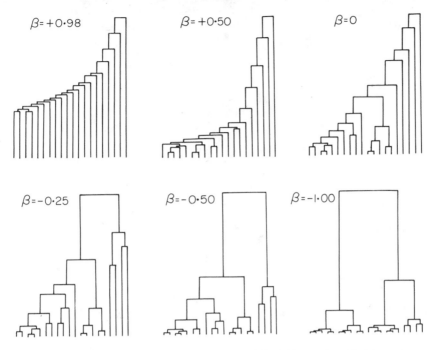

Fig. 37. Flexible sorting, with various values of β, of 20 stands with 76 species (presence/absence data). (From Lance & Williams, 1967a, by courtesy of *Computer Journal*.)

The use of information statistics has previously been discussed in relation to divisive classification (p. 183). It was, however, first introduced for agglomerative classification using the Shannon measure on presence/absence data (Williams *et al.*, 1966; Lambert & Williams, 1966). Unlike the other measures of interstand distance considered, which are all (ij) measures, the information statistic is an (ij, k) measure derived from the information contents before and after fusion as

$$\Delta I_{(ij, k)} = I_k - I_i - I_j.$$

Lance and Williams pointed out that for fusion between individual stands, ΔI reduces to $2(b + c) \log 2$, i.e. into a multiple of squared Euclidean distance, and is indistinguishable from an (ij) measure. Thus for strategies depending at all levels on interstand measures (nearest-neighbour, furthest neighbour, group-average) the use of the information statistic is pointless, giving the same answer as squared Euclidean distance by unnecessarily cumbersome computation (Fig. 38). It is incompatible for median and flexible sorting because stand/group and group/group values would remain (ij) measures and no

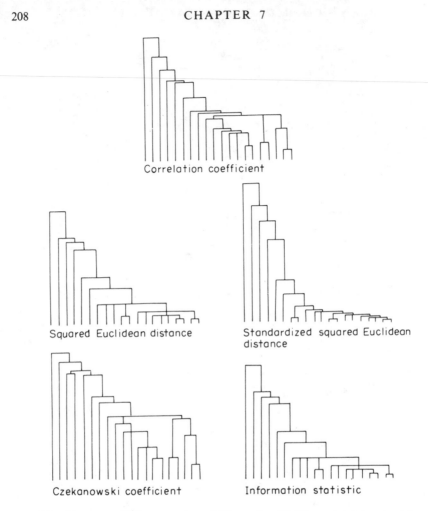

Fig. 38. Nearest-neighbour sorting of 20 stands with 76 species (presence/absence data) with different similarity measures. The hierarchies from squared Euclidean distance and the information statistic are identical. (From Williams *et al.*, 1966, by courtesy of *Journal of Ecology*.)

longer represent information gain. The strategy used is a centroid one, though the centroid is not derived, and the definition of the group under study is used in calculating ΔI. Though in theory reversals in level are possible with the information statistic (Burr, 1970), in practice the strategy is normally monotonic. Though originally introduced in relation to presence/absence data, information analysis can be extended to quantitative data in the manner discussed above. (For an example of practical use of information analysis with quantitative data, see Lloyd, 1972.)

There is thus a wide variety of possible procedures for agglomerative classification available. Although the measure of similarity and the sorting strategy to be used are not independent, in that the measure may restrict possible sorting strategies, the considerations in choosing measure and sorting strategy are fairly distinctive. Choice of measure is dictated by those aspects of the data to which most importance is attached in the context of an investigation.

Choice of sorting strategy is influenced firstly by two practical considerations. For all but very small sets of data there is an overwhelming computational advantage in a combinatorial strategy or one (centroid sorting of information statistics) which requires only the composition of groups as they are formed to be retained. Secondly, reversals in the hierarchy so complicate ecological interpretations that a non-monotonic strategy will normally be excluded from consideration. Once these conditions are satisfied the choice of strategy is more open, and must be made in the light of the kind of variation in composition present and the purpose of the analysis. If the data include quite distinct vegetation types with no intermediates (such as might happen in a broad survey covering a wide geographical range and distinct soil-types), it is important that the classification should recognize these distinctive types as major groups. Under these conditions most strategies will do so and the choice made will be relatively unimportant. It is, however, at least arguable that if there are quite distinct, well isolated, vegetation types, numerical analysis will not be necessary to recognize them.

More usually the data being analysed will show to a greater or lesser extent a continuum of variation and the objective is *dissection* (Kendall, 1966) rather than classification in the strict sense; boundaries between groups will necessarily be somewhat nebulous and the boundary position that is most efficient (in the sense that groups have minimum internal heterogeneity) will depend on the composition of the particular stands that have been included in the sample represented by the data. The purpose of analysis of such data may be to delimit vegetation types, e.g. as a basis for a vegetation map or catalogue of resources, or as a tool in management planning, or as a means of generating hypotheses about the factors determining the composition of the vegetation.

A space-contracting strategy, as seen above, tends to obscure boundaries between types and to give a more or less strongly chained hierarchy (Fig. 38). As Lance and Williams (1967a) suggest, no user is likely to require a classification in which the classificatory boundaries are weakened, and there is therefore no place for space-contracting strategies. Space-dilating strategies sharpen apparent boundaries by giving more intense clustering in the hierarchy, but do so at the cost of a tendency to form '"non-conformist groups", the members of which have little in common beyond the fact that they are rather unlike everything else, including each other' (Lance & Williams, 1967a)*. If interest centres on the classes themselves the formation of non-conformist groups is clearly undesirable, and eventual attachment of isolated stands to the group which they most resemble is preferable. This indicates the use of a space-conserving strategy. If interest centres on hypothesis generation then the advantages of the more clear-cut clustering of space-dilating strategies outweight the 'misclassification' inherent in the formation of non-conformist groups (cf. the space-dilating information statistic in Fig. 39 with the

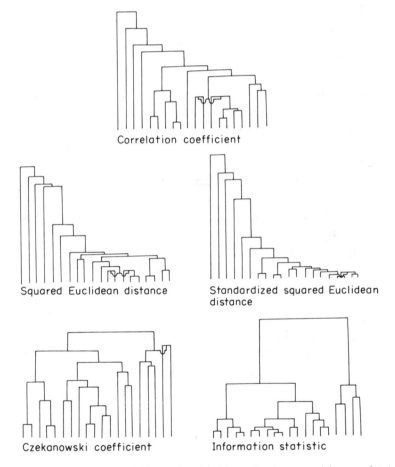

Fig. 39. Centroid sorting of 20 stands with 76 species (presence/absence data) with different similarity measures. (From Williams *et al.*, 1966, by courtesy of *Journal of Ecology*.)

  * See Williams, Clifford and Lance (1971) for discussion of the tendency to form non-conformist groups.

remaining, space-conserving, hierarchies). Flexible sorting has the advantage of allowing the degree of space-dilation, and hence the intensity of clustering, to be manipulated. In practice, flexible sorting with $\beta = -0.25$ appears to be satisfactory with most data sets (Williams, 1976).

If there is no unique smallest interstand distance in the initial matrix, for all except nearest-neighbour sorting the exact form of the hierarchy will depend on which pair of stands with the smallest interstand distance are fused first. Consider, for example, the very simple case of classification by furthest-neighbour sorting of five points in a straight line separated by the following distances:

$$A \ 1.2 \ \ B \ 1.0 \ \ C \ 1.0 \ \ D \ 1.4 \ \ E.$$

If the first fusion is between $B + C$ successive stages will be (A 2·2 BC 2·0 D 1·4 E), (A 2·2 BC 3·4 DE), (ABC 4·6 DE). If, however, the first fusion is between C and D, successive stages will be (A 1·2 B 2·0 CD 2·4 E), (AB 3·2 CD 2·4 E), (AB 4·6 CDE). The groups at the penultimate level in the hierarchy will be different in the two cases. With nearest-neighbour sorting B, C and D fuse at the same level and the following stages are (A 1·2 BCD 1·4 E), (ABCD 1·4 E) so that which pair is fused first is immaterial. It can readily be shown that similar difficulties occur with other strategies.

Principally on account of this uncertainty when there are ambiguities in the initial matrix, it has been argued that only nearest-neighbour sorting can validly be used (Sibson, 1971; cf. Williams, Lance *et al.*, 1971). It has been amply demonstrated that the nearest-neighbour strategy is of little practical use in the context of vegetation and that other strategies can be productive of useful hypotheses. In practice ambiguities are likely to occur only at the lowest levels in the hierarchy, at which ecological interpretation is unlikely to be attempted, and it is immaterial at this level that groups of different composition may result from the arbitrary choice of pair for the first fusion. If there was reason to believe that genuinely distinct classes exist, as is commonly the case in taxonomy, then the uncertainty would have more serious implications, but in most studies of vegetation the data cover gradients of composition or, at the most, types with indistinct boundaries. The analogy between the numerical analysis of vegetation and numerical taxonomy is not as precise as is often assumed and conclusions on methodology appropriate to numerical taxonomy should not be applied uncritically to the study of vegetation.

There remains one further question to be considered in relation to choice of procedure in classifying data. This is whether to apply a standardization to the data and, if so, which standardization to use. Some strategies and coefficients imply a particular standardization but others do not, and this has confused consideration of standardization. Thus among divisive procedures, association-analysis implies standard-deviate standardization (to zero mean and unit variance) of species representation whereas division on an information statistic or between group sum of squares is free of standardization. Some of the coefficients of similarity or dissimilarity used in agglomerative procedures likewise involve standardization. The correlation coefficient implies standard deviate standardization by stands. The Jaccard and Sørensen coefficients (and any empirical coefficients which ignore double negative matches) apply a variable standardization to presence/absence data; since the divisor of the fraction forming the measure of dissimilarity between a pair of stands depends on the total number of species present in the two stands, the weight attached to a single species difference will vary. Other measures e.g. Euclidean distance, are free of standardization.

If no standardization is implicit in the procedure used, then standardization may be applied to the data before analysis. Whether it is desirable to do so will depend in part on the biological assumptions made. The implications of standardization are more straightforward for quantitative data. If attention is focussed entirely on the relative composition of the vegetation of a stand and the total amount of vegetation is considered irrelevant, then a standardization to eliminate interstand differences in standing crop is clearly desirable. The most obvious way to achieve this is to express the amount of each species in a stand as a proportion of the total amount in the stand. Standardization by stand norm (Noy-Meir, 1973a; see p. 248), which on calculation of Euclidean distance between stands gives 'standard distance' instead of 'absolute distance' (Orloci, 1967a), has a similar effect but does not equalize stand totals exactly.

Though the ecological advantages of removing differences in standing crop may be debated, the effect of such standardization is clear cut. The effects of standardizations which alter the relative weighting of different species are less obvious. Amounts of species have sometimes been expressed as proportion or percentage of the maximum amount recorded in the data set for the species in question. The objective is to prevent the more abundant species dominating the analysis. The maximum value recorded for a species in a particular data set will be subject to considerable random variation and though it does approximate to the required result, the standardization will be vague and imprecise. Standard-deviate standardization will achieve the same result in a more satisfactory way, but may make a more drastic alteration in weighting than proportion of species maximum. Consider a species present in small but relatively constant amount; the occasional complete absence or presence of an unusually large, though still small in absolute terms, amount may be weighted far more heavily than large differences in amount of an abundant species.

There is no reason why even such drastic alteration in species weighting should not be made, but the implication for subsequent interpretation must be kept in mind.

A possible alternative approach to species weighting is to assign weights either on the basis of *a priori* decisions about the species concerned (Macnaughton–Smith, 1965) or by calculation in some way from the data being analysed (Williams *et al.*, 1964). As an example of the latter, Hall (1970), assuming that a high abundance value is a more precise indicator of environmental conditions than a low abundance value or complete absence, first calculated the absolute distance contribution for each species from data standardized by species maximum value and then weighted the species contribution by its unstandardized total representation. The effect is to allow similarity in representation of abundant species to carry greater weight than similarity in representation of less abundant species, but to constrain the distance between plots to the limits of zero and unity (thus eliminating the effects of species-richness).

At least with data for ordination a double standardization has sometimes been applied. Species amounts have first been expressed as percentage of maximum and then the data for each stand have been converted to proportion or percentage of the stand total. Austin and Greig-Smith (1968) pointed out that this results in an approximation to presence and absence and there is little advantage (and considerable computational disadvantage) over direct use of presence/absence data.

Standardization may also be applied to presence/absence data. Standardizations which remove the effects of differences in standing crop with quantitative data will remove differences in species-richness with presence/ absence data. It is doubtful whether any circumstances will normally arise in which this is desirable*. Application of standard-deviate standardization will result in greater weighting being given to the occurrence of rare species and the absence of common ones than vice versa.

There appears to be considerable variability in the interpretability of classification following different standardization. Figure 40 shows the hierarchies resulting from optimal agglomeration classification of twenty-one stands of rain forest using six different standardizations. Four of the six classifications were satisfactory, in that they were ecologically interpretable, and two were not. Moreover, standard distance using presence/absence data

---

* In association analysis, in which differences in species-richness are retained, but differences in number of stands occupied are eliminated (Williams & Lambert, 1961a), it is relationships between species rather than stands which are analysed.

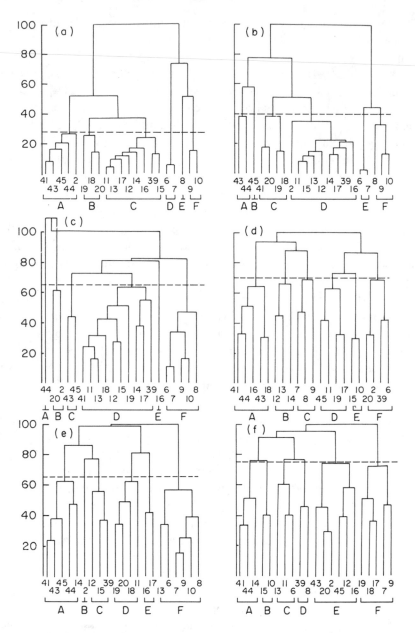

Fig. 40. Optical agglomeration of twenty-one stands of rain forest using different initial standardizations of the data. (a–d) Quantitative data; (a) absolute distance, (b) standard distance, (c) absolute distance after standardization by species standard deviate, (d) standard distance after standardization by species standard deviate. (e, f) Presence/absence data; (e) absolute distance, (f) standard distance. (From Austin & Greig-Smith, 1968, by courtesy of *Journal of Ecology*.)

was not satisfactory although Orloci (1967a) had found it satisfactory with data from a different type of vegetation (successional stages on a flood plain in British Columbia). The value of different standardization evidently varies with the type of data being analysed. Austin and Greig-Smith suggested that it depends on three factors, variations in species-richness between stands, variations in standing crop between stands and degree of species predominance (differences in percentage contribution of different species, especially of the most abundant species). It is still not clear what determines the most profitable standardization to use for different combination of these three factors.

Discussion so far has been concerned with the classification of stands on the basis of their floristic composition. If the data are of presence, they can also be considered as specifying the stands in which different species occur, rather than the species composition of different stands, and analysed to define groups of species with similar stand occurrences. The members of such a group may be presumed to have similar ecological characteristics or, at least, similar characteristics in relation to the controlling factors within the set of stands under consideration. Such a classification of species will allow information from other sources on the ecological behaviour of some species to be used to evaluate the ecological behaviour of other species and to erect hypotheses about controlling factors of major importance in particular stands. If the initial data are quantitative the analysis may still be directed to classification of the species, the amount of a species in a stand being regarded as a measure of the suitability of the stand for that species.

Williams and Lambert (1961a) extended the use of association analysis to produce species groups in this way, introducing the useful terms 'normal analysis' for classification of stands and 'inverse analysis' for classification of species. In inverse association analysis the effect of the standardization involved is to eliminate the differences in species richness but retain differences in species abundance. Thus species groups tend to reflect the relative abundance of species; this is not, however, necessarily misleading as the most abundant species in this sense are likely to be those with the widest range of tolerance in respect of the environmental range of the stands analysed. Similarly, differences in richness in normal analysis are likely to reflect real differences in habitat conditions of stands. In principle, any classificatory procedure may be used to give an inverse analysis, though the dimensions of the data matrix may make a particular procedure impractical, e.g. Webb *et al.* (1967a) estimated that inverse information analysis of a matrix of 18 stands × 818 species would have taken between 6 and 7 hours on the computer used, compared with $2\frac{1}{2}$ minutes for inverse association analysis.

If both normal and inverse analyses have been carried out it is worthwhile to examine a two-way table (Fig. 41), to assess how far particular species groups are characteristic of particular stand groups. To a certain extent this may be obvious on inspection. Thus in Fig. 41, species of group A represent a common element of the vegetation, but species of group B are most characteristic of stands of group 3. Nevertheless, not all the species and stands of a more or less fully occupied cell of the table contribute to a defined species/stand unit. Thus *Quercus robur* clearly does not contribute to the unit defined by stand group 1 and species group C; its assignment to group C results from its similarity in ecological behaviour to the other species of the group over the data as a whole. Williams and Lambert (Williams & Lambert, 1961b; Lambert & Williams, 1962) proposed a procedure, 'nodal analysis' aimed at 'extracting the genuinely central species-in-habitat coincidences round which the population under examination may be considered to be varying'. Nodal analysis determines 'coincidence parameters'. When a set of data (either the original set or a group resulting from a previous division) is subjected to a normal analysis, it is also subjected to inverse analysis and the stand in each subgroup which has the highest sum of χ^2 (or other measure being used) is termed the coincidence parameter. Coincidence parameters are similarly determined for species groups. Noda of different degrees of importance are determined according to the presence of one or both the coincidence parameters appropriate to the cell. Only if the stand coincidence parameter contains all the species of the group and the species coincidence parameter is present in all the stands of the group will a nodum contain all the entries of the cell.

Feoli and Orloci (1979) proposed testing the number of positive entries in the cells of a two-way table against random expectation, using χ^2 after first adjusting the observed values to a constant cell size. This is not in itself very helpful, but the relative divergence from random expectation can then be estimated as $\chi^2/\{F.\min(t-1), (z-1)\}$, where F is the overall total, t the number of rows and z the number of columns. $(F.\min(t-1), (z-1)$ is the maximum possible χ^2 for given F, t and z.)

If the objective of analysis is to identify species-in-habitat groups a strategy aimed directly at achieving this, rather than the superposition of normal and inverse analyses, is desirable. Moreover, there are theoretical objections to the superposition of two analyses; they may assume different numerical models, not necessarily compatible (Webb *et al.*, 1967a; Dale & Anderson, 1973), e.g. normal and inverse association analyses involve different standardizations.

Tharu and Williams (1966) suggested dividing monothetically at each stage by reference either to species or to stands, according to which gives the most

	Species	Stand ① (3 8 13 19 20)	Stand ② (7 10 11 17)	Stand ③ (2 9 12 14)	Stand ④ (1 4 5 6 15 16 18)
A	1 Calluna vulgaris	+ + + + +	+ + + +	+ + + +	+ + + + + + +
	4 Erica cinerea	+ + + + ·	+ · + +	· · · ·	+ · · + · + +
	3 E. tetralix	· + · + ·	· · + +	+ + + +	+ + + + + + +
	2 Molinia caerulea	+ + + + +	+ + + +	+ + + +	+ + + + + + +
	10 Polygala serpyllifolia	+ + · + ·	· · + +	· · · ·	· + · + + + +
	8 Ulex minor	+ · · + +	· · · +	· · · ·	· · · + + + +
B	23 Drosera intermedia	· · · · ·	· · · ·	+ · + · + ·	· · · · · · ·
	18 D. rotundifolia	· · · · ·	· · · ·	+ · + · + ·	· · · · · · ·
	15 Eriophorum angustifolium	· · · · ·	· · · ·	+ · + · + ·	· · · · · · ·
	27 Juncus squarrosus	· · + · ·	· · · ·	+ · + ·	· · · + · · ·
	32 Pinus sylvestris	· · · · ·	+ · · +	· · + ·	· + · · · + ·
	6 Trichophorum cespitosum	· · · · ·	· · · ·	+ + + +	· + + · + + ·
C	9 Agrostis setacea	+ · · + +	+ · · +	· · · ·	· · · · · · +
	36 Carex pilulifera	+ + + + ·	+ + + +	· · · +	· · · · · · +
	21 Festuca ovina	· + + · ·	+ · + ·	· · · ·	· · · · · · ·
	11 Potentilla erecta	+ + + + +	+ · + +	· · · ·	· · · · · · ·
	5 Pteridium aquilinum	+ + + · ·	+ + + ·	· · · ·	· · · · · · ·
	43 Quercus robur	· · · · ·	+ · + · + ·	· · · ·	· · · · · · ·
D	14 Carex panicea	+ · · + +	· · · ·	· + · +	· · · · · · ·
	19 Juncus acutiflorus	· · · · ·	· · · ·	· + · +	· · · · · · ·
	17 Northecium ossifragum	· · · · ·	· · · ·	· + · +	· · · · · · ·
	13 Pedicularis sylvatica	· + · + ·	· · · +	· + · ·	· · · · · · ·
E	56 Anemone nemorosa	· + · · ·	· · · ·	· · · ·	· · · · · · ·
	58 Anthoxanthum odoratum	· + · · ·	· · · ·	· · · ·	· · · · · · ·
	44 Betula verrucosa	· + + · ·	· + · ·	· · · ·	· · · · · · ·
	57 Campanula rotundifolia	· + · · ·	· · · ·	· · · ·	· · · · · · ·
	12 Carex binervis	· + · · +	· · · ·	· · · ·	· · · · · · ·
	61 Castanea sativa	· + · · ·	· · · ·	· · · ·	· · · · · · ·
	46 Cerastium vulgatum	· + + + ·	· · · ·	· · · ·	· · · · · · ·
	35 Galium hercynicum	+ + · + ·	· · + ·	· · · ·	· · · · · · +
	37 Hieracium pilosella	+ · + · + ·	· · · ·	· · · ·	· · · · · · ·
	41 Hypericum humifusum	+ + · · ·	· · · ·	· · · ·	· · · · · · ·
	38 Hypochaeris radicata	· + · + +	· · · ·	· · · ·	· · · · · · ·
	50 Lathyrus montanus	· + · · ·	· · · ·	· · · ·	· · · · · · ·
	48 Lonicera periclymenum	· + · · ·	· · + ·	· · · ·	· · · · · · ·
	54 Lotus corniculatus	· + · · ·	· · · ·	· · · ·	· · · · · · ·
	55 L. uliginosus	· · · · ·	· · · ·	· · · ·	· · · · · · ·
	52 Luzula multiflora	· + + + ·	· · · ·	· · · ·	· · · · · · ·
	45 Orchis ?ericetorum	· + · · ·	· · · ·	· · · ·	· · · · · · ·
	39 Sieglingia decumbens	+ + · + +	· · · ·	· · · ·	· · · · · · ·
	34 Succisa pratensis	+ + · · ·	· · · ·	· · · ·	· · · · · · ·
	49 Teucrium scorodonia	· + · · ·	· · · ·	· · · ·	· · · · · · ·
	53 Veronica chamaedrys	· + · + ·	· · · ·	· · · ·	· · · · · · ·
F	+ 33 other species				

Fig. 41. Two-table showing normal and inverse analyses. (Modified from Lambert & Williams, 1962, by courtesy of *Journal of Ecology*.)

effective division in terms of concentration of entries in cells of the two-way table. Practical tests of this 'concentration analysis' have been disappointing, giving complex results not readily interpretable (Webb *et al.*, 1967a; Holland, 1969; Dale & Anderson, 1973).

Another approach to the problem of combining normal and inverse analysis, 'inosculate analysis' (Dale & Anderson, 1973; Dale & Webb, 1975), adopts a different model for classification. Most of the methods considered in this chapter so far assume that the members of a 'perfect' group have all species present at all sites and, additionally if quantitative data are used, each species is present in the same amount at each site. Strategies thus aim, explicitly or implicitly, to minimize the variance within groups (Austin, 1972), i.e. to make them as homogeneous as possible. The minimizing of variance is applicable to data after standardization, whether the standardization is necessarily involved in the analytical procedure, as with association analysis, or is made before analysis, so that in terms of the original data, minimizing variance is not necessarily achieved, certain aspects of heterogeneity being ignored, e.g. as a result of standardization that removes the effect of standing crop. Dale and Anderson, following what they took to be the objective of classical phyto-sociology, assumed that a useful grouping is one within which there is a consistent relationship between the performance of different species and the productivity of different stands, so that the least productive stands (having the fewest species, or the least amount of vegetation by whatever measure of species quantity is being used) will contain only the best performing species (those occurring in most stands or contributing the greatest amount to the set of stands) and least well performing species will be found only in the most productive stands.

In the more usual, 'one-parameter' models, the probability of occurrence of a species j in a stand i in a final group, P_{ij} is assumed to be a constant for the species. Dale and Anderson adopted a model put forward for presence/absence data by Macnaughton-Smith (1965) (originally suggested in a different context by G. Rasch). The probability P_{ij} is regarded as a function of two parameters α_i, relating to the productivity of stand i, and β_j, relating to the performance of species j. For technical reasons the obvious functions $P_{ij} = \alpha_i \beta_j$ and $P_{ij} = \alpha_i + \beta_j$ are unsuitable (Macnaughton-Smith, 1965) and $\alpha_i \beta_j$ is defined as the ratio of the probability of occurrence to probability of non-occurrence:

$$\frac{P_{ij}}{1 - P_{ij}} = \alpha_i \beta_j \quad \text{or} \quad P_{ij} = \frac{\alpha_i \beta_j}{1 + \alpha_i \beta_j}.$$

For quantitative data the expected amount of species j in stand i

$$x_{ij} = \frac{\alpha_i \beta_j}{1 + \alpha_i \beta_j}.$$

This model may be used as the basis of a monothetic divisive strategy. The initial set of data is fitted to the model, estimates of α's and β's being obtained from the data and goodness of fit assessed by an information statistic. If a satisfactory fit is not obtained, the data are divided on the species (for normal analysis) or the stand (for inverse analysis) which gives the best fit in the resulting subgroups, and the procedure repeated in the usual way until final groups giving an acceptable fit are obtained or until a predetermined number of groups have been formed.

This procedure can thus be used either for a normal or an inverse analysis, but the common basis allows it to be used in what Dale and Anderson term an inosculate strategy, dividing either normally or inversely according to which direction gives the greatest improvement in fit. They discuss an example, using data for 80 species in 68 stands of birch copse and shrub heath in Iceland. The course of the analysis, as divisions on a rearranged two-way table, is shown diagrammatically in Fig. 42. The final grouping shows three stand groups (3, 10, 11), in two of which there is no division of the species present into distinctive groups, and one group with five groups of species behaving differentially in relation to the variation between sites.

The values of α for the stands in a group allow the stands to be arranged in an ordered sequence, those stands with the highest values being the ones in which the vegetation has its fullest expression*. Similarly, the values of β can be regarded as indicating the relative importance ('dominance' in a loose sense) of different species within the vegetation.

As Dale and Anderson point out, the two-parameter model, although not overtly designed to do so, involves an interaction between species. If the amounts of two species j and k in stand i are compared, these are not in direct proportion to their β values but

$$\frac{x_{ij}}{x_{ik}} = \frac{\beta_j + \alpha_i \beta_j \beta_k}{\beta_k + \alpha_i \beta_j \beta_k}$$

so that as the productivity of a stand tends to zero the comparison of the amounts of the species tends to the simple ratio of the species performance

---

* For this reason Dale and Anderson refer to inosculate analysis as combining ordination with classification. This terminology is somewhat misleading; 'ordination' is normally used to refer to the approach to analysis of data discussed in Chapter 8.

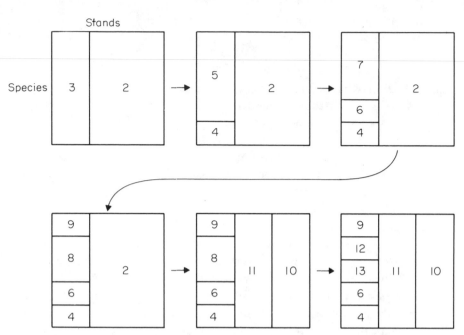

Fig. 42. The course of an inosculate analysis, from an example quoted by Dale and Anderson (1973).

parameters. Conversely, as the stand productivity increases, the performance of the species becomes more nearly independent of the species performance parameters. There is a comparable interaction when the amounts of the same species in two different stands are compared, with stand productivity becoming less important as β increases.

This interaction may be interpreted as taking account of competition between species for the available resources of a site, a potentially valuable feature for classification used for hypothesis generation. With most techniques of classification, interpretation must assume that species performance in a stand is determined by the environmental characteristics of that stand, although which of two species of equivalent tolerance is present, or in greater amount, may depend on chances of initial establishment and competition between the two.

Gilbert and Wells (1966) proposed a model explicitly based on the interaction between species in the context of detecting patchiness along transects, rather than for formal classification, but Dale (quoted by Austin, 1972) has developed a divisive technique using it in a similar way to Macnaughton-Smith's model.

If the frequencies of species i and k in N stands, expressed as proportions, are P_j and P_k respectively, then, if the species were distributed among the stands independently of each other, the expected number of stands containing both species would be

$$n_{jk} = N P_j P_k$$

and so $\log n_{jk} = b_j + b_k$, where $b_j = \frac{1}{2}\log N + \log P_j$.

Gilbert and Wells therefore fitted the model

$$\log n_{jk} = b_j + b_k + c_j c_k$$

where the c's are 'purely statistical parameters describing, as far as possible, the two-species interactions'. They reflect both the differing responses of species to environment and any direct competitive (or stimulatory) interactions between species, although the two cannot be distinguished. As Austin (1972) suggested, the utility of the model in classification is worth further examination.

The validity of using a classification of a limited set of stands, commonly by a technique that treats the data as a universe rather than a set of samples, as a more general classification in which further stands can be placed, was queried earlier in this chapter. Nevertheless, workers desiring a comprehensive classificatory scheme, and some using classification only for hypothesis generation, need to assign further stands to the most appropriate group, and, indeed, their choice of technique of classification may be influenced by the facility with which this may be done.

Divisive monothetic procedures provide a direct 'key' for allocation of further stands. How readily a further stand can be allocated in a divisive polythetic procedure, will depend on the procedure used. If it is based on splitting of ordinational axes, and the ordination procedure used involves the calculation of species loadings and their use to determine axis positions (see Chapter 8) allocation is again straightforward—the position of a new stand on the axes, and hence the group to which it belongs, can be calculated from its species composition. Indicator species analysis (Hill et al., 1975) specifically provides for ready allocation of further stands by identifying for each division the five species which have the highest indicator value (p. 189). If the procedure does not provide a direct means, allocation of further stands can be made in the same way as for agglomerative classification, to be considered next.

In principle, the allocation of further stands to groups in an agglomerative classification is straightforward. The similarity of an additional stand to each of the groups of the classification is calculated and the stand is assigned to the group to which it is most similar. Clearly the same measure of similarity or distance, and the same standardization, if any, as were used in constructing the classification must be used. How computationally demanding doing so will be, will depend not only on the number of groups involved but also on the sorting strategy used for deriving the classification. With centroid, median or optimal agglomeration strategies, only the similarity of the new stand to the centroids of each of the groups need be calculated. If, however, nearest or furthest neighbour, or group average strategies, have been used, strictly the similarities of the new stand to all stands contributing to the classification will be needed. It should be remembered, however, that the addition of any further stands to a set of data is liable to alter the classification. Allocation of further stands into an existing classification is thus necessarily suboptimal and it may be sufficient for practical purposes to use the centroids of groups for allocation of further stands even if single linkages are theoretically more appropriate.

An alternative approach, applicable however the groups have been derived initially, is to select as the most efficient discriminatory species for groups those having the largest ratios of between-group to within-group variance (Jancey, 1979). Multiple discriminant analysis (p. 288) may also be available, depending on the classificatory procedure used.

So far in this chapter, classification has been discussed in terms of an initial matrix of data recording either the presence or some measure of the amount of all the species recorded in each stand. Brief mention should be made of other possible initial data sets.

In very species-rich situations, e.g. tropical rain forest, stands may be 'overdefined' in that a part only of the species complement is sufficient to draw conclusions about the interrelationships of the stands (Webb et al., 1967a; Greig-Smith, 1971b). Thus Webb et al. (1967b), examining a set of data with 818 species from 18 stands, found that classification (by agglomerative information analysis) of the whole data was reproduced exactly by one based on the 269 species of 'big trees' (species capable of reaching the canopy) and substantially by one based on 65 species of big trees, selected on the basis of an inverse analysis. There are clear computational advantages with very species-rich data in reducing the number of redundant species before analysis. Austin and Greig-Smith (1968) showed that reduction in species number by considering only the more abundant species in rain forest data is effective in ordination and this is almost certainly true of classification also*. Webb et al. also pointed out that certain biological groupings of species, e.g. epiphytes and herbs, gave classifications of the stands that were interpretable although markedly different from that based on all species. Such synusiae presumably

---

* Theoretically, species present in all or nearly all stands could also make little useful contribution to analysis but such species would necessarily be few in number. Such species are, in any case, unlikely to occur in very species-rich situations.

respond to different controlling factors in their environment from those determining the main bulk of the vegetation. It may thus be useful to analyse in terms of such species only. Equally their elimination may give a more satisfactory overall classification.

The major divisions of the vegetation of the world have long been based on physiognomy (cf. Beard, 1973). Webb *et al.* (1970) considered the use of non-taxonomic criteria for classification at a much more detailed level, and showed that classifications of rain forest data based on such criteria can be as satisfactory as those based on floristic criteria. There are evident advantages in using non-taxonomic data in species-rich vegetation, the flora of which is poorly known. Collection of data in the field is very much quicker and can be done by observers lacking detailed knowledge of the flora, and, at least in theory, it is possible to extend a classification to data covering very wide geographical areas. Such data do, however, raise problems of analysis. Floristic data are well defined in such terms as 'all vascular species' or 'all woody species' and the decision on what data to collect is fairly well defined. Possible non-taxonomic data are open-ended and the choice of characteristics to be recorded may substantially affect the classification that results. Further, data are of several kinds. They may be (a) quantitative, involving measures or counts values of which have absolute meaning, e.g. height of canopy, (b) qualitative, e.g. presence or absence of lianes, (c) ordered multistate (Lance & Williams, 1967b), in which a character has several states which can be ranked, e.g. predominant leaf-size class or (d) disordered multistate, in which a character has several states which cannot be ranked, e.g. predominant bark type. These mixed data place constraints on possible measures of similarity and appropriate standardizations. Lance, Williams and Milne give a useful discussion (Lance & Williams, 1967b, 1968b; Lance *et al.*, 1968). Lambert (1972) referred briefly to unpublished work by S. L. Attanapola on the use of non-taxonomic characteristics of individual organs ('eco-organs') rather than whole plants. Satisfactory classifications of test communities, in most cases at least as good as those based on floristic criteria, were obtained from presence/absence data for a set of some 80–90 eco-organ categories.

If floristic data are quantitative, there are likely to be a large number of zero entries in the data matrix. Although a classification will ostensibly be based on quantitative data, the differentiation between stands and the resulting grouping may be regarded as based partly on the presence or absence of species and partly on differences in the amount of species when they are present. Though the difference between these two elements is in a sense arbitrary, it is ecologically meaningful in that whether a species can establish in a given site at all may be dependent on different factors from those that determine its performance if it does establish. Possible techniques of determining the relative importance of the qualitative and quantitative elements in a data matrix have aroused interest. Williams and Dale (1962) suggested that the mean value for a species within stands in which it is present, M_x, represents the qualitative element (L) and that the deviation of the actual value in non-zero stands from this mean $X - M_x$, represents the quantitative element (N). The total hetero-geneity among stands can then be partitioned into four parts, a qualitative part (L/L), a quantitative part (N/N) and two interaction parts $(L/N, N/L)$ representing the relationship between presence and absence of one species and quantity of another. Noy-Meir (1971a) carried out this partition for stand groups formed at successive stages of association analysis using different stand sizes (measures of cover were available as well as the presence/absence records used for association analysis). He showed that the qualitative element usually predominated at the level of the whole population and when small stands were used, but the quantitative and interaction elements were always important and the quantitative element became predominant after a few sub-divisions particularly with larger stand sizes. This helps to explain the less satisfactory nature of the lower levels often noted with association analysis, and found previously by Noy-Meir *et al.* (1970) for the same data. The partition into qualitative and quantitative elements is more complicated when information statistics are used, but analogous partitions can be made (Orloci, 1968; Williams, 1972, 1973). A somewhat similar possible partition of total hetero-geneity, into species ignoring size of individual, and sizes within species was suggested by Williams *et al.* (1973) for forest data, where sizes of individual trees are commonly recorded; they presented a solution for the information statistic case.

   If it is desired to reduce the overall influence of the quantitative element in the data (rather than primarily to alter the relative weighting of different species (see p. 212), a common transformation may be applied to all positive entries, e.g. square root or logarithmic transformation*.

   With increased use of numerical methods of classification, there has been interest in speedier approximate procedures for preliminary analysis of very large data sets (e.g. Janssen, 1975, Loupen & van der Maarel, 1979; Gauch, 1980). These depend on taking either a stand, or a randomly derived 'pseudostand' made up from the total species list, as the centre of a cluster. All stands within a declared distance of this centre are assigned to the cluster. The

---

* Initial data transformation may also be necessary if the data are in the form of values on an abundance scale rather than a measure. For useful discussions of initial data transformations see, for example, Clymo (1980), Jensen (1978), van der Maarel (1979), Nichols (1977).

process is repeated until all stands are included in clusters or remain isolated. The procedure is non-hierarchical and will normally be a preliminary to further analysis after removal of redundant and outlier stands, or to hierarchical treatment of the clusters.

Almost all discussions of classification of vegetation assume that the objective is the production of mutually exclusive groups—a stand may be assigned to only one group in a normal analysis and a species to only one group in an inverse analysis. Yarranton *et al.* (1972) suggested consideration of the advantages of non-exclusive groups in certain cases, on the grounds that a stand may include more than one element of vegetation and that a species of wide ecological amplitude may be associated with more than one group of species of narrower ecological amplitude. They used a modification of centroid sorting of the correlation coefficient by which one group was allowed to build up until no further individuals could be added at a predetermined minimum value of the correlation coefficient and then starting the next group from the largest correlation coefficient not involving a member of the first group. (After this starting point of the second group, the group may draw in stands already included in the first group.) Hierarchical structure is then imposed by similar treatment at a lower termination level of the centroids of the groups and any remaining single stands. If production of non-exclusive groups is thought desirable other measures of similarity could be used.

With the increasing availability of computing facilities, it is not surprising that attempts were made to adapt numerical methods to the procedure of the Braun-Blanquet system of classification (see for example, Benninghoff & Southworth, 1964; Moore *et al.*, 1967, 1970; Lieth & Moore, 1971; Češka & Roemer, 1971; Spatz & Siegmund, 1973; Dale & Quadraccia, 1973; Westhoff & van der Maarel, 1973). Procedures are based more or less explicitly on the methods used manually for the Braun-Blanquet system*. The most developed numerical programme for Braun-Blanquet classification is TABORD (van der Maarel *et al.*, 1978), which starts either from arbitrary groups of stands or, preferably, from groups based on the judgement of the investigator. All stands are then tested individually for their similarity to each group. After each stand has been checked and, if necessary, relocated, the process is repeated until stability is achieved. Groups below a stated distance from one another are fused. Various options are then available to order the groups of stands and to order the species, starting with species constant in stand group 1 but not in stand group 2, then those constant in stand groups 1 and 2, and so on.

---

* For a clear account of the manual operation of the Braun-Blanquet system, see Mueller-Dombois and Ellenberg (1974).

Provision is made for further visual ordering of species and stands within their groups.

There is an evident similarity with the preparation of a two-way table entirely by numerical methods but with the additional step that the two-way table is then rearranged to concentrate the positive entries along the diagonal. Hill's (1979) TWINSPAN uses reciprocal averaging, which does concentrate positive entries along the diagonal, to achieve this. It dichotomizes an initial ordination of stands, identifies 'differential species' and uses these to refine the ordination. This procedure is then repeated for each branch of the dichotomy.

CHAPTER 8

Plant Communities—III. Ordination

The objective of ordinating sets of vegetation data is normally to generate hypotheses about the relationships between composition of vegetation and the environmental or other factors which determine it. In principle, as with classification, the resulting presentation of the data might be used as a tool in, for example, management of the vegetation, but in practice a classification is nearly always more convenient for such purposes.

Since the objective is to explore relationships between vegetation and influencing factors, there are two obvious approaches. If the range of composition is simplified into an ordination, this can be used as a framework on which values of any environmental factor considered as possibly important can be plotted and the resulting distribution of values examined. If these values show a degree of order when so plotted, there is evidence of correlation between the factor and the overall composition of the vegetation. It does not follow that the environmental factor is therefore controlling, even in part, the vegetational composition; quite the reverse may be the case, e.g. the type of vegetation may in fact determine such features of the environment as type of soil humus or level of some nutrients in the soil, or both composition and environmental factor may be determined by some other environmental factor. The value of ordination is basically in selecting from the indefinite number of possibly important influencing factors those which it is worth investigating further, though there may also be useful clues from the interaction of different factors as they appear on the ordination diagram. Figures 43 and 44 show examples of environmental factors plotted on vegetational ordinations. Trends in values of an influencing factor will not necessarily run parallel to an axis, and attempts to interpret axes as representing influencing factors are, particularly with less efficient ordination techniques, liable to confuse interpretation.

It may also be informative to plot the occurrence of species onto a vegetational ordination as a means of comparing tolerances and performances (Fig. 45). Caution is needed in interpreting such species plots, especially if the total number of species is relatively small, because each species will normally have contributed to the ordination and will therefore have a distinctive pattern on it.

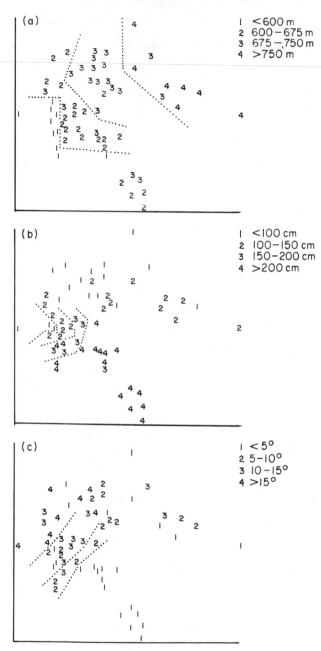

Fig. 43. Distribution of values of environmental factors on an ordination of stands of blanket bog vegetation. (a) Altitude, (b) peat depth, (c) angle of slope. (From Tallis, 1969, by courtesy of *Journal of Ecology*.)

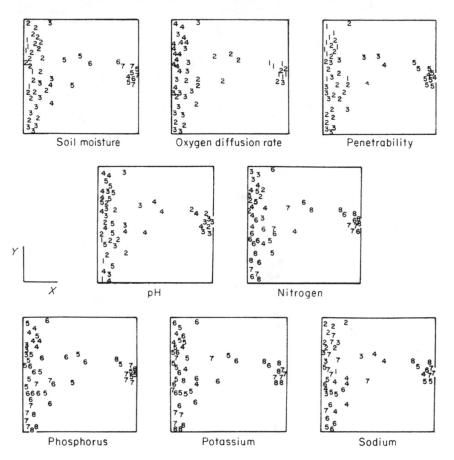

Fig. 44. Distribution of environmental factors on an ordination of stands of sand dune vegetation. Scores for each environmental factor represent equal segments of its range. (From Pemadasa *et al.*, 1974, by courtesy of *Journal of Ecology*.)

The alternative strategy is to simplify the available environmental data by an ordination and plot onto it information on vegetational composition. There are two obvious difficulties. Unlike vegetational ordination, where either the whole species complement or a clearly defined part ('all vascular plants', 'all woody plants' etc.) can be used, the list of possible environmental features is open-ended, and the choice of features to be included may greatly affect the resulting ordination. The other problem is that only individual species can be plotted onto the environmental ordination, unless some previous simplification of the vegetational data has been made (see below, p. 278).

The idea of simplifying and presenting vegetational data in terms of axes of

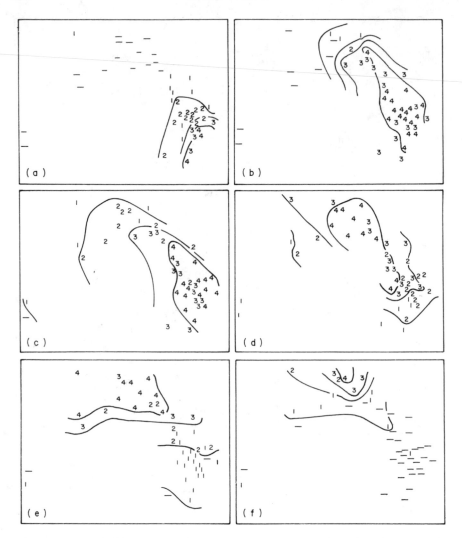

Fig. 45. Distribution of selected species on an ordination of stands of limestone grassland vegetation. 1–4, Quartiles of maximum abundance; −, species absent. (a) *Cladonia impexa*, (b) *Helianthemum chamaecistus*, (c) *Briza media*, (d) *Poterium sanguisorba*, (e) *Agrostis tenuis*, (f) *Trifolium pratense*. (From Gittins, 1965a, by courtesy of *Journal of Ecology*.)

variation rather than classes was initially developed in relation to the concept of vegetation as showing a continuum of variation. It is not surprising that ordination initially appealed especially to workers holding this view, but it must be emphasized that ordination techniques are not dependent on variation being continuous, any more than classification techniques are dependent on the existence of discontinuities in the data.

Dale (1975), in a useful discussion of approaches to ordination, distinguished three major sources of ordination methodology. The first, associated especially with the work of Whittaker (e.g. Whittaker, 1956, 1967), arranged stands directly in relation to specified environmental gradients*. The second aimed to order stands in a linear sequence such that a given stand is closely similar in composition to its neighbours on either side, an approach which Dale described as 'path-seeking'. The sequence was assumed to reflect a gradient of environmental factors (e.g. Curtis & McIntosh, 1951; Curtis, 1959)[†]. The third approach aimed primarily at reduction of dimensionality. The composition of a stand can be regarded as specified in 'species space', i.e. by coordinates on species axes. Goodall (1954c) pointed out that, if species are correlated in their occurrences in stands, a description in a space of much lower dimensionality will convey most of the information in the data. He used a principal component analysis (see below, p. 242) to derive a reduced number of axes. Many later techniques have aimed primarily at reduction in dimensionality, but this and the path-seeking approach have, as Dale pointed out, tended to converge in practice.

It is useful to consider simple approaches of each of the three kinds, before turning to more powerful techniques. Whittaker's direct gradient analysis[‡] was developed initially in relation to mountainous areas bearing mainly forest vegetation (Whittaker, 1952, 1954, 1956, 1960, 1967, 1973b). Initial arrangement was in relation to a 'moisture gradient' defined in terms of a physiographic series from deep ravines with flowing streams to open south-facing slopes, i.e. from the most mesic to the most xeric sites. As Whittaker pointed out, this 'moisture gradient' is not a gradient of moisture alone but includes a complex of correlated environmental gradients. On the results of placing stands in relation to this gradient, species were selected which had a clear mode in representation either towards one end of the gradient or around the middle. These species were given weightings according to their position on the gradient and the weightings used with the amounts of the species in a stand to obtain a

* The first practical use of this approach appears to have been by L. G. Ramensky in the 1920's, but his publications were little known outside Russia (see Sobolov & Utekhin, 1973).

† The first use of this approach appears to have been by J. Paczoski in the 1930's (see Maycock, 1967).

‡ Whittaker originally termed this simply 'gradient analysis' but in later publications designated it 'direct gradient analysis' in contrast to 'indirect gradient analysis', used for dimensionality reduction methods.

more precise placing of that stand on the gradient*. Successive altitudinal belts were examined in terms of the moisture gradient, giving a two-dimensional ordination (Fig. 46). In his study of the Siskiyou Mountains Whittaker (1960) found it necessary to use separate two-dimensional ordinations for the vegetation on the main rock types present, giving, in effect, a three-dimensional arrangement. The success of this technique clearly depends on the choice of environmental axes to be used. It was developed by Whittaker in relation to vegetation of mountainous areas where the main controlling environmental

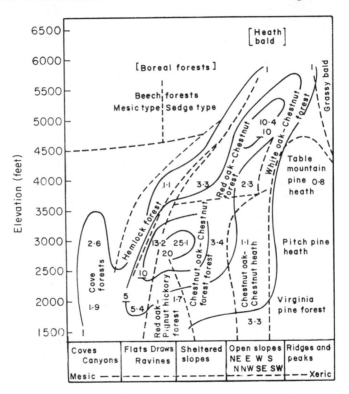

Fig. 46. An example of direct gradient analysis. Stems of *Hamamelis virginiana* as percentage of total stems in stands in the Great Smoky Mountains in relation to gradients of elevation and moisture conditions. Figures are values for percentage of stems. Heavy lines are 'isorithms' (lines of equal percentage contribution to stands). Broken lines indicate the approximate boundaries of vegetation types in relation to the gradients studied. (From Whittaker, 1956, by courtesy of *Ecological Monographs*).

  * King (1962) used a similar procedure to place stands more precisely along a gradient of soil types.

gradients are pronounced and obvious, and in these circumstances it gave informative ordinations. With many data sets, where the important environmental gradients are less readily determined, the requirement for initial choice of axes is a serious limitation.

Curtis and McIntosh (1951; see also Brown & Curtis, 1952) developed a technique based on arranging stands in an order such that, as far as possible, the amount of each species showed a single peak only, or increased or decreased monotonically along the series. Each species was then assigned a weighting according to the position of its peak, and these weightings used to give a more precise position of stands on the axis corresponding to the initial ordering*. Curtis and McIntosh derived the weightings by first grouping the stands according to the 'leading dominant' species (the species present in greatest amount) and arranging these groups in order (Table 19). The average occurrence of other species within these groups (Table 20) were then used to assign weightings for these species (Table 21). The resulting single-axis ordination can be used to examine the occurrence of species, including those

Table 19. Upland hardwood forests in Wisconsin: average importance value (IV) and constancy (%) of trees in stands with given species as the leading dominant (From Curtis & McIntosh, 1951, by courtesy of *Ecology*) (For species with highest importance potential only—80 stands)

Species	Leading dominant in stand			
	Q. velutina	*Q. alba*	*Q. rubra*	*A. saccharum*
Q. velutina				
Average IV	165·1	39·6	13·6	0
Constancy (%)	100·0	72·3	38·3	0
Q. alba				
Average IV	69·9	126·8	52·7	13·7
Constancy (%)	100·0	100·0	97·1	66·7
Q. rubra				
Average IV	3·6	39·2	152·3	37·2
Constancy (%)	25·0	94·5	100·0	76·3
A. saccharum				
Average IV	0	0·8	11·7	127·0
Constancy (%)	0	5·6	29·4	100·0

* The main variation in the data examined by Curtis and McIntosh was related to successional differences and they therefore used the term 'climax adaptation number' for the species weighting. The term 'continuum index' (cf. Fig. 47) was replaced by 'compositional index' in later publications (Curtis, 1959).

Table 20. Upland hardwood forests in Wisconsin: average importance value (IV) and constancy (%) of trees in stands with given species as the leading dominant (From Curtis & McIntosh, 1951, by courtesy of *Ecology*) (Eleven species of intermediate importance potential—80 stands)

Species		Leading dominant in stand			
		Q. velutina	*Q. alba*	*Q. rubra*	*A. saccharum*
Q. macrocarpa	IV	15·6	3·5	4·2	0·1
	C (%)	50·0	38·9	20·6	4·8
Prunus serotina	IV	21·4	21·8	5·9	1·4
	C (%)	87·5	89·0	64·8	19·0
Carya ovata	IV	0·3	8·8	5·2	5·9
	C (%)	12·5	61·2	38·3	33·3
Juglans nigra	IV	1·5	1·2	2·2	1·9
	C (%)	12·5	11·1	20·6	23·8
Acer rubrum	IV	3·9	2·3	2·4	1·0
	C (%)	12·5	33·3	23·5	4·8
Juglans cinerea	IV	0	2·7	1·7	4·8
	C (%)	0	11·1	20·6	47·6
Fraxinus americana	IV	0	1·9	5·1	7·6
	C (%)	0	11·1	20·6	42·8
Ulmus rubra	IV	4·6	7·7	8·3	32·5
	C (%)	25·0	27·8	53·3	85·7
Tilia americana	IV	0·3	5·9	19·0	33·0
	C (%)	12·5	16·7	73·5	100·0
Carya cordiformis	IV	2·5	5·8	4·1	8·2
	C (%)	12·5	33·3	41·2	66·7
Ostrya virginiana	IV	0	2·4	5·5	16·2
	C (%)	0	22·2	41·2	95·3

not used in its construction (Fig. 47), or the values of environmental factors, in relation to the ordination.

Curtis and McIntosh's technique has two major limitations. It can give only a single-axis ordination and, though the ranking of species weightings may be clear-cut, the precise weighting assigned to a species is, to a considerable extent, a matter of judgement (cf. Tables 20 and 21). Thus, two workers, analysing the same set of data, will not necessarily produce the same ordination. Although the emphasis of this approach and that of Whittaker are different, the derivation of Curtis and McIntosh's single axis is very similar to that of Whittaker's moisture relations axis.

Goff and Cottam (1967) removed the need for subjective assignment of species weightings by first calculating a matrix of species similarities. The two

Table 21. Upland hardwood forests in Wisconsin: tree species found in stands studied, with the climax adaptation numbers of each (From Curtis & McIntosh, 1951, by courtesy of *Ecology*)

	Climax adaptation number
Quercus macrocarpa Michx.	1·0
Populus tremuloides Michx.	1·0
* *Acer negundo* L.	1·0
Populus grandidentata Michx.	1·5
Quercus velutina Lam.	2·0
Carya ovata (Mill.) K. Koch.	3·5
Prunus serotina Ehrh.	3·5
Quercus alba L.	4·0
Juglans nigra L.	5·0
Quercus rubra L.	6·0
Juglans cinerea L.	7·0
* *Ulmus thomasi* Sarg.	7·0
* *Acer rubrum* L.	7·0
Fraxinus americana L.	7·5
* *Gymnocladus dioica* (L.) Koch.	7·5
Tilia americana L.	8·0
Ulmus rubra Muhl.	8·0
* *Carpinus caroliniana* Walt.	8·0
* *Celtis occidentalis* L.	8·0
Carya cordiformis (Wang) K. Koch.	8·5
Ostrya virginiana (Mill.) K. Koch.	9·0
Acer saccharum March.	10·0

* Climax adaptation number of these species is tentative, because of their low frequency of occurrence in this study.

most dissimilar species were assigned weightings of 1 and 10 respectively and weightings for the remaining species determined by their similarities to them, e.g. the weighting for a species with 30% similarity to the species with weighting 1 and 50% similarity to the species with weighting 10 is $((30 \times 1) + (50 \times 10))/(30 + 50) = 6.63$. New estimates of species weightings were then calculated by reference to similarities to, and weightings of, all other species, the process being repeated until stability of weightings was achieved.

Bray and Curtis (1957) put forward a multi-axis technique which, though not explicitly described as such, is a simple dimensionality reduction technique. They calculated a matrix of similarities between stands which were then converted to interstand distances. Bray and Curtis themselves applied a preliminary double standardization to their data (species representation

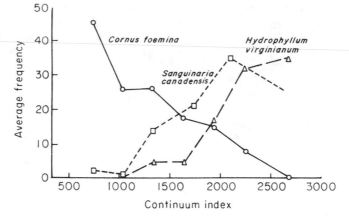

Fig. 47. Upland hardwood forests in Wisconsin. Average frequency values of three herbs, arranged in order of continuum index. (From Curtis & McIntosh, 1951, by courtesy of *Ecology*.)

expressed as percentage of maximum observed and the standardized scores for species in each stand then placed on a relative basis). They used the quantitative form of the Sørensen coefficient (Czekanowski coefficient), expressed as a percentage, as the measure of similarity, and converted the percentage similarity to distance by subtracting not from 100% but from 80%, a value regarded as representing 'identity' on the basis of replicate sampling of the same stands. Neither standardization nor the particular coefficient used are essential to the technique. Conversion to distance by subtraction from a value itself subject to a sampling error is undesirable if for no other reason than the practical one that it may give rise to negative distances; it is likewise not an essential feature of the technique.

The construction of axes used by Bray and Curtis may be illustrated from the hypothetical example they quote. The upper right hand portion of Table 22 shows the similarities between five stands and the lower left hand portion the corresponding interstand distances (obtained by subtraction from 100). The maximum interstand distance is 99·9 between stands 1 and 2, which are therefore placed at the opposite ends of a first axis. The distances of the remaining stands are known and they can be placed along the first axis either by geometric construction as Bray and Curtis did (Fig. 48) or by solving the two right angled triangles formed by the projection of the stand position on to the axis (Beals, 1960)*. A second axis is formed by selecting two stands with similar

* Gauch and Scruggs (1979) examined several alternative ways of defining the position of stands on the axis.

Table 22. Matrix of hypothetical exact interpoint distances (From Bray & Curtis, 1957, by courtesy of *Ecological Monographs*)

Stand No.	1	2	3	4	5
1	—	0·1	30	30	30
2	99·9	—	30	50	50
3	70	70	—	17·8	79·6
4	70	50	82·2	—	35·2
5	70	50	20·4	64·8	—

The upper right portion of the table shows hypothetical data on point similarity for an exact spatial system. The lower left portion shows data on similarity, inverted to give interpoint distances.

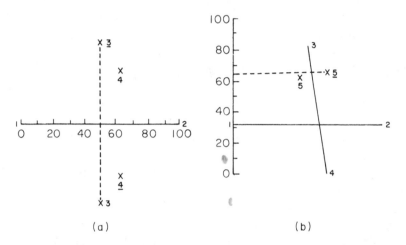

(a)   (b)

Fig. 48. Method of stand location used by Bray and Curtis (see text). (From Bray & Curtis, 1957, by courtesy of *Ecological Monographs*.)

positions on the first axis but having a relatively large interstand distance (stands 3 and 4 in Fig. 48). The remaining stands are placed in relation to this axis in the same way. If there are stands with similar position on both axes, but having appreciable interstand distances, a third (and further axes) may be derived in a similar manner.

The two techniques developed by Curtis and his associates are important because they have been used as practical tools in studying both widely varying vegetation (e.g. Curtis, 1959; Ashton, 1964) and variation on a much smaller scale (e.g. Anderson, 1963; Gittins, 1965a).

The technique introduced by Bray and Curtis, which, together with various modifications of it, has come to be known as 'polar ordination' (Cottam *et al.*,

1973), proved attractive for several reasons. It is simple in conception and the mechanisms of axis construction are readily understood; at least in its original form, it requires relatively little computation and limited data sets can be ordinated 'by hand'; and it allows choice of axis directions, one of the advantages of Whittaker's simple environmental ordination. In its original form, however, the technique has been criticised on three grounds, the distance measure used, the relationship of the axes derived and the choice of end-points for the axes.

A spatial representation of interstand distances which reproduces them exactly can only be made if the distance measure is a metric (and if as many dimensions are retained as are present in the original data). If spatial representation is attempted with a non-metric measure a point will be reached where, after a stand has been fixed by reference to its distance from other stands, other of the interstand distances in which it is involved are not satisfied. Put in another way, the position assigned to a stand will vary according to which of its interstand distances are used to fix its position. If a small number of dimensions only are being retained these effects may not become apparent, but distortion of stand relationships still occurs. This is clearly undesirable.

The one-complement of the Czekanowski coefficient used by Bray and Curtis is a metric only if the stand totals are the same, when it is equivalent to a Manhattan metric. Bray and Curtis expressed the data for each stand as relative values, but then eliminated the metric property by subtracting the calculated similarity from a lesser value than one; if a similarity is to be converted to a distance it must be subtracted from the value of similarity corresponding to complete identity if the distance is to be metric (though not all similarity measures are capable of giving metric distances).

A variety of similarity and distance measures could be used. Orloci (1974, 1978) suggested using Euclidean distance on data standardized by stand vectors ('standard distance', p. 194) or the Manhattan metric on data standardized by stand total (i.e. expressed as relative values). He rejected the use of these measures on unstandardized data because they are then sensitive to differences in total amount in stands and have no upper limits to their values. Whether these features are desirable or not depends in part on the nature of the data. If there are stands with no species in common, it may be thought useful to differentiate between a pair of species-poor stands with no species in common and a pair of species-rich stands with no species in common; use of unstandardized data will do so.

In the original technique, the axes, except by chance, neither intersect nor are perpendicular to one another. Bray and Curtis noted that they are normally oblique but accepted the distortion produced by then plotting stands as if the axes were perpendicular. Provided the axes intersect, stands can readily be plotted onto oblique axes (Beals, 1965). Equally, provided the axes are perpendicular, although they do not intersect, coordinates remain the same if the axes are projected orthogonally onto a plane parallel to both and an ordination diagram can be plotted directly from the calculated coordinates. When the axes are oblique and do not intersect, the basis of a diagram must be a plane containing one of the original axes and parallel to the other. Coordinates of stands will not then be those on the original axes. Orloci (1974, 1978) provided a method for determining the coordinates. Similar considerations apply if further axes are being constructed.

Orloci (1966) earlier suggested an alternative method of deriving directly axes which are perpendicular (Fig. 49)*. The first axis is taken as the line joining the pair of stands (R_1, R_2) most distant from one another, as in the original technique. The position of any other stand, P_j, on this axes (X_{1j}) is obtained (by solution of right-angled triangles) as

$$X_{1j} = (D_{1j}^2 + D_{12}^2 - D_{2j}^2)/2D_{12}.$$

The distance of P_j from the axis is $a_j = \sqrt{(D^2{}_{1j} - X^2{}_{1j})}.$

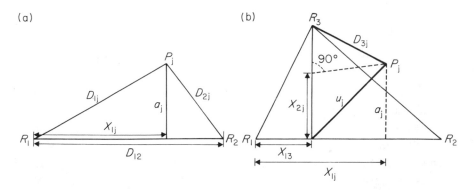

Fig. 49. Derivation of stand positions on the first two axes of perpendicular ordination. Note that P_j in (b) is not in the plane $R_1 R_2 R_3$. (See text).

Orloci took as the reference stand (R_3) for a second axis that stand which has the maximum distance from the first axis (a_{max}) and defined the second axis as the perpendicular from R_3 onto the first axis. Position on the second axis is

* The procedure proposed by Swan et al. (1969) is identical but the expressions for axis positions are presented in a slightly different form.

then readily obtained (by solution of right-angled triangles) as

$$X_{2j} = (u^2_j + a^2_{max} - D^2_{3j})/2a_{max},$$

where $u^2_j = a^2_j + (X_{1j} - X_{13})^2$.

The distances of P_j from the plane defined by the first two axes is b_j $= \sqrt{(D^2_{1j} - X^2_{1j} - X^2_{2j})}$. The stand which has the maximum distance from this plane (b_{max}) may be taken as the reference stand (R_4) for a third axis. The position on the third axis is then

$$X_{3j} = (v^2_j + b^2_{max} - D^2_{4j})/2b_{max},$$

where $v^2_j = b^2_j + (X_{1j} - X_{14})^2 + (X_{2j} - X_{24})^2 *$.

As Swan *et al.* (1969) pointed out, if the relationship of a stand to the first axis is split into orthogonal components a represents the residuum unaccounted for by position of the first axis. Similarly b represents the residuum unaccounted for by positions on the first and second axis. After determining position on the third axis, a residuum $c = \sqrt{(D^2_{1j} - X^2_{1j} - X^2_{2j} - X^2_{3j})}$, will remain. From successive residua, further orthogonal axes can be derived by expressions analogous to those given above. The position on the fourth axis will be obtained as

$$X_{4j} = (w^2_j + c^2_{max} - D_{5j})/2c_{max},$$

where $w^2_j = c^2_j + (X_{1j} - X_{15})^2 + (X_{2j} - X_{25})^2 + (X_{3j} - X_{35})^2$.

Orloci selected the reference stands on the criteria of maximum interstand distance (for the first two reference stands) or maximum residual distance unaccounted for, but the procedure can clearly be used also if reference stands are chosen on some other basis.

If the stands represented by a set of data form a more or less elongated cluster in unreduced, multidimensional space, as they are likely to do if the stands are responding primarily to a main influencing factor, it is desirable that the first axis should lie, at least approximately, in the direction of the major axis of the cluster. Only if it does will the relative observed interstand distances be retained as nearly as possible on the axis. If the data set include any stands which are markedly different from the rest, and the two most distant stands are chosen as the endpoints of the first axis, it is likely that one, and possibly both, of the end stands will be outside the range of the majority in this way, and the resulting axis may be oblique to the major axis of the data. This effect is evident in many published uses of the Bray and Curtis technique in its original form

---

* The formula for position on the third axis given by Orloci is incorrect (M.O. Hill, personal communication).

(see Fig. 50, where the first axis is based on the average position on axes specified by stands 9 and 43, and by stands 2 and 15 respectively).

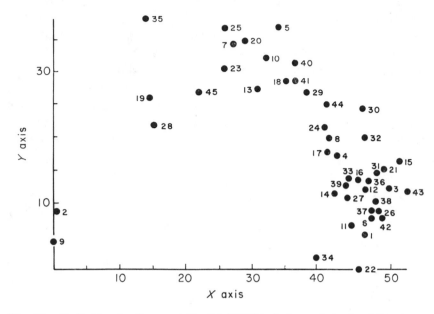

Fig. 50. Ordination by Bray and Curtis' (1957) technique of stands of limestone grassland vegetation. (From Gittins, 1965a, by courtesy of *Journal of Ecology.*)

Van der Maarel (1969) noted that stands defining an axis which is efficient in this sense will show strong negative correlation in their distances to other stands and suggested taking as the endpoints of the first axis the pair of stands showing the greatest negative correlation. To reduce computation he adopted as an approximate estimate of negative correlation the sum of the absolute differences for two stands between corresponding cells of the matrix of interstand distances.

Swan *et al.* (1969) approached the problem of selecting the stand pair accounting for the maximum amount of total interstand distance more directly. They calculated the total squared interstand distances along the axes defined by all possible stand pairs and accepted as the first axis that pair of stands for which this is greatest. Since total squared interstand distance is a multiple of the sum of squares of stand position*, it is sufficient to determine the latter. After fitting the first axis, the second axis is based on that reference

* Total squared interstand distance is n times the sum of squares not $2n$ times, as Swan *et al.* state. Their formula includes each interstand distance twice.

stand which gives a maximum sum of squares of positions on the corresponding axis and similarly for further axes.

　　Orloci (1966) suggested a different approach to selection of stand-defined
axes. Using data centred by species, he took that stand vector (i.e. the line
joining the centroid of the stands to the stand position) on which the projection
of other stand vectors was maximum as the first axis.

　　It is thus possible to increase the efficiency of stand-defined axes over the
original form of Bray and Curtis's technique or Orloci's modification of it, but
this can only be done at the cost of increased computation. With other than
small sets of data computation by hand will be unacceptably lengthy and
recourse must be made to electronic computing. The technique of principal
component analysis, which determines the axis or component which accounts
for the maximum variance at each stage, is then available.

　　Principal component analysis has often been described as a form of 'factor
analysis' and is so referred to in the earlier literature (e.g. by Goodall (1954c),
who first used it for vegetation analysis, and in previous editions of this book).
Although the algebraic manipulations involved in principal component
analysis and in factor analysis in the strict sense are closely similar, the concepts
and assumptions are different and it is important to distinguish the two groups
of techniques. Accounts of principal component analysis are available in
general texts on multivariate analysis, e.g. Kendall (1957), Lawley and Maxwell
(1963), Seal (1964). Useful discussions include those by van Groenewoud
(1965), Gittins (1969), Jeffers (1978), Noy-Meir (1971b) and Williams (1976) in
a specifically ecological context and that by Blackith and Reyment (1971) in a
more general biological context.

　　The basis of principal component analysis is most readily introduced in
terms of the very simple situation of a number of stands containing varying
amounts of two species only. Such stands may be represented in relation to
orthogonal species axes, the coordinates of a stand representing the amounts it
contains of species 1 and species 2 respectively (Fig. 51(a)). The relationship of
the stands could be reproduced exactly by any other two axes lying in the same
plane. Principal component analysis in its original form operates on data
centred by species, i.e. the origin of the species axes is moved to the centroid of
the set of stands (Fig. 51(b)). It then projects the stands onto a line through the
origin so orientated that the sum of squared distances of the stands from the
line is minimized, i.e. the sum of squares of distances of the projected positions
of the stands along the line is maximized. This line is the first component or
axis and the coordinates of stands on it will be of the form

$$z_1 = a_{11}y_1 + a_{12}y_2,$$

where y_1 and y_2 are the coordinates on the original axes (the amounts of species 1 and 2 respectively) expressed as deviation from species mean and a_{11}, a_{12} are constants calculated to maximize stand distances along the line.

In the simple two-dimensional case the distances of stands from the first axis will give coordinates on a second axis, orthogonal to the first, $z_2 = a_{21} y_1 + a_{22} y_2$. The stands have been referred to two new axes, describing exactly their positions relative to one another, but with as much as possible of their variability in position accounted for by the first axis, whereas with the original species axes in Fig. 51b, the variability was expressed more or less equally on the two axes*.

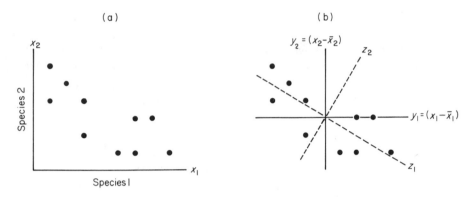

Fig. 51.   Relation of stands containing varying amounts of two species (see text).

If there are three species in the data, the second axis could be placed in any orientation perpendicular to the first axis. It is so orientated that a maximum amount of the variability not accounted for by the first axis is expressed by projection of the stands onto it, leaving the residual variability to be expressed on the third axis. Stands will now have coordinates

$$z_1 = a_{11} y_1 + a_{12} y_2 + a_{13} y_3$$
$$z_2 = a_{21} y_1 + a_{22} y_2 + a_{23} y_3$$
$$z_3 = a_{31} y_1 + a_{32} y_2 + a_{33} y_3.$$

* It will be seen that the effect of this derivation of axes is to partition the sum of squares of distances of stands from their centroid into two orthogonal components. As Williamson (1972) pointed out, this is similar to the partition of sum of squares in an analysis of variance, the major difference being that in analysis of variance the criteria on which the partition is based are external, but in principal component analysis they are internal.

In the more realistic case of n stands containing m species, the new coordinates will be of the form

$$z_1 = a_{11}y_1 + a_{12}y_2 + a_{13}y_3 + \ldots\ldots + a_{1m}y_m$$
$$z_2 = a_{21}y_1 + a_{22}y_2 + a_{23}y_3 + \ldots\ldots + a_{2m}y_m \text{ etc.}$$

To describe relative stand positions completely will require as many axes as there are dimensions in the original data. The maximum number required is m or $(n-1)$, whichever is less. Principal component analysis does not in itself produce any reduction in dimensionality. Its value is that in changing the reference axes to a new orthogonal set, it concentrates the variability in the successive axes derived, so that the information of value in interpreting the data is likely all to be accounted for by the first few axes and the later ones may be ignored*.

It is worth noting that the first principal component is that derived variable which has maximum correlation with the data variables and similarly the next principal component has maximum correlation with the data variables subject to the constraint that it is uncorrelated with the first component (Hill & Smith, 1976).

If the first few axes extracted are accepted as adequate to display the information of interest in the original data, then a reduction in dimensionality has been achieved and the z's, or component scores, provide the coordinates of stands in an ordination. The a's, or component loadings, are also of interest. Each a is characteristic of a particular species in relation to a particular axis or component. Thus species which have similar distributions among the stands being analysed will have similar component loadings, and the component loadings provide the coordinates of an ordination of species in relation to the same components. If information is available about the ecological tolerances of at least some species such a species ordination will assist the interpretation of the stand ordination.

For the simple two-dimensional case presented above the derivation of the first axis could be made from the first principles, in a manner analogous to the derivation of a simple regression. In the multiple-dimensional case recourse must be made to a matrix algebra approach and analysis of more than a small initial data matrix is only practical if a computer is available.

---

* The initial species axes are equally orthogonal, in that change of coordinate on one axis does not necessarily imply change of coordinate on other axes. This does *not* mean that species occurrences or amounts are uncorrelated. Indeed, it is the correlations between species in vegetation that are of interest. The meaning of orthogonality in this context appears sometimes to have been misunderstood (e.g. Beals, 1973).

The starting point is a square matrix of similarities between species (correlation coefficients or covariances/variances in the usual form of principal component analysis) (see, for example, Williams, 1976). For such a matrix there are number (equal to the dimension of the matrix) of *eigenvalues* or *latent roots*, which are the solutions of the determinant equation

$$\begin{vmatrix} (r_{11}-\lambda) & r_{12} & r_{13} & r_{14}\cdots\cdots r_{1m} \\ r_{21} & (r_{22}-\lambda) & r_{23} & r_{24}\cdots\cdots r_{2m} \\ r_{m1} & r_{m2} & r_{m3} & r_{m4}\cdots\cdots (r_{mm}-\lambda) \end{vmatrix} = 0$$

or, in matrix notation,

$$|\mathbf{R} - \lambda\mathbf{I}| = 0$$

Associated with each value of λ is an *eigenvector* or *latent vector*, $\mathbf{V}$, obtained by solving the equation

$$\begin{pmatrix} (r_{11}-\lambda_p) & r_{12} & r_{13} & r_{14}\cdots\cdots r_{1m} \\ r_{21} & (r_{22}-\lambda_p) & r_{23} & r_{24}\cdots\cdots r_{2m} \\ \vdots & \vdots & \vdots & \vdots \\ r_{m1} & r_{m2} & r_{m3} & r_{m4}\cdots\cdots (r_{mm}-\lambda_p) \end{pmatrix} \begin{pmatrix} v_{1p} \\ v_{2p} \\ \vdots \\ v_{mp} \end{pmatrix} = \begin{pmatrix} 0 \\ 0 \\ \vdots \\ 0 \end{pmatrix}$$

or, in matrix notation,

$$(\mathbf{R} - \lambda\mathbf{I})\mathbf{v}_p = 0.$$

The latent vector $\mathbf{v}_p$ then gives the species loadings (a's) on the pth axis. The value of λ for a component is proportional to the variability accounted for by that component. For analysis of covariances $\Sigma\lambda$ is equal to the total variance and for analysis of correlation coefficients, where each species has unit variance, $\Sigma\lambda$ is equal to the number of species. The 'efficiency' of each axis can thus be defined as a percentage of the total variability in the data accounted for.

For the conditions of principal component analysis that the components are orthogonal to one another and that the component accounts for the maximum amount of the variability not accounted for by previous components,

$$\lambda = \Sigma a^2 \Sigma z^2.$$

This defines the relationship between the a's and z's for a component but does not determine their absolute values. Two alternative conventions are commonly used. With R-scaling $\Sigma a^2 = \lambda$, so that $\Sigma z^2 = 1$ and with Q-scaling $\Sigma a^2 = 1$ so that $\Sigma z^2 = \lambda$. It is helpful to use R-scaling for species ordinations and Q-scaling for stand ordinations; the spread of points in a graphical representation is then greater on axes which account for more of the total variability.

The data matrix has so far been considered as representing stands in relation to species axes, and principal component analysis as giving an ordination of species directly (component loadings or species loadings) from which an ordination of stands can be obtained (component scores or stand scores). This has commonly been referred to as R-analysis. The same data matrix can equally be regarded as defining species positions in relation to stand axes. The same procedure applied now to similarities between stands will give an ordination of stands directly (component loadings, now stand loadings) from which an ordination of species may be obtained (component scores, now species scores). This is a Q-analysis.

It can be shown (e.g. Gower, 1966; Orloci, 1967b) that, provided the same transformation of the data matrix are used, R and Q analyses result in identical components. Thus if covariance is used in the R-analysis, implying centring but not standardization of the data ($\Sigma_i(x_{ih} - \bar{x}_h)(x_{ik} - \bar{x}_k)$ between species h and k), the corresponding similarity coefficient between stands i and j is $\Sigma_k(x_{ik} - \bar{x}_k)$ $(x_{jk} - \bar{x}_k)$, the 'weighted similarity coefficient' of Orloci (1966), and not the covariance between stands, $\Sigma_k(x_{ik} - \bar{x}_i)(x_{jk} - \bar{x}_j)$. Since R and Q analyses give identical results, choice of the method to use is one of convenience, R being more economical of computing if there are fewer species than stands and Q more economical if the reverse applies. This R/Q duality and its dependence on using the same data transformation have not always been realized and this has led to argument about the basis of selection of R and Q approaches*.

Principal component analysis has been considered above in terms of change of axes in a 'species space' (R) or 'stand space' (Q). The equations

$$z_{ip} = a_{p1}y_{i1} + a_{p2}y_{i2} + \ldots \ldots + a_{pm}y_{im}$$

for a particular stand i, can be rearranged in the form

$$y_{ik} = a_{1k}z_{i1} + a_{2k}z_{i2} + \ldots \ldots + a_{pk}z_{ip},$$

which is the underlying model of principal component analysis, expressing the performance of species k in stand i in terms of independent 'properties' of the site (z's) and the 'responses' of the species to these properties (a's). Expressed in this way the value and the limitations of the method become clearer (Noy-Meir, 1970). The dependence on the two parameters is assumed to be multiplicative or interactive and only when a site is 'suitable' for a species (high values of z's) and the species has a large 'capacity' of responding to the suitability of the site (high values of a's) will the performance of the species be high. Further, several

* For earlier confusion about the definition of R and Q approaches, see Ivimey-Cook et al. (1969).

kinds of suitability and capability give rise to the different components, which are assumed to contribute additively to the total performance. Noy-Meir (1970) concludes 'such a general model seems ecologically acceptable, if "capability" and "suitability" are interpreted in a wide enough sense. A too narrow interpretation of these terms as related to a specific environmental factor can, however, lead to improper use of the model . . . . . '.

Though the extraction of principal components from a matrix of measures of similarity is a fixed technique, the results of principal component analysis differ greatly according to the transformation, if any, applied to the initial data. Principal component analysis was developed largely in relation to the analysis of psychological tests before it was applied to vegetation data, and for this purpose centring and, often, standard deviate standardization of the data are appropriate and may be necessary. Perhaps because of this it appears to have been widely assumed by both users and critics of principal component analysis as an ordination technique that it necessarily involves either covariance or the correlation coefficient as the measure of similarity. Some discussions of principal component analysis further fail to realise that covariances and correlation coefficients imply particular data transformations. Noy-Meir (1970, 1971b, 1973a; Noy-Meir et al., 1975) has greatly clarified the application of principal components analysis to vegetational data and the following discussion of data transformation is based on his account.

Principal component analysis operates on similarities which are in the form of sums of cross-products between the measures of species representation in stands, i.e. similarity between species h and k is $s_{hk} = \Sigma_i y_{ih} y_{ik}$, and similarity between stands i and j is $s_{ij} = \Sigma_k y_{ik} y_{jk}$, where y_{ik} is the transformed amount of species k in stand i. The difference between different analyses is in the transformation from x_{ik}, the amount of species k actually recorded in stand i, to y_{ik}. Thus, if the correlation coefficient between species h and k,

$$s_{hk} = \Sigma_i (x_{ih} - \bar{x}_h)(x_{ik} - \bar{x}_{ik}) / \sqrt{\{\Sigma_i (x_{ih} - \bar{x}_h)^2 \Sigma_i (x_{ik} - \bar{x}_h)^2\}},$$

is used, the transformation

$$y_{ik} = (x_{ik} - \bar{x}_k) / \sqrt{\Sigma_i (x_{ik} - \bar{x}_k)^2}$$

has been applied, i.e. the data have been centred by species and standardized by species standard deviate. Before applying principal component analysis to a set of data, the decisions must be made whether to centre the data and whether to apply a standardization. These decisions have to be based on ecological not statistical considerations. A range, not exhaustive, of possible transformations is given in Table 23.

If a variable has an arbitrary zero-point, e.g. temperature in °C, centring is

Table 23. Transformations in Euclidean space (modified from Noy-Meir, 1973a)

	Transformed scores y_{ik}		Similarity between sites (Q)	s_{ij} presence	Similarity between species (R)	s_{hk} presence
	Quantity	Presence				
Untransformed	x_{ik}	0 , 1	Product-moment between stands	m_{ij}	Product-moment between species	n_{hk}
By stand centred	$x_{ik} - \bar{x}_i$	$-\dfrac{m_i}{m}$, $1 - \dfrac{m_i}{m}$	Covariance between stands	$m_{ij} - \dfrac{m_i m_j}{m}$		
standardized by norm	x_{ik}/q_i	0 , $\dfrac{1}{\sqrt{m_i}}$	Congruence between stands	$\dfrac{m_{ij}}{\sqrt{(m_i m_j)}}$		
standardized by total	$x_{ik}/m\bar{x}_i$	0 , $\dfrac{1}{m_i}$		$\dfrac{m_{ij}}{m_i m_j}$		
centred + standardized by standard deviation	$(x_{ik} - \bar{x}_i)/s_i$	$\sqrt{\left(\dfrac{m_i/m}{m - m_i}\right)}$, $\sqrt{\left(\dfrac{m - m_i}{m m_i}\right)}$	Correlation between stands	$\sqrt{(\chi^2_{ij}/m)}$		

By species

				Weighted similarity coefficient		Covariance between species	$n_{hk} - \dfrac{n_h n_k}{n}$
centred	$x_{ik} - \bar{x}_k$	$\dfrac{-n_k}{n}$	$1 - \dfrac{n_k}{n}$				
standardized by norm	x_{ik}/q_k	0	$\dfrac{1}{\sqrt{n_k}}$			Congruence between species	$\dfrac{n_{hk}}{\sqrt{(n_h n_k)}}$
standardized by total	$x_{ik}/n\bar{x}_k$	0	$\dfrac{1}{n_k}$				$\dfrac{n_{hk}}{n_h n_k}$
centred + standardized by standard deviation	$(x_{ik} - x_k)/s_k$	$\dfrac{\sqrt{(n_k/n)}}{\sqrt{(n - n_k)}}$	$\sqrt{\left(\dfrac{n - n_k}{mn_k}\right)}$			Correlation between species	$\sqrt{(\chi^2_{hk}/n)}$

Doubly

centred	$x_{ik} - \bar{x}_i - \bar{x}_k + \bar{\bar{x}}$		
standardized by total	$x_{ik}/\sqrt{(mn\bar{x}_k\bar{x}_i)}$	0	$\dfrac{1}{\sqrt{(m_i n_k)}}$

---

i, j = stand indices; h, k = species indices; x_{ik} = score for species k in stand i; n = number of stands; m = number of species; $\bar{x}_i$ = mean score for stand i, $\sqrt{\Sigma_i^n x_{ik}^2}$; $\bar{x}_k$ = mean score for species k; $\bar{\bar{x}}$ = mean of all scores; q_k = species norm, $\sqrt{\Sigma_k^m x_{ik}^2}$; q_i = stand norm, $\sqrt{\Sigma_i^n x_{ik}^2}$; $s_i = \sqrt{\Sigma_k^m (x_{ik} - \bar{x}_i)^2}$; $s_k = \sqrt{\Sigma_i^n (x_{ik} - \bar{x}_k)^2}$.

logically necessary. Most measures of species representation, however, have a natural zero-point, the absence of a species from a stand, and the option of using non-centred data is available. It has commonly been assumed that the first component of a non-centred ordination is an uninformative 'general' one passing from the origin to the centroid of the set of points (Dagnelie, 1960; Orloci, 1966). Noy-Meir (1973a) showed that if the data include two discrete groups or series of stands, with no species in common, (so that there are correspondingly two discrete groups of species with no stands containing representatives of both groups) each group will have a general or unipolar component with positive values for all the species and stands for that group; other components are bipolar but with non-zero values for members of one group only. Table 24 is an artificial matrix including two distinct groups and Fig. 52 shows the results of stand ordinations with and without centring. Only the non-centred ordination (Fig. 52(a)) brings out the distinctiveness of the two groups, with stands 1–7 spread on the first two axes but having zero values on the third axis and, conversely, stands 8–10 spread on the third axis but with zero values on the first two.

Table 24.   Artificial 10×10 matrix (Noy-Meir, 1973a)

Stands	A	B	C	D	E	F	G	H	I	J	m_i
					Species						
1	+	+									2
2		+	+	+							3
3				+							1
4		+		+	+	+					4
5				+	+						2
6				+	+	+					3
7	+	+	+	+	+	+	+				7
8								+	+		2
9									+	+	2
10										+	1
n_k	2	3	3	6	4	3	1	1	2	2	27

With real data there are unlikely to be completely distinct groups, but quasi-disjunct groups with only a small proportion of intermediate stands or of species occurring in stands of more than one group may occur, especially in data from a wide range of stands. Figure 53 shows the direction of the first two axes in non-centred ordinations and the first axis in centred ordinations for a series of artificial 10×10 two-cluster matrices ranging from homogeneous to

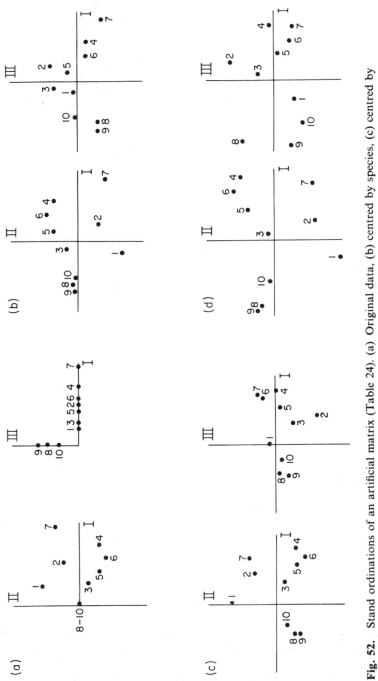

Fig. 52. Stand ordinations of an artificial matrix (Table 24). (a) Original data, (b) centred by species, (c) centred by stand, (d) doubly centred. (From Noy-Meir, 1973a, by courtesy of *Journal of Ecology*.)

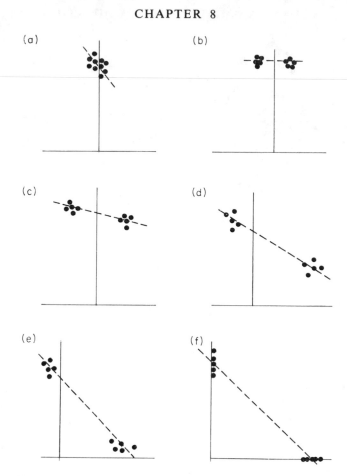

Fig. 53. The direction of (———) the first two non-centred and (---) the first centred principal components for artificial data sets from (a) homogeneous to (f) disjunct. (From Noy-Meir, 1973a, by courtesy of *Journal of Ecology*.)

disjunct. The first axis is naturally unipolar for all non-centred ordinations. The second axis becomes increasingly asymmetric with increasing disjunction in the data*. The centred ordination, on the other hand, scarcely distinguishes between different degrees of disjunction (Fig. 53 (b – f)). There are evident

* Noy-Meir suggested as a measure of asymmetry a coefficient

$$\alpha = 1 - (\overline{\Sigma}a^2 / \overset{+}{\Sigma}a^2)$$

where $\overline{\Sigma}a^2$ and $\overset{+}{\Sigma}a^2$ are the sum of squares of negative and positive loadings respectively (the larger sum being taken by convention as $\overset{+}{\Sigma}a^2$). All axes that are unipolar or nearly so (say $\alpha > 0.9$) may be taken as defining distinctive vegetational series, the remaining bipolar axes describing aspects of variation within the series.

advantages in non-centred ordinations if there is any considerable degree of disjunction in the data (see also Feoli, 1977; Noy-Meir & Whittaker, 1977; Carleton, 1979, 1980; Carleton & Maycock, 1980). If the data are homogeneous or at least continuous, however, there is no advantage in abstention from centring; the first axis conveys no useful information and may be disadvantageous in that its direction constrains that of the second axis which may then not be in the direction of maximum variability within the single cluster.

If centring is desirable, the alternatives of centring by species, centring by site and double centring are open. Centring involves specification of the origin or point of reference of the multivariate model. 'It is the "point of zero information"; anything that is at it, is trivial and uninteresting; anything that deviates from it, is "information"' (Noy-Meir). When data are not centred the point of reference is the all-zero record and an interest in the absolute occurrences and co-occurrences of species, and in absolute composition and joint composition of stands is implied. Centring by species transfers the reference point to a hypothetical 'average stand' containing the mean amount of each species, so that stands contribute information in so far as they depart from this average composition, and species in so far as they depart from uniform distribution over all stands. Centring by stand transfers the point of reference to an 'average species', which contributes the same proportion $(1/m)$ of the total vegetation in all stands, so that only species whose distribution differs from that of total vegetation and stands whose composition departs from equal proportions of all species contribute information. As Noy-Meir said, the effect of double centring is difficult to interpret. He suggested that ecological problems are rarely stated in a form that requires centring by stand, or double centring. The choice will generally be between centring by species and no centring.

A further important consideration is that if the data are centred, the absence of a species influences the analysis, whereas it does not do so with uncentred data. The desirability of allowing absence of species to influence the analysis will depend on the kind of data and the purpose of the analysis.

Just as variables without a natural zero-point must be centred, so variables in one analysis measured on different scales must be standardized to eliminate differences due solely to different scales of measurement. Two standardizations are commonly used, by norm and by total. The norm is $\sqrt{\Sigma x^2}$ so that $y = x/(\sqrt{\Sigma x^2})$. If the data have previously been centred (so that $y' = (x - \bar{x})$ subsequent standardization by norm is $y = \dfrac{y'}{\sqrt{\Sigma y'^2}} = \dfrac{(x - \bar{x})}{\sqrt{\Sigma(x - \bar{x})^2}}$, i.e. what is generally referred to as 'standardization by standard deviate'. For standardization by total, $y = x/\Sigma x$. Standardization by total cannot be applied after centring ($\Sigma y = 0$). Vegetation data are normally in the same form throughout a set and, except for densities of species of widely different size and growth form, directly comparable (Noy-Meir et al., 1975). Standardization is therefore optional and its desirability must be assessed on ecological grounds.

Standardization may greatly affect the presentation of relationships between stands and between species (Fig. 54), and the resulting ordination (Fig. 55). Noy-Meir et al. (1975) have elucidated the effects of standardization in ecological terms. They have shown theoretically, and demonstrated by analysis of varied sets of data, that $\Sigma_i y^2_{ki}$, the sum of squares of all values for species k (squared norm or length of position vector of the species in stand-space) is a measure of the 'weight' or 'importance' of the species in the configuration of the points around the origin, regardless of its relationships with other species. Similarly, $\Sigma_k y^2_{ki}$, is a measure of the a priori importance of stand i in the configuration of points around the origin and the ordination derived from it. Choice of standardization thus depends on the relative weighting of species and stands of different kinds considered desirable.

With untransformed data, stands will clearly be weighted by their richness (number of species occurring for presence data, total amount of vegetation for quantitative data) and species by their abundance (number of stands in which they occur for presence data, total amount of the species in all stands for quantitative data). Centring will in itself alter the weightings. Centring by species will weight species according to their variance, so that for presence data species with 50% frequency will contribute most. Centring by stand will have the same effect on stands and for presence data stands containing 50% of species will contribute most. With quantitative data the most abundant species generally have the highest variance and will therefore contribute most; the relation between stand total and variance of species contributions to that total is less predictable.

Standardizing by stand norm (or by standard deviate after centring) equalizes the weighting of stands. The weight of a species still depends on its abundance but also on the 'poorness' of sites in which it occurs so that the species which are the only or the dominant species in the largest number of stands will contribute most. Conversely, standardizing by species norm equalizes the weighting of species, while the weighting of stands depends jointly on their richness and the rarity of the species which occur in them. Standardization by total gives weights proportional to $\Sigma x^2/(\Sigma x)^2$, so that for presence data rare species and poor sites are weighted more heavily by standardizing by species and stand totals respectively. 'The expressions for Σy^2 of doubly standardized data are not so simple to interpret, but they indicate

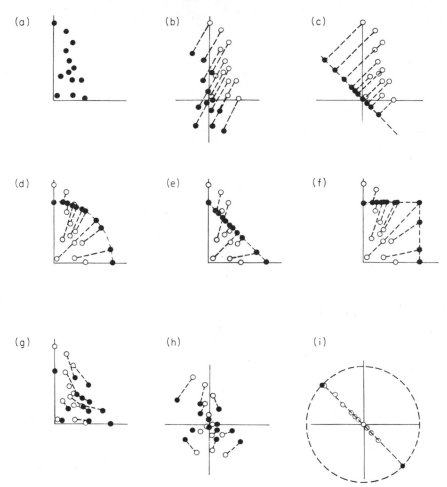

Fig. 54. The effects of data transformations on the representation of a sample of stands in two-dimensional species-space. (a) The original space (raw data). (b) Centring by species mean; translation of the centroid to the origin. (c) Centring by stand mean; projection on diagonal (diagonal hyperplane in multidimensional case). (d–i) Standardizations. (d) By stand norm; radial projection to unit circle (surface of hypersphere). (e) By stand total; radial projection to unit hypotenuse (hyperplane). (f) By stand maximum; radial projection to unit square (hypercube). (g) By species norm; rescaling to isotropic dispersion around the origin. (h) By species standard deviation after centring by species (species correlation); rescaling to isotropic dispersion around the centroid. (i) By stand standard deviation after centring by stand; radial projection to intersection of diagonal hyperplane with unit hypersphere (reduction to two points in the two-dimensional case only). (From Noy-Meir *et al.*, 1975, by courtesy of *Journal of Ecology*.)

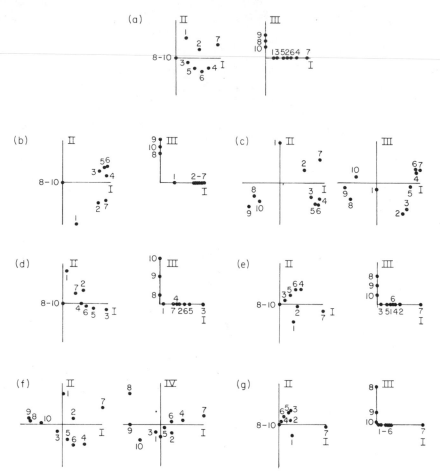

Fig. 55. Stand ordination of an artificial matrix (Table 24) with different transformations. (a) Original data, (b) standardized by stand norm, (c) centred by stand and standardized by stand standard deviation, (d) standardized by stand total, (e) standardized by species norm, (f) centred by species and standardized by species standard deviation, (g) standardized by species total. (From Noy-Meir *et al.*, 1975, by courtesy of *Journal of Ecology*.)

some overall weighting in favour of both rare species and poor sites, and particularly coincidences of the two' (Noy-Meir *et al.*, 1975).

Noy-Meir's clarification of cross-product similarity coefficients is also relevant to the use of such coefficients in agglomerative classification (Chapter 7). The only cross-product coefficient that has commonly been considered for direct use in classification is the correlation coefficient between stands (i.e.

centring by stands and standardizing by stand norm) which has been rejected empirically as not giving satisfactory classifications, but Euclidean distances are related to cross-product similarity coefficients*. The same considerations of centring and standardization can thus be applied to the use of Euclidean distance in classification. Orloci (1967a) used standardization by stand norm (when squared distance $= 2(1 - \Sigma y_i y_j)$ but otherwise little attention has been paid to transformation of data before calculating Euclidean distance. It is interesting, however, to note that some element of standardization by stand enters into many of the 'empirical' similarity coefficients that have been used in classification.

Hill (1973b) introduced, under the name 'reciprocal averaging', a technique designated 'l'analyse factorielle des correspondances' ('correspondence analysis') by Benzecri (1973), which had earlier been discussed in various other contexts (Hill, 1974). Although reciprocal averaging is a particular form of principal component analysis, uncentred with simultaneous double standardization by total, Hill derived it by a different and simple rationale from Whittaker's direct gradient analysis approach.

Consider the simple case of presence/absence data from stands occupying a range of positions along a well-defined floristic gradient, corresponding to a single environmental gradient. Whittaker's approach (pp. 231) is to assign approximate weightings to species according to their optimum positions on the gradient and use these weightings to derive a position along the gradient for each stand. There are two limitations in this procedure; the gradient has to be recognized initially, and the weightings assigned to species are approximate and subjectively determined. However, the stand scores can be used to give corrected species scores, as the average score for the stands in which a species occurs. These new species scores, after rescaling to the same range as originally used (Hill used a range of 0–100) can then be used to recalculate stand scores. Hill showed that successive cross-calibrations in this way converge to a unique set of values, providing a simultaneous one-dimensional ordination of species and stands, even if the initial species scores are assigned arbitrarily. Thus not only is the subjective element in the determination of species scores eliminated, but it is no longer necessary to recognize the gradient (floristic or environmental) involved; a good initial guess at species scores will reduce the number of iterations necessary to achieve stability, but it is not essential.

The one-dimensional ordination obtained in this way is the equivalent to the first axis of the corresponding principal component analysis. It may also be

* Squared Euclidean distance between stands i and j is

$$\Sigma (y_i - y_j)^2 = \Sigma y^2_i + \Sigma y^2_j - 2\Sigma y_i y_j.$$

interpreted as a rearrangement of the initial data matrix in such a way that positive entries are concentrated as nearly as possible around a diagonal (Hill, 1974), an approach which is intuitively attractive in interpreting vegetational data (Fig. 56).

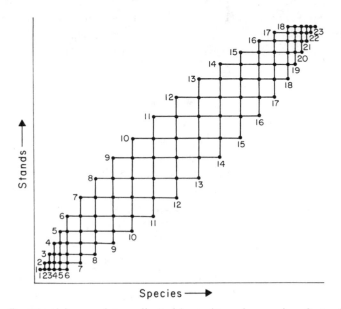

Fig. 56. Presence/absence data ordinated by reciprocal averaging, first axis. (From Hill & Gauch, 1980, by courtesy of *Vegetatio.*)

Although two-way successive calibration will converge to a unique solution, there are other 'solutions' which have the same property (species score = average of scores for stands in which the species occurs and stand score = average of scores of species occurring in the stand, rescaled to constant total range in each case). These are equivalent to the axes other than the first of the corresponding principal component axes.

Reciprocal averaging has been discussed above in terms of presence/absence. It is equally applicable to quantitative data, when the contribution of species to a stand score depends not only on the score for the species but also on the amount of each species (and similarly for the determination of a species score from stand scores). It represents a form of principal component analysis with considerable advantages for 'general purpose' vegetational ordination; being non-centred, it is efficient with heterogeneous data, and the double standardization gives some emphasis on rarer species and poorer sites and especially coincidences between them representing distinctive noda (Noy-Meir *et al.*, 1975). It has the further advantage that it is possible, though somewhat laborious, to extract at least the first two axes by hand, and this may be important to workers in the field in remote situations, for whom a preliminary analysis is often valuable before planning further work. Hill (1973b) provides a worked example.

Reciprocal averaging has two faults (Hill, 1979b; Hill & Gauch, 1980). It shows the 'arch' or 'horseshoe' effect; a linear gradient of composition is expressed as an arch in two dimensions of the ordination (Fig. 57). This is an artefact of the method not reflecting any real feature of the data and results because the second axis is derived in such a way that it is not correlated with the first axis, but it is not necessarily independent of the first axis. The second fault is that equivalent differences in composition are not represented by the same differences in first axis position, e.g. in Fig. 56 stands 15 and 18 differ in composition by exactly the same amount as stands 9 and 12 but are closer together on the first axis.

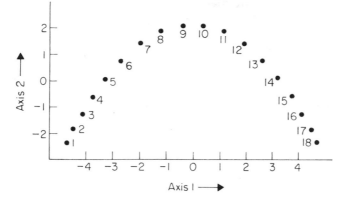

Fig. 57. First two axes of reciprocal averaging ordination of the data of Fig. 56, showing the 'arch' effect. (From Hill & Gauch, 1980, by courtesy of *Vegetatio*.)

Hill (1979b; Hill & Gauch, 1980) proposed a modified version, 'detrended correspondence analysis' (DECORANA), to overcome these faults. The arch effect is removed by adjusting the values on the second axis in successive segments by centring them to zero mean (Fig. 58). The variable scaling on the axes is corrected by adjusting the variance of species scores within stands to a constant value.

Principal component analysis is a flexible and powerful technique of exploring vegetational data, but it has certain limitations. Some authors (e.g. Seal, 1964) have argued that since principal component analysis assumes that

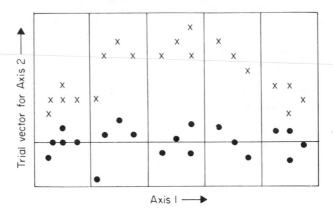

Fig. 58. Method of detrending used in detrended correspondence analysis. X, Initial trial positions on second axis; ●, adjusted positions on second axis. (From Hill & Gauch, 1980, by courtesy of *Vegetatio*.)

the data being analysed are drawn from a single multinormal population, which is clearly very far from true for vegetational data, it cannot validly be used. This criticism involves a misunderstanding of the way it is used. As Crovello (1970), who gives a useful discussion of statistical criticism of the use of the technique, pointed out, the assumption of multinormal distribution is only important if the user desires to draw statistical inferences from the analysis, e.g. to assess the numbers of significant components or, as Crovello suggested, to draw probabilistic frequency ellipses round a stand or species. The ecologist is not usually concerned with such precise interpretation but is needing rather a rearrangement of the data, the interpretation of which is made more subjectively. For this purpose principal component analysis may validly be used, provided the basic model (p. 246) is accepted*.

A more important limitation is that principal component analysis can only validly be used on cross-product coefficients of similarity and it may be argued that the relationships between stands are more validly expressed by distance measurements. Some distance measures can be expressed in terms of cross-products, e.g. Euclidean distance, but others cannot, e.g. the Manhattan metric $\Sigma |y_i - y_j|$. Gower (1966, 1967a) developed a technique, principal coordinate analysis[†], which derives from a matrix of interstand distances coordinates, referred to principal axes, of the corresponding stand positions. From the

---

* There are limitations in the model, the importance of which is discussed below (pp. 264).

[†] Also referred to as 'principal axes analysis' (e.g. Orloci, 1978) or included under principal component analysis in a broad sense (e.g. Anderson, 1971).

matrix of interstand distances, d_{ij}, between stands i and j, a matrix $\mathbf{E}$ with elements $e_{ij} = -\frac{1}{2}d^2_{ij}$ is formed. A further matrix $\mathbf{F}$, with elements

$$f_{ij} = e_{ij} - e_{i.} - e_{.j} + e_{..}$$

is formed. The latent vectors of $\mathbf{F}$, each normalized so that its sum of squares is equal to the corresponding latent root of $\mathbf{F}$, define the coordinates of stands in relation to principal axes. This follows because the distance between points the coordinates of which are the ith and jth rows of the matrix of latent vectors, normalized in this way, of a matrix $\mathbf{F}$ is

$$f_{ii} + f_{jj} - 2f_{ij} = e_{ii} + e_{jj} - 2e_{ij} = -\tfrac{1}{2}d^2_{ii} - \tfrac{1}{2}d^2_{jj} + d^2_{ij} = d^2_{ij}.$$

(The distance between a point and itself, d_{ii}, is zero.)

If the appropriate distance measure is used, the results of principal coordinate analysis will be the same as those of the corresponding principal component analysis, and may in some cases be computationally more convenient (Orloci, 1973). More important, however, is its ability to analyse a matrix of any distance measure between stands, which not only allows a wider choice of resemblance function, but permits the inclusion of mixed data (Dale, 1975)*, such as may be involved in ordination based on stands described by structural features instead of floristic composition. Williams, Dale and Lance (1971) have proposed extensions which would allow asymmetric measures of relationship to be used. Such measures have been little used, but are potentially important in ordinations of species based on nearest-neighbour relationship where if an individual of species A has as its nearest neighbour an individual of species B, it is not necessarily the nearest neighbour of the latter.

If principal coordinate analysis is applied to stands with such a wide range of composition that some pairs of stands have no species in common, the arch effect (p. 259) will occur. The contribution to interstand distance from a species, once it is absent from one of a pair of stands, can be no greater however far apart along a gradient the stands are, so that the complete absence of species in common gives only a lower limit for the appropriate distances. Williamson (1978) suggested a method, 'step across', for presence/absence data which avoids this difficulty. The distance between two stands with no species in common is derived as the shortest distance obtained by moving through one or more stands with intermediate composition. Using the Manhattan metric,

* Use of principal coordinate analysis is clearly more satisfactory than the transformation of a distance to a similarity not directly corresponding to it, e.g. van Groenewoud's (1965) use of $(1 + D^2)^{-1}$ and of e^{-D^2} to analyse a matrix of the generalized distance, D^2.

which is additive provided there are species in common, the arch effect is eliminated.

Factor analysis, in the strict sense, has been much less used in vegetational analysis than principal component analysis, although attention was directed to it by Dagnelie at an early stage in the development of ordination techniques (Dagnelie, 1960; see also Dagnelie, 1965a,b, 1973). It has certain potential advantages for fairly homogeneous sets of data but requires more constrained assumptions than principal component analysis and, as Dale (1975) pointed out, presents numerical problems*.

Principal component analysis relates the total variability in the data to a number of orthogonal axes, extracted in order of the amount of variability they account for; it is essentially an altered presentation of the original data, and does not involve any prior hypothesis on the structure of the data. Factor analysis is concerned with that part of the variability which is accounted for by correlation between the variables. The underlying model of principal component analysis is

$$y_{ik} = a_{1k}z_{i1} + a_{2k}z_{i2} + \dots\dots + a_{pk}z_{ip}$$

relating the amount y_{ik} of species k in stand i, to the 'properties' of the site, z_{i1}, z_{i2} etc. by means of 'responses' a_{1k}, a_{2k} etc. of the species to these properties (p. 246). After analysis, the investigator may, and normally does, decide to interest himself in a limited number, n, of the components only, i.e. to regard the model as

$$y_{ik} = a_{1k}z_{i1} + a_{2k}z_{i2} + \dots\dots + a_{nk}z_{in} + E,$$

where E is regarded as an error, or unexplained, part of the representation of the species. Factor analysis assumes that there are only a limited number, m, of 'common factors' each affecting the representation of two or more species together with a 'unique factor', u_{ik}, peculiar to each species at each site:

$$y_{ik} = a_{1k}z_{i1} + a_{2k}z_{i2} + \dots\dots + a_{mk}z_{im} + u_{ik}.$$

The unique factor includes both a response of the species to the site and an 'error' element, but these cannot in practice be distinguished. The extent to which the representation of a species is determined by the common factors is termed its 'communality'. When factor analysis is performed on the matrix of correlation coefficients between species, as it usually is, so that the diagonal elements (the correlation coefficients between species and themselves) are all unity, these elements are replaced by the corresponding communalities before

* Useful discussions of factor analysis include those of Blackith and Reyment (1971), Dagnelie (1960, 1965a, b, 1973), Dale (1975), Ferrari *et al.* (1957), Gower (1966, 1967a) and Williams (1976).

determining the species loadings. With the correlation matrix the communalities

$$h^2_k = a^2_k + a^2_{2k} + \ldots\ldots + a^2_{mk}.$$

Factor analysis solutions depend either on a specified number of common
factors or on a specified set of communalities. In principle, the number of
common factors can be determined objectively by testing the significance of the
residual correlations after removing the effect of each successive common
factor, but the test assumes that each variable is normally distributed (Lawley
& Maxwell, 1963); this assumption is practically never valid for species values
in vegetation data. Equally, there are no grounds with vegetation data for
initial assignment of communalities. Unlike principal component analysis
there is no unique solution; either an arbitrary decision must be made about the
number of common factors to be extracted, or an unjustified assumption made
that the species are normally distributed. There is a further difficulty in the use
of factor analysis that the calculations of factor scores, to give a stand
ordination, is not straightforward, they can only be estimated 'by methods of
doubtful utility' (Blackith & Rayment, 1971).

The concentration in factor analysis on interrelationships between species
is attractive; it is these interrelationships that are normally of greatest interest,
the proportion of variability assigned to unique factors representing 'background noise' in the context of most sets of data. However, the methodological
difficulties are clearly serious; Dale (1975) emphasized that factor analysis is not
a single method, but has a large number of variants, which differ in their
procedures of estimating communalities and factor scores. Dale suggested that
if the numerical problems of factor analysis can be solved, it may become a
more attractive approach. In practice, in the few cases where factor analysis has
been applied to relatively heterogeneous data, the results have not been
markedly different from those obtained by the simpler technique of principal
component analysis. There seems little advantage in using the more complex
technique, especially as it lacks the flexibility of choice of coefficient available
with principal component analysis.

In concluding discussion of factor analysis it cannot be emphasized too
strongly that the common factors are not necessarily related to factors in the
ecological sense. A common factor represents a set of correlations between
species. It may subsequently be interpreted as being determined by an
identifiable ecological factor but this will not necessarily be so. The same
applies to the components of a principal component analysis used in an
ordination.

In applications of principal component and factor analyses in some fields it is usual to consider rotation of the axes as an aid to interpretation (see, for example, Harman, 1976). Once a limited number of axes have been derived, they can be replaced by the same number of new axes without loss of information. In the ecological context, interpretation may be simplified if each axis is so orientated from the origin that as far as possible each species has a high loading on one axis and zero or low loadings on other axes. Each axis will then represent a type of vegetation and the species characteristic of it will have high loadings on that axis (Ivimey-Cook & Proctor, 1967). Rotation may be to orthogonal or to oblique axes. Noy-Meir (1970) considered that there was little advantage in oblique rotation, but Carleton (1980) found it useful to examine the results of both orthogonal and oblique rotation. Noy-Meir (1971b) used Kaiser's (1958) varimax orthogonal rotation after non-centred component analysis ('nodal ordination') to distinguish vegetational noda* in heterogeneous vegetation (see also Carleton, 1979; Carleton & Maycock, 1980). Varimax rotation maximizes the squared loadings on each axis; this tends to bring individual loadings nearer to unity or zero. Carleton (1979) took nodal ordination a stage further by making a principal component analysis of the resulting noda.

Stand-defined ordination techniques, principal component analysis and factor analysis all assume a monotonic, linear model of species interactions along a single compositional gradient, i.e. if the amount of one species is plotted in a series of stands against the amount of another species the relationship is linear (Fig. 59(b)). Such a relationship will result if both species are responding to an environmental gradient linearly (Fig. 59(a)). It is possible for there to be a linear relationship between species although their relationship with the environmental gradient is curvilinear (Fig. 60), but more frequently curvilinear response to the environmental gradient will result in curvilinear relationships between species (Fig. 61). If one species has zero values in some stands in which the other is varying, the species relationship will again not be linear (Fig. 62). Clearly species relationships are unlikely to be perfectly linear even in a relatively homogeneous set of stands, though they may be monotonic. Abundant evidence from the known response of species to environmental factors in experiments and from the results of direct gradient analysis indicates that species commonly show a peaked response curve in relation to an environmental gradient. Species relationships are then complex (Figs 63, 64) (van Groenewoud, 1976). Thus, with greater heterogeneity among stands, the majority of species relationships are likely to be neither linear nor monotonic.

* This use of 'nodum' corresponds approximately with the use of the term by Poore (1955a, b) for an 'abstract vegetation unit of any category' and by Lambert and Williams (1962) for 'species-*cum*-habitat units'.

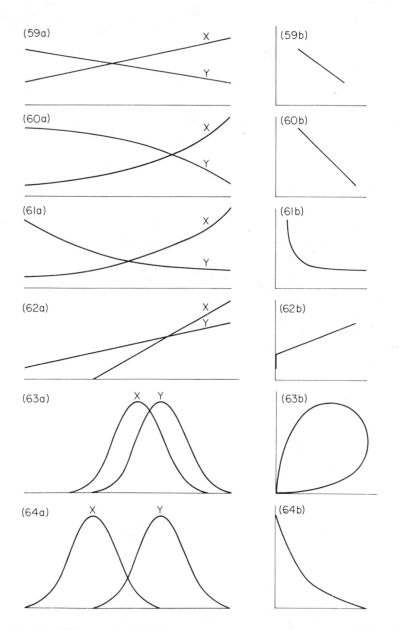

(59a) (59b) (60a) (60b) (61a) (61b) (62a) (62b) (63a) (63b) (64a) (64b)

Figs 59–64. Different relationships between two species, X and Y, along a single environmental gradient. (a) Amount of species at different positions along the gradient, (b) amount of species Y plotted against species X for different positions along the gradient.

The distorting effect of non-linearity of species relationships on ordination techniques that assume linearity has received considerable attention and the problem was reviewed by Austin (1976a, b). The non-linearity may be expected to be manifest in an ordination as a failure of a single gradient of composition, if composed of successive peaked curves of species contributions, to be accounted for by a single axis. Goodall (1954c), in a pioneering application of principal component analysis, interpreted the second component extracted from his data as a non-linear component of species responses to the gradient represented by the first component and emphasized the difficulties that arise from non-linearity. Many published ordinations show a curvilinear arrangement of stands that might intuitively have been expected to form a single gradient.

Harberd (1962) appears to have been the first to use artificial data to examine the problem. Using presence and absence in successive stands each differing from its neighbours by having one species not in its neighbour and lacking one species present in its neighbour, he demonstrated that the end stands in the series occupied less extreme positions on the first axis than the penultimate ones*.

Swan (1970) prepared sets of artificial data representing successive bell-shaped species curves along single gradient and analysed them by the stand-defined technique of Swan *et al.* (1969); the resulting ordinations represented the gradient as a complex curve in several dimensions. Noy-Meir and Austin (1970) analysed the same data by principal component analysis of species-centred data and obtained similar results (Fig. 65). Austin and Noy-Meir (1971) constructed sets of artificial data with the representation of each species having a bell-shaped surface relative to a two-dimensional grid, representing two independent environmental gradients. Principal component analysis of species-centred data gave complex curved surfaces which could not be adequately represented in two or even three dimensions (Fig. 66). Austin and Noy-Meir also examined the effects of several standardizations (by stand norm, successive double standardization and simultaneous double standardization); standardization resulted in marked improvement in the reproduction of the original plane in some cases, but the effects of any one standardization were not consistent over the range of data examined.

Gauch and Whittaker (1972b; Whittaker & Gauch, 1973) analysed artificial 'coenoclines' or 'community gradients' composed of curves of species represen-

* Harberd ostensibly examined the first canonical variate of a matrix of generalized distances, but under the particular conditions that he regarded each stand as a group within which species variances are unity and covariances zero, this was equivalent to extraction of the first axis of a principle coordinate analysis of Euclidean distances.

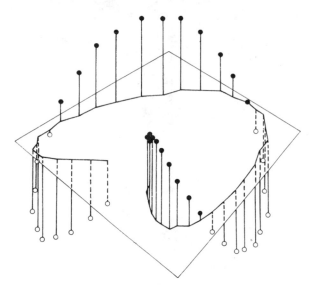

Fig. 65. Principal component ordination of sequence of stands along a single environmental gradient to which species have bell-shaped response curves. (From Noy-Meir & Austin, 1970, by courtesy of *Ecology*.)

tation of normal (Gaussian) form, with varying degrees of 'beta diversity' (species change along the coenocline) and of background noise (Gauch & Whittaker, 1972a, 1976). They concluded that the several variants of stand-defined ordination used produced less distortion, expressed as an arching of the gradient, than principal component analysis, which produced not only arching but incurving of the extreme stands comparable to that found by Noy-Meir and Austin. Only one, unspecified, form of principal component analysis was used. In view of Noy-Meir and Austin's demonstration of the marked effect of standardization in some circumstances, and later demonstrations of the efficiency of reciprocal averaging (Gauch *et al.*, 1977; Austin, 1976b), Gauch and Whittaker's rejection of principal component analysis cannot be accepted uncritically.

The distortions that result from the use of essentially linear techniques of ordination clearly involve problems in the interpretation of ordinations of real data. If the analysis is confined to relatively homogeneous sets of data, these may approximate sufficiently to linearity for distortion to be relatively unimportant. With more heterogeneous sets of data, interpretation must be made with awareness of the likely effects of non-linearity, especially that a second axis may represent largely the non-linear parts of interspecific relationships represented primarily by the first axis, and similarly for later axes.

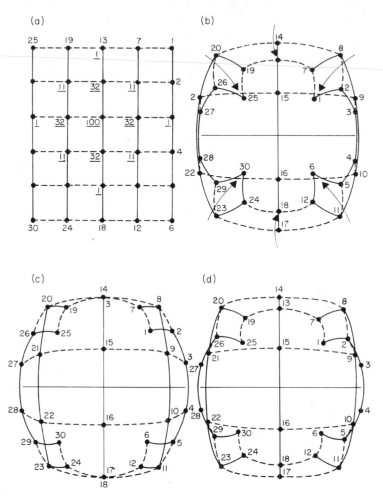

Fig. 66. Principal component ordinations of artificial vegetation data from a two-dimensional environmental grid. (a) Positions of stands in original grid showing the distribution of 'quantities' 1, 11, etc. of species with optimum in stand 15, (b) species with low overlap, (c) species with greater overlap, (d) with additional species outside the grid. (From Austin & Noy-Meir, 1971, by courtesy of *Journal of Ecology*.)

The importance of non-linearity depends to some extent on the objective of ordination. If this is to reproduce a coenocline exactly, as a basis of studying individual species responses (and this seems to be implicit in the discussions of Whittaker and his associates, but see Noy-Meir and Whittaker (1977)), then any failure to reproduce the exact form is a serious shortcoming. If the objective is exploration of the correlation between species composition of stands and the levels of environmental factors (and this is the objective of most practical uses of ordination) as a basis of hypotheses about controlling factors, then what is important is that the ordering of stands should be monotonic with the order on a coenocline; the relationship need not be linear. These different objectives will be considered in turn.

It has commonly been assumed that the response of a species to an environmental gradient has the form of a Gaussian or normal curve (e.g. Gauch & Whittaker, 1972a). While general understanding of the physiological response of individual species to environmental gradients indicates that a unimodal response curve is to be expected (cf. Mueller-Dombois & Ellenberg, 1974, pp. 344 et seq.), there are no theoretical or practical reasons for thinking that such a curve should be of Gaussian rather than of some other broadly 'bell-shaped' form. Further, competitive effects of other species will modify the ideal physiological response curve (Mueller-Dombois & Ellenberg, 1974). Austin (1976a, 1980a) pointed out that the species curves obtained in direct gradient analyses are often far from Gaussian, not infrequently being asymmetric or even bimodal, and has used other, more variable, functions in preparing artificial coenoclines.

There is a further difficulty in developing an adequate model of a coenocline. The shape of response curve will be markedly affected by the scale used for the environmental gradient. Thus Whittaker and Fairbanks (1958) found that planktonic copepod species showed markedly skewed response curves to a linear salinity gradient but apparently symmetrical ones to a logarithmic salinity gradient. An artificial coenocline may be produced by relating whatever form of species response curve is considered appropriate to an axis, but with real data definition of distance along a coenocline is difficult. Most ecological gradients involve a number of environmental factors which vary monotonically but not linearly with one another; consider, for example, salt-marsh stands with such factors as exposure, humus content, salinity, mechanical composition of mud, etc. varying monotonically but not linearly with height above a datum level. These considerations make the concept of 'ecological distance' (Beals, 1973; Gauch, 1973), distance along the environmental gradient, an unrealistic one in any exact sense.

If the objective of analysis is to identify a postulated coenocline, then that ordering and spacing of stands which gives the best fit of the values for individual species to the assumed form of species response curves is required*.

* Noy-Meir (1974a, b) suggested the term 'catenation' for methods which order stands or species in a way which optimally accounts for local similarities, rather than overall similarities.

There are various techniques available for fitting Gaussian curves to data. Gauch and Chase (1974) developed an approach suited to ecological data, which present the difficulties that the sampling points are not usually uniformly spaced along the abscissa (ordination axis) and that the sampling range may not include the entire Gaussian curve. They take the highest observed value for a species as a first estimate of its modal value, Y_0, and the corresponding axis position as a first estimate of the position of the mode, μ. The observed data points are then used to obtain values of σ from the Gaussian equation $y = Y_0 e^{-(x-\mu)^2/2\sigma^2}$, and a weighted average of these values is taken as a first estimate of the standard deviation, σ. Small changes in the values of the three parameters are then made iteratively until a satisfactory fit, assessed by sum of squared deviations, is obtained.

Gauch *et al.* (1974) used this procedure to derive an ordination. The fitting of a curve to the data for a species requires the axis positions to be stated. These may be obtained from a direct environmental measure, e.g. altitude, or as positions on the first axis of a preliminary ordination. Gaussian curves are calculated for all species and then small changes in axis position made iteratively (with consequent changes in the parameters for individual species) until the best possible overall fit to Gaussian curves is obtained.

Johnson and Goodall (1979) suggested a similar approach, but fitting two Gaussian curves for each species, one based on quantitative values where the species is present and the other on probabilities of absence. These are used, as appropriate, in relation to stands and the best approximations for stand positions obtained by an iterative procedure as in Gauch *et al.*'s method.

These techniques can derive one axis only and are only applicable to situations where there is a single overriding gradient represented in the data. Ihm and van Groenewoud (1975) adopted a similar approach, aiming at the optimal overall fit of species values to Gaussian curves, but used a mathematically more complex method to derive stand and species mode positions on the axis directly. Their technique does provide for simultaneous fit to further axes, i.e. multivariate normal surfaces are fitted to individual species values. The assumptions made, that all species have the same standard deviation and that the number of species present in stands is constant, are, however, unrealistic in relation to field data. The method has the further disadvantage that, because the calculations depend on the summed cross-products between species for stands, it becomes inefficient if some pairs of stands have no species in common, as is likely to be the case if an extensive range of a gradient is included in the data. By contrast, the iterative procedure used by Gauch *et al.* is not limited by the range of gradient included and, indeed, is likely to be more efficient for greater ranges because a smaller proportion of the species involved will be represented by truncated curves.

Derivation of an ordination by these methods has the evident limitation that a response curve of Gaussian form is assumed. Moreover, the fitting procedure will tend to distort a non-Gaussian response curve into a more nearly Gaussian form, as Austin (1976b) demonstrated by the analysis of artificial data. Curtis and McIntosh's (1951) continuum analysis was based on species response curves that were unimodal, but not otherwise defined. Their pioneer approach depended on hand-sorting of stand order and subjectively assessed species weightings. A more formalized procedure based on ordering stands in such a way that individual species show a smooth rise and fall in representation is an appropriate basis for detecting and elucidating a coenocline.

Noy-Meir (1974a, b) considered the application to vegetation data of 'parametric mapping', a method developed in psychology by Carroll (Shepard & Carroll, 1966; Kruskal & Carroll, 1969)*. The method seeks an ordering such that changes in the variables being considered (generally amounts of each species, but alternatively a derived entity e.g. axis position in a preliminary ordination) are as smooth or continuous as possible. Continuity is interpreted as requiring that a very small change in axis position should involve only a small change in each variable. Maximal continuity for one variable is obtained when

$$q = \sum_{i=1}^{n} \sum_{j=1}^{n} \left(\frac{\Delta y}{\Delta z}\right)^2 W_{ij}$$

is minimal, where Δy is the change in variable y associated with a change Δz in axis position and summation is over all pairs of points i, j. W_{ij} is a weighting reflecting the adjacency of the points. Carroll initially suggested

$$W_{ij} = 1/(\Delta z)^2$$

so that the quantity to be minimized is

$$q = \sum_{i=1}^{n} \sum_{j=1}^{n} \frac{(\Delta y)^2}{(\Delta z)^4}.$$

In the real case the requirement is the best overall fit for a number of variables to several axes and Δy is replaced by the Euclidean distance, d_{ij}, between points in the original variable space and Δz by the Euclidean distance, D_{ij}, between points in the space defined by the new axes, so that the expression

---

* In ecological usage, Noy-Meir suggested the term 'continuity analysis' as preferable to 'parametric mapping'.

to be minimized is

$$Q = \sum_{i=1}^{n} \sum_{j=1}^{n} \frac{d^2_{ij}}{D_{ij}^{4}}.$$

There is, however, an indeterminancy in this expression. The scale used for the new axes is arbitrary and Q could be made smaller by increasing all z values and hence the values of D_{ij}. Q must thus be normalized by the same power of D_{ij}

e.g.
$$K = \sum_{i=1}^{n} \sum_{j=1}^{n} \frac{d^2_{ij}}{D_{ij}^{4}} \bigg/ \left\{ \sum_{i=1}^{n} \sum_{j=1}^{n} \frac{1}{D^2_{ij}} \right\}^2.$$

The measure of adjacency, W_{ij}, and the form of normalization are to some extent arbitrary and Carroll suggested a more general index of continuity

$$K^* = \sum_{i=1}^{n} \sum_{j=1}^{n} \frac{(d^2_{ij})^{\alpha}}{(D^2_{ij})^{\beta}} \bigg/ \left\{ \sum_{i=1}^{n} \sum_{j=1}^{n} (D_{ij}^2)^{\gamma} \right\}^{-\frac{\beta}{\gamma}}$$

where $\gamma = \alpha - \beta$. K is a special case where $\alpha = 1$, $\beta = 2$, $\gamma = -1$. Noy-Meir (1974a) pointed out that the larger β is set, the stronger is the weighting of pairs of points which are close together on the derived axes ($\equiv$ putative coenoclines) relative to pairs of points which are more distant from one another and considers this to be desirable because Euclidean distance between sites distorts coenocline distances most strongly for large distances.

Derivation of the new axes is achieved by an iterative procedure to minimize K, starting from initial guessed positions on a specified number of axes. Unfortunately, this is subject to technical problems; Kruskal and Carroll (1969) found that unsatisfactory solutions could occur if samples were randomly distributed along the gradients, a situation that is likely to apply to many sets of vegetation data (cf. Austin, 1976b).

Since the unsatisfactory features of principal component analysis and principal coordinate analysis result to a considerable extent from the non-linear relationship between the similarity or dissimilarity measure used and interpoint distance along the underlying gradients, the possibility of basing an ordination on the ranking of similarities or dissimilarities, rather than their actual values, is attractive. A method of such 'non-metric multidimensional scaling' or 'non-parametric multidimensional scaling' (Clymo, 1980) was proposed by Kruskal (1964a, b; see also Kruskal & Carmone, 1971; Fasham, 1977; Prentice, 1980) and was apparently first applied to ecological data by Anderson (1971). This seeks an arrangement in which interstand distances are monotonic with the distances or dissimilarities in the original data, i.e.

$$d_{ij} > d_{hl} \rightarrow D_{ij} \geqslant D_{hl}$$

for all i, j, k, l where d_{ij}, d_{kl} are the original dissimilarities between points i, j and k, l and D_{ij}, D_{hl} are the corresponding distances in the ordination. Ties are ignored as no information about relative habitat differences is conveyed by dissimilarities between stands with no species in common (Prentice, 1977).

The starting point of non-metric multidimensional scaling is an initial placing of stands in relation to a specified number of ordination axes*, either arbitrarily or on the basis of a preliminary ordination. A regression of D on d is calculated and the axis positions are then changed slightly and the regression recalculated, the procedure being continued iteratively to minimize the deviation from the regression. Kruskal proposed as a measure of 'stress', the failure of fit to the regression,

$$S = \sum_{i<j} (D_{ij} - \hat{D}_{ij})^2 \left| \sum_{i<j} (D_{ij} - \bar{D})^2 \right.$$

where $\hat{D}_{ij}$ is the regression estimate of d_{ij} and $\bar{D}$ is the mean of all D_{ij}. Stress is thus the sum of the squared deviations from regression estimates normalized to allow for the arbitrary scale of ordination axes, as in the measure of continuity in parametric mapping.

Sibson (1972) suggested a slightly different criterion,

$$d_{ij} > d_{ik} \rightarrow D_{ij} \geqslant D_{ik}$$

for all i, j, k. The effect of this, with a long gradient, is to place more emphasis on local relationships within the gradient and less on overall relationships[†]. Prentice (1977) and Clymo (1980) applied Sibson's technique with satisfactory results. Anderson (1971) suggested that with a limited number of stands (say < 30) much information may be lost in non-metric multidimensional scaling, especially in the case of two more or less discrete clusters, which will appear as a continuum, but Prentice (1977) found this not to be so in the data he examined. Gordon and Birks (1974) pointed out that if a single gradient solution is used,

---

* Fasham (1977) suggested using the *minimum spanning tree* for the data as a guide to the dimensionality of the data. The minimum spanning tree (Gower & Ross, 1969) is a set of straight lines such that no closed loops occur, each point is visited by at least one line and the tree is connected and of minimum overall length. If the tree branches, at least two dimensions are indicated. If, when the tree is plotted on a two-dimensional ordination, a substantial number of branches cross over each other, higher dimensionality is indicated.

[†] M. O. Hill (personal communication) has pointed out that the importance of Sibson's modification is in part that distances may not be comparable in different parts of the gradient. If one end is markedly more species-rich, the distances there may (depending on the distance measure used) appear larger than at the other end. If comparisons are confined to distances from a particular stand, this distortion is unimportant.

the procedure of minimization of stress used may result in a local minimum rather than an overall one; this was confirmed for artificial data by Austin (1976b).

Anderson (1971) proposed a metric 'minimization of quadratic loss functions' method, rather similar in approach to non-metric multidimensional scaling. The latter, by minimizing deviation from a (normally linear)* regression of ordination distances on original distances, takes note of ranking of distances only. Anderson's technique minimizes deviation of ordination distances from original distances, minimizing

$$\frac{\Sigma g_{ij}(D_{ij} - d_{ij})^2}{\Sigma g_{ij} d^2_{ij}}.$$

The g_{ij} are coefficients permitting unequal weighting of the comparisons. Anderson suggested, for example, that if it is desired to attach greater weighting to larger distances, g_{ij} might be set equal to d_{ij}. In view of the type of non-linearity in data from coenoclines, however, this is unlikely to be helpful and normally $g_{ij} = 1$. Anderson suggested starting with a principal component ordination and then minimizing the expression above by successive approximation. This technique is thus a method of improving on an ordination already derived. Its limitation is that it aims to relate ordination distances as nearly as possible to original distances, precisely the relationship that causes the difficulty in reproducing a coenocline.

Discussion of ordination so far has mostly assumed that the data to be analysed are of the form of records of some measure or estimate of the amount of each species in each stand. Quantitative data may be laborious to gather and the question arises whether such data add sufficiently to the information conveyed by lists of species present in stands to justify the additional labour involved in the field. Alternatively it may be more useful to spend the same time in the field obtaining species lists from more stands rather than quantitative data from fewer stands.

In relatively homogeneous sets of data, in which the majority of species are present in all stands, quantitative measures are clearly necessary for an informative ordination. It is with more heterogeneous data, with a considerable number of zero entries in the original data matrix, that the question is more open. Some investigations have included the analysis of both quantitative data and the corresponding presence/absence matrix. Orloci (1966) analysed

---

* Orloci (1973, 1978) suggested that one of the potential advantages of non-metric multidimensional scaling is that, by using a curvilinear regression of ordination distances on original distances, it permits a specified non-linear model of species response to an environmental gradient to be used.

frequency for a range of sand-dune stands and Noy-Meir (1971b) analysed cover in semi-arid vegetation and both found the resulting ordination more informative than the corresponding ordination of presence/absence data. Though the inclusion of the quantitative information may increase the effectiveness of an ordination, qualitative data only have frequently been used effectively, e.g. by Hall and Swaine (1976) for data for all species in 155 stands of closed-canopy forest in Ghana, where the time required to make quantitative estimates for non-tree species would have been prohibitive.

Direct analysis of quantitative data includes both the presence and absence elements of the data and the variation in amount of a species in those stands in which it occurs. Williams and Dale (1962) showed that the qualitative and quantitative elements in a data matrix can be separated by expressing amount of a species as the deviation from its mean within those stands in which it occurs (see p. 224). Norris and Barkham (1970) in a multiple discriminant analysis (p. 288) of the ground flora of a number of beechwoods found that the analysis of total and qualitative data differed little, and concluded that the information on between-wood differences lay mainly in the qualitative element; the quantitative element was, however, important in within-wood differences.

Description on stands and of vegetation types has traditionally been in terms of species, though vegetation types have often subsequently been characterised further by their having a high representation of higher order taxa, e.g. lowland dipterocarp forest. Van der Maarel (1972) suggested analysing data in the first place in terms of representation of different genera, families or orders. Based on the assumption that the morphological similarities between members of a higher order taxon are likely to be associated with physiological similarities, this is a possible approach to the analysis of data drawn from a wide geographical or habitat range. Van der Maarel analysed composite tables for 51 communities representing 49 alliances of the Braun-Blanquet system by principal component analysis. Analysis was in terms of representation of orders, with only species having a high constancy contributing to the representation of their orders, and the data being standardized by community ($\equiv$ stand) total. The resulting ordination was interpretable in terms of environmental features. Sarmiento (1972), Dale and Clifford (1976) and del Moral and Denton (1977) have also used higher order taxa successfully in analyses.

Although variation in ecological behaviour at the sub-specific level is well-known, and ecotypic differentiation has been much studied (e.g. Heslop-Harrison, 1964; Langlet, 1971), no attempt appears to have been made to use sub-specific categories as the basis of ordinational analyses. Such analysis would be expected to be well-worthwhile in studying the control of the proportions of different genotypes in closely similar situations, but the practical difficulties are formidable; in this situation quantitative data would be essential and it would thus be necessary to be able to assign every plant encountered to its genotype.

The use of non-taxonomic data in the characterization of stands has been considered in relation to classification and Webb *et al.* (1970) have shown that such data can be used successfully to classify rain forest (p. 223). Similarly, Knight and Loucks used simple techniques applied to non-taxonomic data to derive ordinations of prairie (Knight, 1965) and temperate forest (Knight & Loucks, 1969). Hall and Okali (1979) used principal component analysis of stem density of different girth classes to ordinate woody fallow vegetation in Nigeria. If principal component analysis is used, the problem arises of disordered multistate characters such as predominant bark type. Hill and Smith (1976) showed that such characters can be accommodated in a principal components analysis by, when calculating a cross-product with a quantitative character, assigning to each state a numerical value corresponding to the average value of the quantitative variable in the stands having that state.

The problem of choice of non-taxonomic characters to use, arising from the open-ended nature of the possible characters, has been referred to earlier. If the objective of the ordination is to provide a framework against which to examine the behaviour of other organisms, the open choice may be turned to advantage. Thus James (1971) successfully used a principal component ordination based on ten structural characters, selected as being likely to be significant in choice of habitat by birds (percentage ground cover, percentage canopy, density of shrubs, density of tree trunks of different size classes, etc.), as a means of comparing their habitat preferences.

In the analysis of vegetational data, interest more usually centres on the composition of stands and often on the environmental factors controlling the composition. Sometimes, however, the interest is rather in the ecological behaviour of species. If classification is used to analyse data, the distinction between 'normal' analysis of stands into groups with similar specific composition and 'inverse' analysis of species into groups with a similar pattern of occurrence in stands is clear-cut and with nearly all techniques separate analyses are needed. Some ordination techniques produce a species ordination concurrently with a stand ordination, e.g. principal component analysis (though not principal coordinate analysis), Gaussian ordination, and, with less precision, Whittaker's direct gradient analysis and Curtis and McIntosh's technique. Polar ordination (Gittins, 1965b) and non-metric multidimensional scaling (Matthews, 1978) have been used to obtain a species ordination;

principal coordinate analysis and parametric mapping, given a suitable measure of dissimilarity between species, could also be used.

Yarranton (1966, 1967a,b,c) used principal component analysis of interspecific contacts (p. 126), rather than data for stands, to produce a species ordination. The data may be represented in alternative forms (Fig. 67). Type A is the unreduced data matrix, Type B the corresponding contingency matrix, in which entries represent the number of joint occurrences of pairs of species or, in the principal diagonal, the total number of occurrences of a species. Correlation coefficients (or other cross-products) between species can be calculated for both data types and principal components derived. Yarranton pointed out that if the area being examined is fragmented into microhabitats comparable in size to the patches of an individual species, so that two species in contact may often have little ecologically in common, Type A coefficients will in general be higher than the corresponding Type B coefficients. In analysis of Type A data, the species loadings can be used to derive an ordination of individual points, allowing an examination of the relationship of any environmental factors measured at sample points to species composition. Although the analysis of Type B data proved generally more informative on species interrelationships, it does not permit direct comparison with environmental variables.

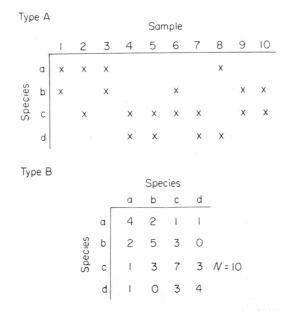

Fig. 67. Types of data for interspecific contacts (see text). A, Original data; B, contingency matrix. (From Yarranton, 1966, by courtesy of *Journal of Ecology*.)

Two possible approaches to the exploration of the relationship between vegetation and environment were outlined at the beginning of this chapter. The ordination of vegetational composition to provide a framework against which to examine levels of environmental factors has evident advantages over the alternative of ordinating environmental information (p. 227) and has been much more widely used. Although environmental ordination has the serious limitations of the open-endedness of possible variables and the need for previous simplification of the vegetational data, if more than individual species are to be considered, it does present fewer technical problems, particularly in the use of principal component analysis. An environmental data matrix, unlike most vegetational data matrices, will contain few zero entries, if any; most environmental variables are measurable and cannot be 'absent'. Moreover, at least over the range of sites that are likely to be included in a single analysis, environmental factors usually vary monotonically in relation to one another. Thus the data are much more appropriate than typical vegetation data to principal component analysis and problems of incurving and distortion less likely to occur. They may even approximate to multivariate normal distribution directly or after suitable transformation, and thus permit the testing of significance of components. Since different variables will be measured in different units, standardization of data will normally be necessary.

The open-ended nature of the variables that may be recorded means that essentially the same feature of the environment may be measured by more than one variable, e.g. a number of soil measures may represent primarily the availability of water to the vegetation. One aspect of the environment may thus become overemphasized in the resulting ordination. The exact effect of this will depend on the scaling convention used (p. 245). The ordering of sites on axes will not be affected, but the order in which components reflecting particular environmental gradients are extracted may be. Certainly the percentage variance accounted for by a component is not as meaningful as in a vegetational ordination. In theory, there is also the possibility that the environmental data may include a component with no influence on the vegetation, leading to difficulty in interpretation when vegetational information is plotted onto the ordination; it is uncertain, however, how far this effect is of practical importance (p. 297). Austin (1968b) and Austin et al. (1972) discussed these limitations. A wide range of procedures are available for environmental ordination, from the straight-forward plotting of species occurrence against a single environmental measure to complex techniques of data simplification. If composition of vegetation is predominantly correlated with a single environmental complex that can be adequately typified by a measure of one of its elements, or a pair of such complexes, the straightforward plotting against environmental measures may be very informative, as Whittaker demonstrated in his original gradient analysis (p. 232), and more complex techniques may have little advantage. This is particularly so if microclimatic differences associated with aspect are important. Though aspect can be treated as a quantitative variable with an arbitrary zero corresponding to north, it is often useful to represent it directly as compass direction, e.g. Perring's (1959) study of species distribution in chalk grassland in relation to aspect and slope (Fig. 68).

Loucks (1962) used synthetic scalars to represent complexes of related environmental factors instead of single representative measures, basing a scalar on understanding of the way different factors interact in their effect on plants. His approach may be illustrated from one stage in his construction of a moisture regime scalar (Fig. 69). The horizontal axis in the figure is based on depth of water table where the latter is less than 7 ft (213 cm) below the surface (right-hand part of scale). Where the water table is deeper the moisture status is expressed in terms of water holding capacity of the soil (left-hand part of the scale). The vertical axis is a previously constructed scalar representing the proportion of heavy rain lost by run-off; this is derived from consideration of angle of slope and position on slope, which affects the amount of water received by run-off from higher up the slope. The boundaries between segments of the moisture regime scalar are such that where the water table is high the value is little affected by the run-off scalar, but where the water table is low, the value is strongly dependent on it.

Loucks successfully used three scalars, moisture regime, nutrient status and local climate, to interpret the composition of forest in an area of varying topography. This approach has apparently not been used since but it is worth considering in situations where there is already considerable understanding of factors controlling vegetation as the basis of more detailed investigations*.

Several of the techniques already discussed have been applied to environmental data. Hole and Hironaka (1960) and Monk (1965) used Bray and Curtis analysis and a number of workers have used some form of principal component analysis e.g. Dagnelie (1960), Austin (1968), Goldsmith (1973).

There remain several procedures that do not relate directly to the main lines of development of technique considered above.

Various workers have found it possible to prepare useful ordinations by simple procedures, involving little or no numerical manipulation of the data (McIntosh, 1973). Thus Gimingham (1961), in an examination of heaths over

---

* See Austin (1972) for a more general discussion of environmental scalars.

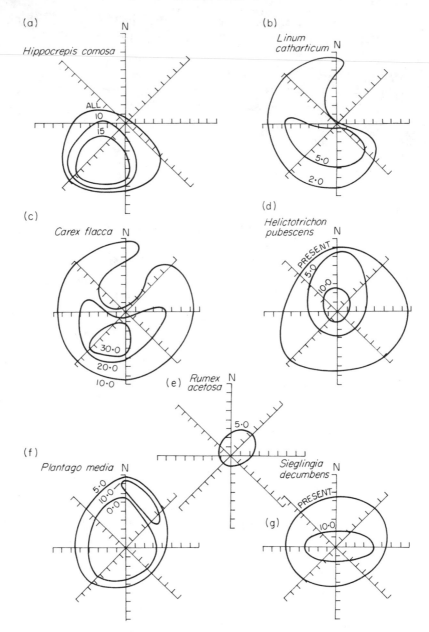

Fig. 68. Distribution of selected species in chalk grassland in north Dorset. Lines represent the eight compass points. Marks on the lines represent 5° intervals in degree of slope from flat ground (at the intersection of the lines) to 40°. Isolines enclose all sites having cover percentage greater than that marked. (From Perring, 1959, by courtesy of *Journal of Ecology*.)

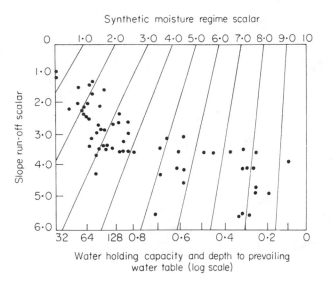

Fig. 69. Nomogram for a synthetic moisture regime scalar (see text). Dots indicate the positions of samples in Loucks's (1962) study of forest communities in New Brunswick. (From Loucks, 1962, by courtesy of *Ecological Monographs.*)

a wide geographical range in Western Europe, used Curtis and McIntosh's approach to arrange stands with similar composition in linear order (but without calculating a precise compositional index) and then placed the resulting series on a network of axes by inspection. Falinski (1960) used a matrix of interstand similarities to produce a comparable network. A matrix of similarities may be rearranged in such a way that high values are placed as near as possible to the diagonal (Fig. 70) to give a one-dimensional ordination (e.g. Clausen, 1957; Dix & Butler, 1960 etc.).

Various workers have prepared plexus or constellation diagrams showing the positive associations between species for a set of data (Fig. 71), e.g. Agnew (1961), Hopkins (1957), Welch (1960), McIntosh (1962). Such diagrams are a form of species ordination, but normally only show what species are associated, and lack the relationship between spacing in the diagram and overall similarity of a typical species ordination. De Vries (1953, 1954a), using $\sin((\chi^2/N) \times 90°)$ as a measure of strength of association, attempted to make the distance between species inversely proportional to the strength of the association between them, but not surprisingly this was not possible in a two-dimensional frame and he was forced to include some species in more than one position.

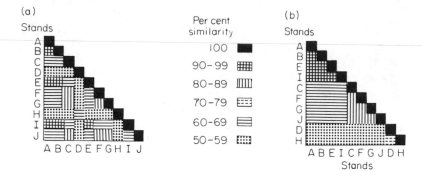

Fig. 70. (a) Schematic diagram showing a matrix of hypothetical indices of similarity between pairs of stands. (b) The same indices arranged with similar stands adjacent to one another. Clearly distinct groups appear as triangles of high values. (Modified from Sneath & Sokal, 1962, by courtesy of *Nature, London.*)

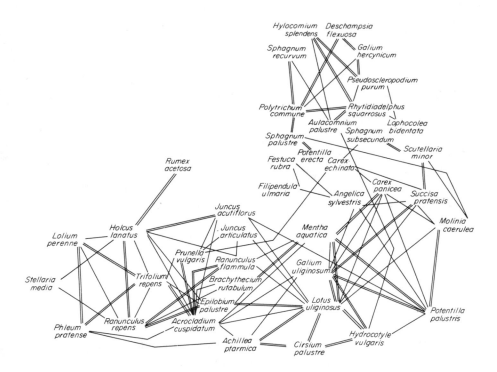

Fig. 71. Positive associations between species in 99 stands containing *Juncus effusus* in North Wales. ——, $P < 0.05$; ══, $P < 0.01$. (From Agnew, 1961, by courtesy of *Journal of Ecology.*)

Austin (1981) demonstrated the value, as an additional method of analysis, of plotting stand positions in relation to axes of diversity properties, including stand richness (number of species per stand), stand abundance (amount of vegetation per stand by some measure) and stand dominance (the largest proportion of stand abundance contributed by a single species).

Interpretation of relationship between vegetation ordination and environment (or vice versa) has commonly been attempted directly, e.g. by plotting the values for an environmental factor onto the vegetational ordination. An alternative approach is to examine the spatial distribution of single or derived variables and compare the resulting maps, i.e. to use the original stand positions in the field as the framework of comparison. This is clearly applicable rather to sets of data obtained from a limited geographical area, sampled reasonably intensively, than to more extensive surveys.

In the first application of principal component analysis to vegetation data, Goodall (1954c) mapped the values of component scores for stands and compared these component maps with a contour map of the area. From this comparison he was able to relate the first component of his data to topography. The axis positions for stands given by any technique of ordination can be treated in this way. The approach is likely to be most profitable when the scale of environmental heterogeneity is reasonably large in relation to the total area examined.

Lange (1968, 1971) devised a technique, 'influence analysis', which identifies 'influences' (effectively, ordinational components) in the vegetation and evaluates them preparatory to mapping them. Association between species is examined and groups of highly associated species ('nodes') identified. Lange used χ^2, taking account of all relationships above a predetermined value, but, in principle, nodes could be determined from any hierarchical species clsssification of the data. One node of Lange's illustrative example, 88 stands of heath vegetation in South Australia, involved four species, designated for convenience A, B, C and D. There was strong positive association between A and B and between C and D, and negative association for both A and B with C and D. Of the possible combinations of species occurrences in a stand AB and CD represent opposite poles of the influence. Denoting AB as having influence rating 1, and other combinations as having rating corresponding to the number of differences from AB, e.g. BCD, differing in the absence of A and presence of C and D, as rating 4, a table of influence ratings can be drawn up (Table 25). The rating of each stand can then be entered on a map of the area (Fig. 72(a), (b)). Comparison with a contour map of the area (Fig. 72(c)) indicates that influence 1 is correlated with altitude and one pole of influence 2 with higher aspects of the windward slope.

Table 25. Arrays of species combinations (Lange, 1968)

	Node 1					Node 2			
Influence rating array:	1	2	3	4	5	1	2	3	4
Corresponding array of	AB	ABC	ABCD	BCD	CD	E	EF	EFG	FG
species combinations		ABD	AC	ACD			EG	F	
		A	AD	C			None	G	
		B	BC	D					
			BD						
			None						

Crawford and Wishart (1968) made a somewhat similar approach. Their group analysis (pp. 185–7) provides an estimate of the affinity of each stand to all other groups as well as the one to which it is assigned. It is thus possible, for a grid of stands, to represent on a map the degree to which each stand approaches the composition of a particular group.

Gittins (1968) investigated the use of 'trend-surface analysis', a technique widely applied to geological data, in examining more critically the patterns that result from mapping a variable in this way. Trend-surface analysis involves the calculation of regressions, of successively higher order, of the variable being considered on the geographical coordinates of the stands, until a satisfactory fit is obtained. Thus if the coordinates of the stands are (U_i, V_j) (U_i east-west coordinate, V_j north-south coordinate) the linear trend is

$$\tau(U_i, V_j) = \alpha_{00} + \alpha_{10} U_i + \alpha_{01} V_j$$

describing a plane surface, α_{00} defining the height at the origin of the coordinates, α_{10} the slope in the east-west direction and α_{01} the slope in the north-south direction.

The second order or quadratic surface is

$$\tau(U_i, V_j) = \beta_{00} + \beta_{10} U_i + \beta_{01} V_j + \beta_{20} U_i^2 + \beta_{11} V_i V_j + \beta_{02} V_j^2,$$

the third order or cubic surface is

$$\tau(U_i, V_j) = \gamma_{00} + \gamma_{10} U_i + \gamma_{01} V_j + \gamma_{20} U_i^2 + \gamma_{11} U_i V_j + \gamma_{02} V_j^2 + \gamma_{30} U_i^3$$
$$+ \gamma_{21} U_i^2 V_j + \gamma_{12} U_i V_j^2 + \gamma_{03} V_j^3$$

and so on.

Successive regressions describe successively more complex surfaces giving successively closer fit to the observed values (Fig. 73). The use of this form of analysis is illustrated in Fig. 74, which shows the trend surfaces for the first three components of a principal component analysis (of correlation coef-

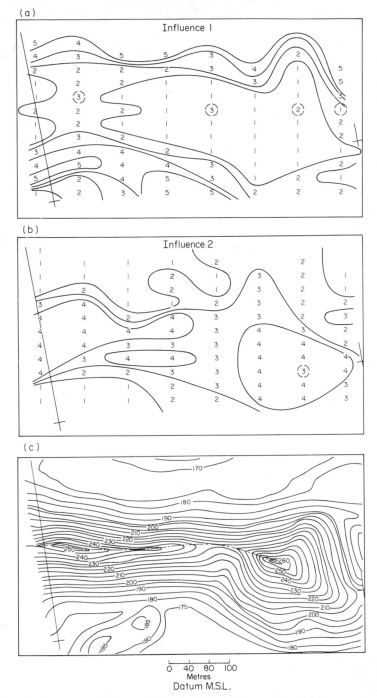

Fig. 72. Influence analysis. (a), (b) Isolines for influences 1 and 2 in a study area, (c) contour map (5 ft., (150 cm) intervals) of the same area. (From Lange, 1968, by courtesy of *Australian Journal of Botany*.)

ficients) of data from grassland vegetation, and for three soil variables. Soil
depth is clearly related to the first component and soil potassium to the third
component, but the relationship of soil phosphate is more complex; it has some
affinity with the trend surfaces of all three components.

Gittins pointed out that when a surface giving a satisfactory fit has been
derived, it may still be profitable to examine the distribution of deviation from
the calculated surface. These will represent very local variations and reflect
pattern in the vegetation, in the sense discussed in Chapter 3.

The need to place further stands into an existing arrangement probably
arises much less frequently with ordination than with classification, but it may
do so with extensive and continuing surveys (e.g. Hall & Swaine, 1976) or in
geographical comparison of floras, as in Lawson's (1978) study of the marine
algal floras of the Atlantic. With some ordinational procedures this can readily

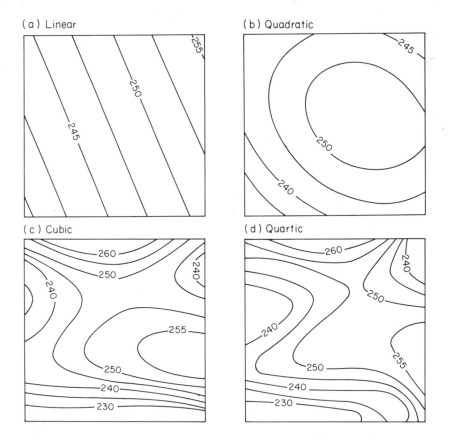

Fig. 73. Trend-surfaces fitted to surface relief for data of Goodall (1954c). (From
Gittins, 1968, by courtesy of *Journal of Ecology*.)

(a) First principal component

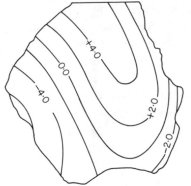

(b) Second principal component

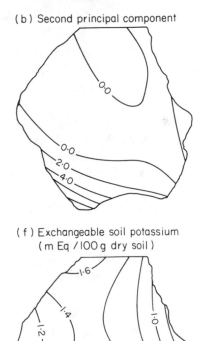

(e) Extractable soil phosphate
     (ppm P)

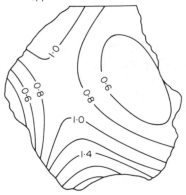

(f) Exchangeable soil potassium
    (m Eq / 100 g dry soil)

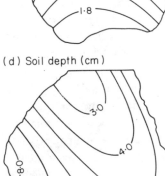

(c) Third principal component

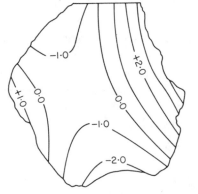

(d) Soil depth (cm)

Fig. 74. Trend surfaces fitted to component scores of stands and to soil variables for data from limestone grassland. (a), (b), (d), (e) Third degree surfaces, (c), (f) second degree surfaces. (From Gittins, 1968, by courtesy of *Journal of Ecology*.)

be done. Distances from the reference stands can be calculated in polar ordination. In most forms of principal component analysis, species loadings can be used to calculate the appropriate stand position. For reciprocal averaging, position for a further stand can be calculated from a regression of stand position on mean score for species occurring in the stand (Swaine & Hall, 1976). Swaine and Hall found, usefully, that a random sub-set of the species in a new stand was normally adequate to given a satisfactory estimate of stand position. With other procedures no straightforward method of placing further stands is available, at least if more than one axis is involved, and recourse may have to be made to estimating distances from a sample of stands of known position to give an approximate position.

When a set of vegetation data have been classified it may be useful to ordinate the resulting groups. Although hierarchical classifications display the degree of dissimilarity between the sub-groups at each node, they do not show the interrelationships of the final groups, nor do procedures normally provide any estimate of similarities between final groups. An ordination will provide information on inter-group relationships and a basis for exploring the correlation between groups and environmental factors additional to that provided by imposing environmental factors on the hierarchy (p. 173).

A straightforward way of ordinating groups of stands is to calculate the mean composition of each group and apply an ordination technique to these means, treating them as if each mean represented a single stand. If the original stand data are for presence only, the mean composition for a group will necessarily be in quantitative form, as frequency of occurrence of species. Even very widely dissimilar vegetation types can be included in one ordination in this way; Curtis (1959) produced a three-dimensional ordination of 28 native terrestrial plant communities. He used the one-complement of the Czekanowski measure of similarity to construct the ordination directly by triangulation from a central type, but other methods could equally well be used, e.g. principal coordinate analysis of a suitable distance measure.

Ordinations of groups should be interpreted with caution. Since each group is represented by its mean, the ordination conveys no information on its range of composition and what appear as two quite distinctive groups may in reality be much less distinctive. If the techniques of classification and ordination used are not strictly compatible, groups may even overlap in the ordination space.

Grigal and Goldstein (1971; Goldstein & Grigal, 1972) used multiple discriminant analysis (canonical variate analysis) to ordinate groups. Blackith and Rayment (1971) give a useful account of this technique (see also Gittins, 1979). The first axis is in the direction which accounts for as much as possible of the variability between the means of the groups, the second is orthogonal to the first, and, within this constraint, in the direction which accounts for as much as possible of the remaining variability, and so on, as in principal component analysis.

Provided that the data are substantially multivariate normal and the dispersion matrices (variances and covariances) of the different groups are the same, statistical tests of the significance of the differences between groups, and of the assignment of a stand to a group, can be made. These assumptions are unlikely to be even approximately satisfied by vegetational data. However, the technique appears to be moderately robust in this respect and it can be used to compare the variability within groups (Matthews, 1979). If an initial classification is compatible with the comparison of within- and between-group variances on which multiple discriminant analysis is based, it also provides a useful method of revising the initial classification (del Moral, 1975).

CHAPTER 9

Vegetation and Environment

The ultimate objective of classification may be and that of ordination commonly is to generate hypotheses about the relationship between vegetation and environment. It is convenient to consider techniques of correlating vegetational and environmental data in relation to classification and ordination together, as some of the problems are common to the two approaches. The subject has been carefully reviewed by Noy-Meir (1970) and the following account largely follows his discussion.

Either or both of the vegetational and environmental data may have been generalized by classification or ordination before their relationship is examined. Vegetational and environmental data may be either qualitative (discontinuous) or quantitative (continuous) or include both. Qualitative variables may be binary (e.g. presence/absence of a species) or multistate (e.g. soil types). There are thus a variety of combinations of data to be considered. For convenience, following Noy-Meir, generalization of vegetation and environment is symbolized by V* and E* respectively, and discontinuous and continuous data by D and C respectively.

The simplest case is V/E, where the occurrence of single species is examined in relation to separate environmental factors. The use of contingency tables for D/D data, of analysis of variance for C/D and D/C data and of regression for C/C data has been outlined in Chapter 5. Regression analysis is not confined to C/C data, but can be modified to use binary or multistate† variables (Noy-Meir et al., 1973).

The earlier uses of regression (e.g. Blackman & Rutter, 1946; Rutter, 1955) involved only a small number of independent variables; the computation involved precluded the inclusion of large numbers. With the present speed and capacity of computers there is now no effective limit on the number of variables that can be included. This does not, however, remove the need for careful prior assessment of the relevance of the variables to be included;

---

† A multistate variable can be treated as several binary variables. Alternatively, if a multistate variable is the only one being considered, the relation between it and quantitative data for a species can be examined by analysis of variance.

unnecessary elaboration of the field and observational programme is as wasteful as is unnecessary computation.

If a number of independent variables are initially included, the question arises how many can be eliminated without seriously reducing the efficiency of the regression (normally assessed as the proportion of the total sum of squares of the dependent variable accounted for by the regression). In some fields, the objective of regression analysis is primarily predictive, to estimate the value of the dependent variable corresponding to stated values of the independent variables. This may be true where the functioning of an ecosystem is being examined (cf. Jeffers, 1978), but in the present context we are concerned rather with detecting those environmental variables with which plant performance is correlated, as a basis for hypotheses about the factors which determine the plant performance. When multiple regression is used as a means of exploring the relationship between variables, rather than to derive a predictive equation for the dependent variable, it is mostly usefully applied in a stepwise manner; the best fit to a single independent variable is selected, and then the best combination of a second independent variable with this one is selected, and further independent variables added to the regression equation until the proportion of the variance of the dependent variable accounted for is considered adequate. For binary multiple regression the environmental attribute which has the highest correlation with the species being examined (measured by $|r|$ for C/D data, or χ^2 for D/D) is the first independent variable to be included in the regression equation. At each subsequent step the one of the remaining environmental attributes with the highest partial correlation with the species is included.

General aspects of the use of regression in ecological contexts have been usefully discussed by, among others, Austin (1971, 1972), Mead (1971) and Yarranton (1967d, 1969b, 1971). Yarranton suggested the inclusion of other species as well as the environmental attributes in the regression equations for species, i.e. a V/(V + E) approach. This makes allowance for interactive effects between species, resulting from modifications of the microenvironment by vegetation. Interpretation in other than species-poor situations is, however, likely to be difficult.

If the relationship is markedly non-linear, the inclusion of linear terms only in a multiple regression on environmental attributes may fail to indicate the possible importance of particular attributes in determining the representation of a species. Difficulty also arises from the strong correlation between environmental attributes that is commonly found. Multiple regression takes account of such correlation between independent variables, so that a multiple regression equation will be satisfactory as an empirical predictor, but may be little help as a means of exploring the importance of environmental attributes. Thus Austin (1971), commenting on Yarranton's (1970) regression analysis of cryptogamic species growing on limestone pavement, pointed out that the independent variables such as light, humidity and temperature are functionally related and form part of a factor-complex determining the evapotranspiration rate which is the primary determinant of plant moisture stress. The use of multiple regression is less straightforward than may appear. As Austin put it, 'judgement is required based on: (a) knowledge of the numerical assumptions and possibilities of the technique, (b) field experience of the vegetation studied, (c) knowledge of the processes influencing plant growth'.

An alternative to multiple regression is predictive attribute analysis (Macnaughton-Smith, 1963, 1965; Noy-Meir *et al.*, 1973). This was developed for binary 'internal' (here, environmental) attributes, but Lance and Williams (1968b) showed that it can be extended to include attributes of mixed C and D form. The first step, as in binary multiple regression, is to identify the environmental attribute showing the maximum correlation with the species, but the data are then divided into two sets, with and without this environmental attribute, and the correlations between the species and other environmental attributes examined in the two sets separately. Correlation of species with environment is thus examined hierarchically. This may be more informative than a single multiple regression equation, at least provided that the original environmental variables are of binary form; if multistate variables have been divided into binary ones, difficulties of interpretation may arise.

For data of C/C form, Westman (1980) suggested, as a means of identifying the environmental factors most influencing the performance of species, fitting Gaussian curves to the response of each species to the values of each environmental factor. The percentage variance accounted for by the fit to a Gaussian curve is determined and the most influential environmental factors are taken to be those for which the mean percentage variance accounted for is greatest. Only actual occurrences of species are taken into account in fitting the Gaussian curves. Westman emphasized that not all, or even most, of the important influencing factors will necessarily be identified because of the multifactorial control of species performance. Nevertheless, this is a promising approach to a preliminary sorting of environmental variables to decide which are worth further investigation, particularly where data are available for a large number of variables.

The V^*/E case, where the vegetation has been generalized but environmental attributes are examined individually, has been the most widely used approach. Commonly, environmental attributes have been plotted onto an ordination of stands or the average values for environmental attributes in the two subgroups at a node of a classification have been examined (Figs 43, 44, 31). Neal and Kershaw (1973) suggested calculating trend-surfaces (p. 284) for a single environmental variable in relation to the coordinates of a vegetational ordination as an improvement on direct plotting. The general failure to examine relationships with the environment any more critically is open to the criticism that the advantages of refined numerical methods of simplifying vegetation data are largely nullified by the subsequent casual approach to correlation with environment. Straightforward plotting onto the ordination does, however, have the advantage that it is unaffected by non-linearity and indeed may expose non-linearity where it has not initially been recognized.

The approaches available for the V/E case are applicable, though not all have been used, and are subject to the same limitations. With vegetational ordination, i.e. C/C or C/D data, the situation is exactly comparable; vegetational components may be regressed on environmental attributes, or mean component scores for discontinuous environmental attributes may be examined by analysis of variance (or t test for binary attributes)†. If vegetation has been classified, though the same analyses can be used as with individual species, interpretation is complicated by the fact that the groups are mutually exclusive in nearly all procedures of classification—a stand belongs to one group only. It is thus not as useful to know for a group of stands as it is for a species what environmental attributes are correlated with its occurrence or non-occurrence; interest lies rather in what environmental attributes differentiate between the groups and it may be useful to ordinate the groups and examine this ordination in relation to environment (e.g. Greig-Smith *et al.*, 1967).

The V/E* approach has not been widely used, and when it has, species correlation with environmental components has been examined by plotting species presence or abundance onto the ordination (e.g. Goldsmith, 1973). The same procedures of more precise analysis as for V*/E are, in principle, available. In view of the limitations of environmental ordinations discussed above (p. 278), it is not surprising that Austin (1968b), applying both V*/E and V/E* approaches to data from chalk grassland, found the vegetational ordination more informative, although the environmental ordination provided some further insight.

Initial simplification of both vegetational and environmental data, a V*/E* approach, is an evident possibility, and several investigations have compared vegetational and environmental ordinations of the same data (e.g.

† In their pioneer ordinational study, Bray and Curtis (1957) calculated correlation coefficients between axis position of stands and environmental attributes.

Dagnelie, 1960; Austin, 1968b; Austin *et al.*, 1972), but the results have been rather disappointing. The most thorough test of this approach is that of Austin (1968b). The highest correlation between any of the first three vegetational and the first three environmental components was only 0·536 (Table 26)*. Plots of environmental components onto the vegetational ordination (Fig. 75) show that the relationships are not linear and suggest that a curvilinear multiple regression would need to be complex and would thus be difficult to interpret usefully.

Table 26. Correlation coefficients between vegetational and environmental components in an analysis of data from chalk grassland (Austin, 1968b)

Vegetational components	Environmental components		
	1	2	3
1	−0·229	−0·109	−0·152
2	0·244	0·536	−0·064
3	−0·191	0·189	−0·149

A vegetational classification could be compared with environmental components in the same ways as comparison with single environmental attributes. Classification of both vegetational and environmental data is unlikely to be used. Environmental data do not generally lend themselves to classification within the range likely to be encountered in a single set of data (but cf. Goldsmith, 1973) and the approach is, in any case, unlikely to be an informative one; procedures are available for estimating the similarity of two hierarchical classifications of the same set of individuals (see, for example, Sneath & Sokal, 1973; Rohlf, 1974; Smartt *et al.*, 1974; Williams, 1976) but it is not very helpful to know merely how similar the vegetational and environmental classifications are.

Instead of separate principal components or factor analysis ordinations of vegetation and environment, both may be included in a single analysis ((V + E)*) (e.g. Ferrari *et al.*, 1957; Dagnelie, 1960; Gittins, 1969; Walker & Wehrhahn, 1971; Schnell *et al.*, 1977; Dye & Walker, 1980). The environmental attributes having high correlation with a particular species may be expected to have similar loadings to those of the species. Examination of the resulting attribute ordination can thus be used to gain an overall impression of

* The highest correlation obtained by Cassie and Michael (1968) in a study of marine fauna and environment was still lower (0·26).

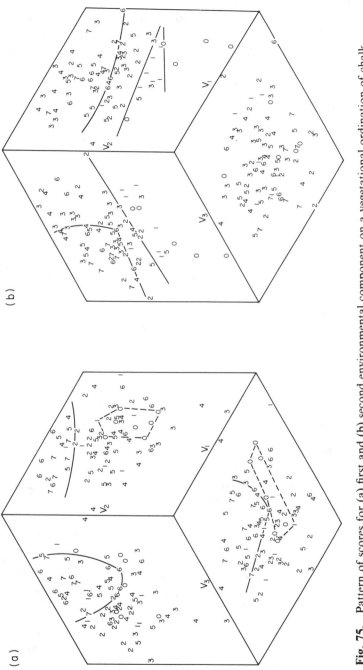

Fig. 75. Pattern of scores for (a) first and (b) second environmental component on a vegetational ordination of chalk grassland. 0–7, range of environmental component scores. (From Austin, 1968b, by courtesy of *Journal of Ecology*.)

the relative importance of environmental attributes and of those most closely correlated with each species. Gittins (1968, 1969) examined the limestone grassland data he had earlier analysed by other procedures (Gittins, 1965a, b, c) (Fig. 76). His earlier recognition of two major trends of variation related to soil depth and soil nutrients respectively was confirmed by the high loadings of soil depth on axis 1 and of soil phosphate and soil potassium on axis 2. The inclusion of environmental attributes in the analysis imposes the constraint that the data must be centred and standardized by attribute*, and this may not

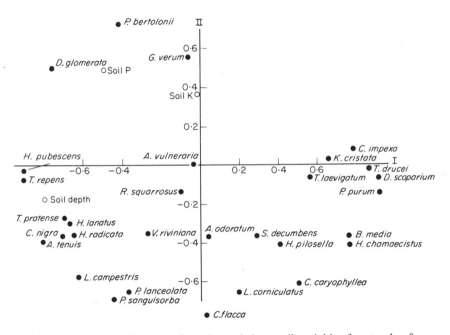

Fig. 76. Attribute ordination of species and three soil variables for stands of limestone grassland. (From Gittins, 1968, by courtesy of *Journal of Ecology*.)

* Walker and Wehrhahn (1971) (see also Jeglum *et al.*, 1971) made the interesting modification of dividing the values for environmental variables by an arbitrary large number before entering them in the data for principal component analysis 'thereby inactivating these attributes in the analysis (but retaining them for calculation of loadings and ensuring that the extracted axes are determined by the vegetation data alone' (Dye & Walker, 1980). The environmental data were scaled before analysis to be of approximately the same order of magnitude. The principal component analysis was of otherwise unstandardized data; the need to standardize the data by attribute was thus avoided.

Table 27. Factor analysis of 50 stands of agricultural grassland based on yield of grass and various environmental variables (Ferrari *et al.*, 1957)

	Loadings on four factors			
	1	2	3	4
1 Soil pH	0·20	0·24	0·41	−0·09
2 Organic matter content of soil	−0·38	0·75	0·02	−0·15
3 Clay content of soil	−0·10	0·80	0·04	−0·20
4 Fine sand content of soil	0·17	0·68	−0·13	0·13
5 'U figure' (soil surface)	−0·04	0·88	−0·14	0·13
6 P status of soil	0·42	0·08	0·45	−0·20
7 K status of soil	0·62	−0·04	0·06	−0·04
8 Mg status of soil	0·27	0·61	0·44	0·02
9 Thickness of humus layer	0·43	−0·11	−0·02	0·22
10 Distance from farm	−0·34	0·05	−0·09	−0·34
11 Water table	0·52	−0·39	−0·09	0·01
12 Fluctuation in water table	−0·29	0·59	−0·05	0·17
13 N fertilizer	0·59	0·05	0·45	0·11
14 P fertilizer	0·37	0·01	0·84	0·00
15 K fertilizer	0·43	0·18	0·71	0·04
16 Pasture quality	0·57	−0·09	−0·04	0·20
17 N content of grass	0·36	−0·02	0·00	0·63
18 P_2O_5 content of grass	0·47	−0·13	0·14	0·58
19 K_2O content of grass	0·68	−0·13	−0·08	0·30
20 MgO content of grass	−0·03	−0·05	0·22	0·15
21 Spring yield of grass (% of annual yield)	0·90	−0·01	−0·09	−0·31
22 Spring yield of grass	0·98	0·03	−0·02	−0·11
23 Autumn yield of grass	0·75	0·07	0·03	0·38
Percentage variance accounted for	24·6	15·1	9·2	6·5

be appropriate to the vegetational data, and introduces the general difficulties of choice of environmental variables. Ferrari *et al.*'s (1957) analysis of agricultural grassland (Table 27) provides an example of a component of environmental variation with little relation to vegetation. The vegetation attributes included (annual yield, spring yield as percentage of annual yield, magnesium, potassium, phosphorus and nitrogen content of grass and pasture quality) all have low loadings on the second axis, which nevertheless accounts for over 60% as much variance as the first axis.

It seems unlikely that a (V + E)* approach will be helpful unless there is already considerable information about which environmental attributes are most relevant or, as Noy-Meir (1970) suggested, integration of vegetation and environment is specifically required in the definition of land systems.

The examination of correlation between components of separate principal component analyses of vegetational and environmental data (V^*/E^*) leads on the theoretically promising approach of canonical correlation analysis. The latter involves the extraction of components in each set of data such that the correlation between equivalent components of the two sets is maximized (($V/E)^*$). Thus species with high loadings on the first vegetational component will be those whose occurrence is most closely correlated with environmental variables having high loadings on the first environmental component. Gittins (1979) has provided a detailed account and assessment of the technique, the application of which to vegetational data was suggested independently by Dagnelie (1961) and Hughes (1961). Trials with field data have mostly not been rewarding (Austin, 1968b; Barkham & Norris, 1970; Gauch & Wentworth, 1976) though Gittins' assessment is more encouraging, at least for data with a broadly linear and continuous structure. Barkham and Norris (1970), examining data from woodland, obtained an interpretable analysis, but concluded that correlation of the components from separate principal component analyses was more informative. Kercher and Goldstein (1977) described a modification, 'canonical group correlation', operating on previously established groups of stands rather than individual stands.

Canonical correlation analysis is subject to the difficulties of non-linearity in vegetational components and of choice of variables and possible inclusion of environmental variables with little or no effect on the vegetation in environmental components, already discussed. Barkham and Norris (1970) further suggested that correlation between vegetational and environmental components may be non-linear. Williams and Lance (1968) suggested a possible strategy of separate ordination of vegetation and environment, followed by a canonical correlation analysis of the relation between them ($V^*/E^*)^*$). As Noy-Meir (1970) suggested, preliminary definition of vegetation and environment in terms of a small number of components could allow easier extension of the canonical analysis to include non-linear relationships.

CHAPTER 10

Practical Considerations

The previous chapters have attempted to summarize the very wide range of techniques available for the analysis of data on the composition and variation of vegetation. Many techniques have not been used in other than very limited trials, often in simplified field situations or with unrealistic simulated data. There has been some discussion of the considerations involved in the choice of appropriate techniques for a particular investigation (e.g. Goodall, 1970; Greig-Smith, 1971a; Lambert, 1972; Lambert & Dale, 1964; Noy-Meir & Whittaker, 1977) but the greatly increased speed and capacity of computers as well as the further development of methodology have made much of the earlier discussion of limited value.

The form of the data to be collected and the type of data analysis to be used in a vegetational study are closely interrelated. The type of data to be collected obviously depends on the question or questions being asked. The method of analysis to be used is equally dependent on the objective of the investigation, but the practicality of collecting appropriate data may place constraints on the choice of method. In contrast to many biological fields there is rarely technical difficulty in obtaining data of the desired form, but the cost in terms of time may be unacceptable, or there may be other objections, for example to destructive sampling such as that involved in biomass determinations. Acceptance of a suboptimal type of data may involve a modification of method of analysis. Because of the heavy investment of time almost always involved in data gathering before any worthwhile analysis can be made, it is important that superfluous gathering of data be avoided. Against this, the mechanics of field work may be such that it may be expensive to return to sites and it may be better to gather data that *may* be required at a later stage, even at the risk that it may not be used.

The nature of vegetation places constraints on the options available. For example, any method of analysing pattern that depends on recognition of discrete individuals is of limited use because many species are not amenable to recording in this way (p. 5), or such density measures may be biologically unrealistic because individuals vary very widely in size. Moreover, in practice, many pattern studies require concomitant assessment of pattern of environmental factors (cf. Greig-Smith, 1979), and the method of analysis used must be capable of handling environmental as well as species data.

Problems of choice of method are greatest in relation to investigations of variation in overall composition of vegetation. Three interconnected decisions are required: the type of data to be collected; the layout of stands; the type of analysis, whether by classification or ordination, or both, and the precise procedure of classification or ordination. These are in turn influenced by the objectives of the investigation. Three principal possible objectives may be identified: simplification of the observed range of vegetation as a basis of mapping or inventory, either as an objective in itself or as a basis of management; search for the existence of discontinuities in the range of composition; detection of correlation between composition and environmental (in the broadest sense) factors as an aid to generation of hypotheses about the causation of differences in composition (Greig-Smith, 1980). Though it is possible for an investigation to serve more than one of these objectives, to do so is likely to require compromise in choice of method, so that a firm decision at the planning stage is important.

Possible forms of data to be collected from each stand range from a simple species list, through estimates of the contribution of each species to the total vegetation of the stand (by one or more of a variety of measures), to records of some aspect of the size or performance of each individual plant (as in some forest enumerations). The cost, in terms of time, of recording a stand varies greatly. Preparation of a species list is, in most vegetation, very much quicker than any form of quantitative measurement, but different forms of quantitative measure themselves vary in the time required.

It is commonly assumed that the recording of quantitative measures adds greatly to the time taken to record a stand (e.g. Lambert & Dale, 1964) but there is little precise information available on how serious this is. Noy-Meir (1971b) estimated, rather than measured, cover, density and height of each species in semi-arid vegetation in south-eastern Australia and found that this approximately doubled recording time for a stand; precise measures, rather than visual estimates, would have greatly increased the time necessary. The extra labour involved in recording quantitative data will clearly vary with the type of vegetation. If large trees only are being recorded in species-rich rain forest, where it is necessary to examine each individual separately for identification, the additional time needed to record density and diameter is negligible; in a moderately species-rich limestone grassland the time needed to note the species present is negligible compared with that needed to obtain quantitative data. The additional labour involved depends also on the measure used, e.g. frequency is normally quicker than a cover measure. If a fixed time is available for gathering data, the time taken to record a stand must be considered in relation to that needed to move between and demarcate stands, which may be considerable in extensive surveys in difficult terrain. If the time saved at each stand by recording presence and absence only is relatively small compared with the time needed to travel to a further stand and demarcate it, the gain in number of stands recorded may not outweigh the loss in total information resulting from the lack of quantitative data.

The data from each stand should be the minimum in terms of time required that will be adequate to answer the questions posed. This will depend on the range of variation being examined and the level of detail of variation that is of interest. The relative contribution of qualitative and quantitative data have been discussed above (p. 224), but it is worth emphasizing that qualitative data only (i.e., a species list from each stand), even when not as efficient as some form of quantitative measure, may lead to an adequate sorting of the stands. Although this may appear surprising, the reason why it should be so is clear. Consider the simplified situation of a single gradient of environmental factors to which species show unimodal response curves (Fig. 77). Although an

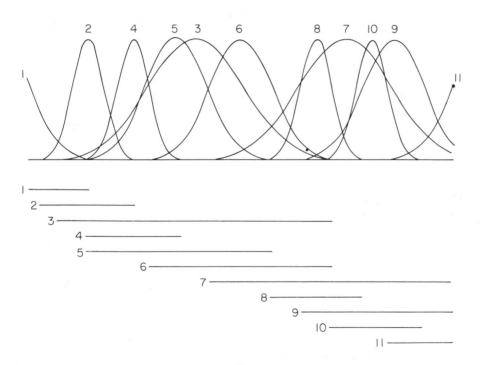

Fig. 77. Unimodal response curves of species along an environmental gradient with corresponding species presence.

observed amount of any one species in a stand gives a more precise indication
of the position of that stand on the gradient than its presence would do, any
particular *combination* of species presences also corresponds to a short
segment of the gradient only (provided a reasonably large number of species
are involved).

If no species have a high representation in any part of the data, analysis of
a quantitative measure is likely to be dominated by the species with the
greatest ranges of representation (unless an appropriate standardization is
used) and a quantitative measure may give a *less* satisfactory analysis than
presence/absence (Austin & Greig-Smith, 1968).

If a quantitative measure is to be used, the choice of measure depends
firstly on the objective. For example, if interest centres on productivity of
different species, either a direct measure of biomass, or one which correlates
well with it, must be used. Only if a proposed measure is satisfactory on this
account, should considerations of economy of time determine the choice. A
further consideration is that partially bounded data (p. 18) may give results
that are so similar to qualitative data that the greater cost is not balanced by
the increase in efficiency (Smartt *et al.*, 1974). This is a disadvantage in the use
of frequency, otherwise attractive on account of its relative ease of recording.
It should also be borne in mind that completely bounded data are, in effect,
standardized by stand and this may or may not be desirable.

In complex vegetation, consideration should be given to whether it is
necessary to record species of all growth forms. It has long been customary to
examine forest composition in terms of the tree layer only, and comparison of
analyses of various sets of growth forms with analysis of the complete
angiosperm complement confirmed the validity of this practice for one set of
data (Webb *et al.*, 1967b). On the other hand, inclusion of a greater range of
life forms increases the information available from a given sample area and
thus allows the use of a smaller stand (Hall & Swaine, 1976). The problem of
whether to record all species has mainly been considered in relation to forest,
especially tropical forest, but similar considerations apply, for instance, to the
inclusion of bryophytes in data from much temperate vegetation, though this
is rarely discussed.

Before leaving the question of type of data to be recorded, it must be
emphasized that quite different considerations apply to description of a
vegetation type as a starting point for understanding structure and functioning. This will always need more information than is needed for efficient
derivation of classification or ordination. If fuller descriptions of the
vegetation are needed, it may be profitable to delay these and examine in more

detail a more limited number of stands selected on the basis of the initial analysis.

If qualitative data only are to be collected, no problems of sampling within the stand arise—the whole stand must be examined. Apart from tree density and basal area, normally enumerated over the whole stand, a quantitative measure will normally be estimated by samples taken within the stand. The placement of samples to ensure an unbiased estimate has been discussed in Chapter 2. Although the estimate must be unbiased, only if tests of significance between estimates for different stands are to be made, so that a standard error of the estimate is required, need the sampling be random. Other sampling schemes, e.g. systematic or stratified unaligned systematic (p. 23), may give a more precise estimate, though at the cost that the precision cannot be estimated. In most survey investigations subsequent significance tests are unlikely to be made. This contrasts with field experiments in vegetation, where subsequent analysis does normally involve significance tests and appropriate random sampling within stands is then necessary.

Stand size is rarely critical, though broad limits can be drawn. Stands must be large enough to cover all phases of any mosaic pattern that may be present, otherwise analysis will reflect this pattern and may be dominated by it. Conversely, if ordination is being used to elucidate pattern within otherwise homogeneous vegetation (p. 103), stands must be small enough to include one phase only of the mosaic. Stands should be large enough to include a reasonable proportion of the total species complement, particularly if subsequent standardization of the data will emphasize rarer species; preliminary examination of a species/area curve may be useful. They should not be so large that a single stand is obviously including markedly different habitats*. It is not even essential that all stands should be of the same size or shape, though this is desirable if the nature of the habitat and vegetation permits. Sometimes, however, this is not possible, e.g. in study of cliff ledge vegetation (cf. Bunce, 1968) or of relict vegetation in a predominantly managed landscape. If stands of varying size are recorded qualitatively the subsequent analysis will be influenced by stand size through the effect of size on species number. With quantitative measures, unless standardization emphasizes rare species, the analysis will be relatively little affected, but any

* Goff and Mitchell (1975) compared ordinations of individual plots or point-centred samples within large stands (up to several hectares) with ordinations based on the lumped data from stands and concluded that they showed little difference. These results should not be taken as a recommendation of such large stands, but do suggest that forest inventory data from large stands may validly be used in vegetation studies.

subsequent interpretation in terms of species-richness must be viewed with caution.

Williams *et al.* (1969) used the nth nearest neighbours to an individual as a 'stand' in a study of pattern, classifying the overlapping 'stands' centred on all individuals in the single plot examined. This approach might be useful in broader studies of forest composition, eliminating the often time-consuming delimitation of plots, and equalizing the amount of information obtained from stands when tree density is variable from stand to stand. Another alternative to the use of plots is to use single transects as stands. Robertson (1977) showed that the results from qualitative data, of species touching a line transect, in sand dune vegetation were comparable to those from plot stands.

If full description of the vegetation is required, rather than the minimal data needed for efficient analysis, a larger stand will generally be necessary both to include the majority of species and to allow reasonably precise quantitative estimates for more of the species. Density estimates present a particular problem in description; since the accuracy of a density estimate, even for random distribution of individuals—the most favourable case—depends on the number of individuals counted (p. 26), even complete enumeration may give low precision for all but the commonest species. The problem is acute in vegetation, such as much tropical rain forest, where no species have high densities, and the size of homogeneous stand theoretically desirable may be unattainable (Greig-Smith, 1965).

The considerations involved in the siting of stands have apparently often been poorly understood. This may, in part, be because the situation is deceptively similar to the design of sampling surveys and censuses aimed at estimates of some parameter for a heterogeneous area (Greig-Smith, 1971c). Whether the situations are really comparable depends on the objectives of the vegetation survey.

If the objective is primarily to simplify the observed range of vegetation, constraints on the siting of stands depend on what is required of the results. If the aim is inventory, e.g. to determine what proportion of the total area is occupied by different vegetation types, the stands must be sited in such a way that an unbiased estimate of amounts of different species or vegetation types in the area as a whole is obtained; the stands are themselves regarded as samples. A form of systematic sampling is likely to give a more precise estimate unless the scale of the sampling grid happens to coincide with that of periodic variation in the vegetation. Any risk of this happening is likely to be evident in the field. Smartt and Grainger (1974) have demonstrated the greater efficiency of stratified systematic unaligned sampling, but in some vegetation types, e.g. forest, the much greater time needed to site stands on this system may make it uneconomic; greater efficiency may result from more stands sited strictly systematically. Only if confidence limits for estimates are required, either to test for differences between parts of the area being investigated, or to compare with other areas, is random siting necessary.

If the primary aim is to produce a vegetation map, stands should be placed as evenly as possible so that the uncertainty of the position of boundaries between one vegetation type and another is more or less constant over the area, i.e. strict systematic siting should be used. If the 'grain' of the vegetation, the average size of patches of different vegetation types, is evidently different in different parts of the whole area, it is useful to stratify the area and site stands systematically at an appropriate spacing in each stratum.

If the primary objective is to search for discontinuities in the range of composition of vegetation, systematic siting is better avoided, because even relatively slight correlation between grid interval and periodic variation will tend to sharpen the boundaries between noda. However, critical delineation of discontinuities in composition is unlikely to be the principal or only objective and appropriate analysis presents problems (see below); useful indications of the discreteness or otherwise of the vegetation types may be obtained from systematically sited stands.

If elucidation of correlations between vegetation and environment is the primary objective, stands should, as far as possible, include all variants of vegetation and an equal representation of all variants. No mechanical system of siting can achieve this dual objective. Both random and, with the usual proviso about periodic variation in the vegetation, systematic siting will, if sufficient stands are examined, ensure that all variants are examined, but unless variants are equally represented in the area, more information will be obtained about some variants than others. There is no complete solution to this dilemma. If there are obvious broad differences in the vegetation, stratification in relation to these will be helpful*, but the same difficulties apply to variation within the broad types. If there is reason to suspect that obvious environmental difference may play an important part in determining the composition of the vegetation, even if corresponding vegetational differences are not obvious, it is sensible to stratify in relation to these and, for example, in undulating topography ensure that equal numbers of stands are sited in valleys, on slopes and on ridges. Note that stratification here aims at equal numbers of stands from different strata, not equal density of stands. There is no objection to including stands because they appear to be in some

---

* In contrast to the undesirability of stratifying samples rather than stands in relation to major vegetational differences (p. 22).

way unlike those already recorded, provided that no attempt is made to draw conclusions from the subsequent analysis about the sharpness of community boundaries or the relative abundance of different vegetation types. A workable approach, if the terrain allows, is to record stands initially by a set scheme, stratifying as above if appropriate, and then add any stands that appear to be different from those already recorded*.

The major decision over analysis of the data is between classification and ordination. This depends partly on the objective of the investigation and partly on the nature of the data. Preparation of a vegetation map or of an inventory of vegetation clearly requires a classification. Classification can be used to identify correlations of vegetation with environment by comparing the average environmental characteristics of the two subgroups at each division in the hierarchy. It does, however, lose information on any correlation between members of a single final group and their environments. This information is, in principle, retained by ordination, but in practice the more detailed correlation may not be detectable if the data represent more than one independent or partially independent gradients of composition. This suggests, for very heterogeneous data, an initial classification down to a level that provides reasonably homogeneous groups, followed both by an ordination of the groups to clarify the relationships between them and by separate ordinations of the members of each group (cf. Greig-Smith et al., 1967), or Noy-Meir's (1971b) nodal ordination (p. 264). In practice, classification and ordination, considered separately, often both contribute to understanding of a data set.

Choice of technique of classification or ordination is not easy; there is no 'best' method. Every simplification of a complex set of data involves ignoring some of the information in the data, and different techniques preserve different information, as is evident from the analysis of the same data set by different techniques (e.g. Austin & Greig-Smith, 1968; Crawford et al., 1970; Ivimey-Cook et al., 1975; Robertson, 1978; Clymo, 1980; del Moral, 1980). Which technique is most appropriate to a particular investigation will depend on which retains the most relevant information, but it is often precisely this which is not known in advance. There is therefore much to be said for analysing data by several techniques.

There has been considerable discussion of the efficiency of different techniques†. Though an understanding of the assumptions and procedures of

---

* Useful discussions of sampling schemes actually used in extensive surveys have been given by, among others, Bunce and Shaw (1973), Curtis (1959, pp. 63–79), Hall and Swaine (1976) and Noy-Meir (1970, 1971b).

† Noy-Meir and Whittaker (1977) gave an excellent account of the merits and limitations of techniques up to 1976.

a technique may indicate its suitability or otherwise in a particular case, there can be no objective assessment of efficiency (Greig-Smith, 1980). How far the results confirm preconceptions is no test of efficiency, though many discussions implicitly adopt this criterion. Blackith and Reyment (1971) put it well '. . . . . there are no objective criteria against which the classifications can be judged. There is therefore a tendency for multivariate techniques to be condemned when they disagree with conventional methods, and regarded as superfluous when they agree'.

The investigator is, perhaps, less likely to have preconceptions about the ordination of a set of data. Efficiency of an ordination has sometimes been judged by the extent to which the variation in the initial data is 'accounted for' in the ordination, but the information retained is not necessarily the most useful; even a low percentage retention of the initial variation may be very informative if there is a large amount of irrelevant information in the original data. An alternative approach to assessment of efficiency is to apply a technique to a set of simulated data and see how far the original structure of the data is retrieved. It is commonly assumed that species response curves to an environmental gradient have a Gaussian form, but there is good reason to doubt whether this is generally so (p. 269). It is thus difficult to produce realistic simulated data related to a single environmental gradient, and more so for response to two or more at least partially independent gradients, a situation more like that commonly found in the field. A common misconception about ordination techniques is that interpretation must be in terms of the axes, i.e. that an axis represents the response of the vegetation to an environmental gradient. This is unnecessarily restrictive; if a pattern of an environmental factor is apparent on a vegetational ordination either by a straight line at an angle to the axes, or by a curved line, there is evidence of correlation between composition of the vegetation and that factor. With only slight exaggeration, it may be said that the axes of an ordination are mere constructional artefacts—it is the display of relative stand positions that is valuable (Greig-Smith, 1971c).

Although direct tests of the efficiency of techniques cannot be made*, they can be judged, albeit subjectively, by how satisfactory they prove to be as working tools and how productive they prove to be in generating hypotheses which can then be tested by further field observation or by experiment, e.g. if analysis of survey data suggests a relationship between composition of vegetation and available level of an inorganic ion in the soil, an obvious

* Wilson (1981) suggested tests for the consistency of ordinations of random subsets of a data set. These test whether an ordination represents a real property of the data, but not whether it is retaining the most relevant information.

experiment is to examine the performance of selected species grown in different concentrations of that ion.

In the earlier applications of numerical methods, the amount of computation required was a serious obstacle to the use of some techniques. With the rapid increase in speed and capacity of computers this is much less often a constraint. Thus association analysis was put forward not as the most satisfactory technique that could be proposed at the time, but as one that was computationally feasible. Now, with its inherent disadvantage of built-in standardization, liability to reversals in the hierarchy and tendency to chain, its use would not be seriously considered.

Although the computational load is now a less serious consideration, the simplest technique that will meet the requirements of the investigation should be used; just as the gathering of data in the field has a cost in time and expense, so has the subsequent data analysis. The question also arises whether field data, at least in extensive surveys, are accurate enough to justify very refined methods of analysis (Greig-Smith, 1980); Hall and Okali (1978) demonstrated the considerable extent to which data in a complex vegetation type are affected by season and by the experience of the observers.

Little has been said above specifically about analysis of data when the objective is the detection of 'real' entities. In the strict sense of detecting complete discontinuities in the range of composition, if they exist, recognition of such entities is not possible by numerical analysis. Where the expressed aim is detection of real entities, the objective is a less demanding one, to identify species combinations of much commoner occurrence, intermediates between which are very rare. This implies the erection of a classification which will have general validity for other stands of comparable vegetation and is more demanding than the extraction of noda to provide convenient reference points in mapping or inventory (and still more so than the use of classification on a particular set of data as a means of examining correlation of composition with environment). Superficially it might appear that this could be achieved by examining the degree of clustering. In a classification those stands which fuse at a low level in a hierarchy and are only linked to others at a much higher level would be recognized as forming a real entity. Alternatively, such an entity would be recognized on an ordination as a cluster of points more or less isolated from others. Extreme caution must, however, be exercised. It is abundantly clear that the degree of clustering depends on the exact technique of analysis used and is a property of the technique rather than of the data; compare, for instance, the effect of nearest-neighbour and furthest-neighbour sorting in agglomerative classification (p. 205–6).

If a classification of general validity is needed, a considerable number of stands should be included to minimize the effects of chance in producing groups, and classes represented by a small number of stands only should be regarded as tentative. If at all practicable, the analysis should be repeated on a separate set of similar number of stands, ideally from a different area, and only if the two results are comparable should the classification be accepted as of general validity.

Two particular situations, not covered above, call for brief comment. The most direct approach to elucidation of change in vegetation with time is to record the composition of permanent plots at successive intervals. The information that results is limited to the length of time over which observations are made. If plots initially at different stages in a succession are available and are recorded over one or more time intervals ordination can be used to summarize the information on change. Each recording of a plot is regarded as a separate 'stand' and a suitable ordination is produced. If the course of change is comparable for different plots, their trajectories in the ordination display will be closely parallel for plots at the same stage and form a sequence for plots at different stages. It is thus possible to reconstruct the whole sequence of changes from shorter term observations and to detect any tendency for plots to converge or diverge in composition (Van der Maarel, 1969; Austin, 1977) (Fig. 78). In the special case of forest vegetation, where tree diameter for a particular species can be regarded as a measure of age, even one set of observations on stands at different successional stages can be used to elucidate the course of change (Goff, 1968; Goff & Zedler, 1968, 1972; Carleton & Maycock, 1978).

Swaine and Greig-Smith (1980) suggested an alternative ordination approach where a series of observations on the same plots are available. From a matrix of species × time for each stand an appropriate cross-product (covariance or correlation coefficient) is calculated for all pairs of species. The values for each species-pair are summed and used as input to principal component analysis. Both species loadings and stand scores can be calculated in the usual way and differential behaviour of stands can be detected. The method has the advantage that initial differences between stands are ignored and analysis concentrates on change, but is only applicable to relatively homogeneous sets of sites, on which the correlations between species may be expected to be the same.

Another approach based on observed short term changes is to estimate transition probabilities for change either from an individual of one species to an individual of another species (as for forest trees, e.g. Enright & Ogden, 1979) or from one vegetation type, defined by a preliminary classification, to another (e.g. Austin, 1980b; Usher, 1981). From the resulting transition

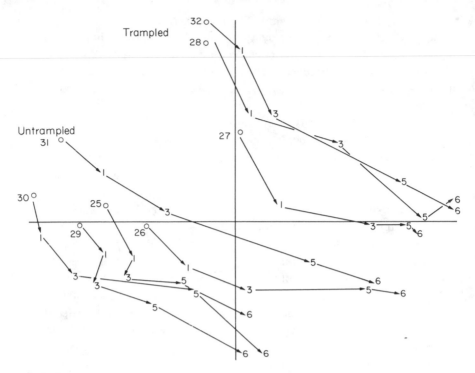

Fig. 78. Ordination of eight individual quadrats on a lawn, recorded on six successive occasions. (From Austin, 1977, by courtesy of *Vegetatio.*)

matrix alternative pathways of change can be elucidated (Fig. 79). Successive multiplication of the initial composition of the vegetation of an area by the transition matrix can then, on the somewhat limiting assumption that the transition probabilities remain constant, be used to make predictions of the future composition. Usher (1981) has a useful critical assessment of this Markovian approach. It is potentially useful especially for elucidating the later stages of succession, where change at any one point is not necessarily unidirectional, and the cyclic changes in vegetation having a mosaic structure of different phases (Watt, 1947) (cf. Legg, 1980).

If the objective of analysis is to identify optimal boundaries between vegetation types in the field (as opposed to discontinuities in the range of composition of the vegetation) a form of agglomerative classification in which fusion is only permitted between stands which are adjacent on the ground might be used. Such methods have been used in the analysis of pollen-analytical and other palaeoecological data from profiles (Gordon & Birks, 1972; Birks & Birks, 1980). These methods could be applied directly to

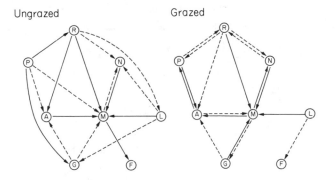

Ungrazed   Grazed

Fig. 79. Diagrams showing main transition probabilities between eight vegetation types, based on a preliminary classification, for grazed and ungrazed grassland. (———) Largest 15% of probabilities; (---) next 15% of probabilities. (From Usher, 1981, by courtesy of *Vegetatio*.)

transect data and comparable methods allow fusion of adjacent stands on a grid (see Gordon, 1981).

Little or no mention has been made in this chapter of tests of significance or probability statements. Tests of significance are appropriate when a hypothesis is being tested, but this is rarely the case with survey data. The objective is much more commonly either the erection of an empirical framework of description, to be judged on its practical utility, as with some uses of classification, or data exploration as a means of generating hypotheses. In neither case is there a null hypothesis to be tested and questions of significance testing and probability do not arise.

Appendix: Tables

Table 1. Variance of binomial distribution after angular transformation. (a) For various values of n (In part from Robinson, 1955)

			p			Approximate value
		0·4	0·3	0·2	0·1	$\left(\dfrac{820·7}{n}\right)$
n	0·5	0·6	0·7	0·8	0·9	
2	1012·5	972·0	850·5	648·0	364·5	410·3
3	575·7	566·4	530·8	449·3	288·8	273·6
4	365·6	369·4	372·6	348·8	250·1	205·2
5	253·6	261·6	279·2	282·8	222·1	164·2
9	106·7	110·6	123·3	148·1	148·9	91·2
10	94·0	96·6	106·4	125·9	136·0	82·1
16	55·2	55·7	57·8	65·8	81·6	51·3
20	43·4	43·5	44·5	48·8	63·5	41·0
25	34·3	34·6	35·3	36·7	47·3	32·8
30	28·2	28·4	28·7	29·9	36·7	27·4

Table 2. Mean range of samples of different sizes (after Pearson & Hartley, 1954). Range expressed in units of population standard deviation

Sample size	Range	Sample size	Range	Sample size	Range	Sample size	Range
2	1·128	9	2·970	16	3·532	23	3·858
3	1·693	10	3·078	17	3·588	24	3·895
4	2·059	11	3·173	18	3·640	25	3·931
5	2·326	12	3·258	19	3·689	50	4·498
6	2·534	13	3·336	20	3·735	100	5·015
7	2·704	14	3·407	21	3·778		
8	2·847	15	3·472	22	3·819		

(Appendix Table 1 (b))
(b) Extended table for $n = 5$ (A. J. Morton, unpublished)

P	0·01	0·02	0·03	0·04	0·05	0·06	0·07	0·08	0·09	0·10
0	33·7	64·3	92·1	117·2	139·9	160·3	178·6	194·8	209·3	222·1
0·1	233·3	243·2	251·7	259·0	265·2	270·3	274·6	278·1	280·8	282·8
0·2	284·2	285·1	285·5	285·4	285·0	284·3	283·3	282·1	280·7	279·2
0·3	277·5	275·7	273·9	272·0	270·2	268·4	266·6	264·8	263·2	261·6
0·4	260·2	258·8	257·6	256·6	255·7	254·9	254·3	253·9	253·6	253·6

Table 3. e^{-m}

m	0	0·01	0·02	0·03	0·04	0·05	0·06	0·07	0·08	0·09
0	1·0000	·9900	·9802	·9704	·9608	·9512	·9418	·9324	·9231	·9139
0·1	·9048	·8958	·8869	·8781	·8694	·8607	·8521	·8437	·8353	·8270
0·2	·8187	·8106	·8025	·7945	·7866	·7788	·7711	·7634	·7558	·7483

m	0·0	0·1	0·2	0·3	0·4	0·5	0·6	0·7	0·8	0·9
0	1·0000	·9048	·8187	·7408	·6703	·6065	·5488	·4966	·4493	·4066
1	·3679	·3329	·3012	·2725	·2466	·2231	·2019	·1827	·1653	·1496
2	·1353	·1225	·1108	·1003	·0907	·0821	·0743	·0672	·0608	·0550
3	·0498	·0450	·0408	·0369	·0334	·0302	·0273	·0247	·0224	·0202
4	·0183	·0166	·0150	·0136	·0123	·0111	·0101	·0091	·0082	·0074
5	·0067	·0061	·0055	·0050	·0045	·0041	·0037	·0033	·0030	·0027
6	·0025	·0022	·0020	·0018	·0017	·0015	·0014	·0012	·0011	·0010
7	·00091	·00083	·00075	·00068	·00061	·00055	·00050	·00045	·00041	·00037
8	·00034	·00030	·00027	·00025	·00022	·00020	·00018	·00017	·00015	·00014
9	·00012	·00011	·00010	·00009	·00008	·00007	·00007	·00006	·00006	·00005

Table 4.

$$\sqrt{\frac{2}{N-1}}$$

N	$\sqrt{\dfrac{2}{N-1}}$	N	$\sqrt{\dfrac{2}{N-1}}$
10	·4714	100	·1421
20	·3244	150	·1159
30	·2626	200	·1003
40	·2265	250	·08962
50	·2020	300	·08179
60	·1841	400	·07080
70	·1703	500	·06331
80	·1591	1000	·04474
90	·1499		

Table 5. Significant points for $\phi = 2n_0 n_2/n_1^2$ (n_0, n_1, n_2 = number of quadrats containing 0, 1, 2, individuals respectively) (from Moore, 1953, by courtesy of *Annals of Botany*)

$$N = \text{total number of quadrats}$$

$$R = \frac{n_0 + n_1 + n_2}{N} \times 100$$

The figure given is the mean of ϕ plus twice its standard error, i.e. approximately the 5% point

Mean number- per quadrat	0·5	1·0	1·5	2·0	2·5
R	99	92	81	68	54
N 50	2·70	2·40	2·46	2·66	2·98
100	2·16	1·95	1·99	2·13	2·34
200	1·80	1·66	1·68	1·77	1·92
300	1·65	1·53	1·55	1·62	1·74
400	1·55	1·46	1·47	1·54	1·63
500	1·49	1·41	1·42	1·48	1·56

Table 6. Densities corresponding to different percentage frequencies in random distributions

F%	0	1	2	3	4	5	6	7	8	9
0	0·	·0101	·0202	·0305	·0408	·0513	·0619	·0726	·0834	·0943
10	·1054	·1165	·1278	·1393	·1508	·1625	·1744	·1863	·1985	·2107
20	·2231	·2357	·2485	·2614	·2744	·2877	·3011	·3147	·3285	·3425
30	·3567	·3711	·3857	·4005	·4155	·4308	·4463	·4620	·4780	·4943
40	·5108	·5276	·5447	·5621	·5798	·5978	·6162	·6349	·6539	·6733
50	·6931	·7133	·7340	·7550	·7765	·7985	·8210	·8440	·8675	·8916
60	·9163	·9416	·9676	·9943	1·0217	1·0498	1·0788	1·1087	1·1394	1·1712
70	1·2040	1·2379	1·2730	1·3093	1·3471	1·3863	1·4271	1·4697	1·5141	1·5606
80	1·6094	1·6607	1·7148	1·7720	1·8326	1·8971	1·9661	2·0402	2·1203	2·2073
90	2·3026	2·4079	2·5257	2·6593	2·8134	2·9957	3·2189	3·5066	3·9120	4·6052

F%	99·1	99·2	99·3	99·4	99·5	99·6	99·7	99·8	99·9
	4·7105	4·8283	4·9618	5·1160	5·2983	5·5215	5·8091	6·2146	6·9078

Table 7. 95% confidence limits for random distribution of the ratio of observed to expected variance in analysis of pattern (from Greig-Smith, 1961b, by courtesy of *Journal of Ecology*)

Degrees of freedom	1	2	3	4	5	6	7	8	9	10	12	14
Lower limit	0·00	0·03	0·07	0·12	0·17	0·21	0·24	0·27	0·30	0·32	0·35	0·40
Upper limit	5·02	3·69	3·12	2·79	2·57	2·41	2·29	2·19	2·11	2·05	1·94	1·87

Degrees of freedom	15	16	18	20	24	28	30	32	36	40	48	56
Lower limit	0·42	0·43	0·45	0·48	0·52	0·55	0·56	0·57	0·59	0·61	0·64	0·66
Upper limit	1·83	1·80	1·75	1·71	1·64	1·59	1·57	1·55	1·51	1·48	1·44	1·40

Degrees of freedom	60	64	72	80	96	112	120	128	144	160	192	224
Lower limit	0·67	0·68	0·70	0·71	0·74	0·75	0·75	0·77	0·78	0·79	0·81	0·82
Upper limit	1·39	1·38	1·35	1·33	1·30	1·27	1·26	1·26	1·24	1·23	1·21	1·19

Degrees of freedom	256	288	320	384	448	480	512	576	640	768	896	960
Lower limit	0·83	0·84	0·85	0·86	0·87	0·88	0·88	0·89	0·89	0·90	0·91	0·91
Upper limit	1·18	1·17	1·16	1·15	1·13	1·13	1·13	1·12	1·11	1·10	1·09	1·09

Degrees of freedom	1024	1152	1280	1536
Lower limit	0·91	0·92	0·92	0·93
Upper limit	1·09	1·08	1·08	1·07

Table 8. Probability that S (for τ) attains or exceeds a specified value. (Shown only for positive values. Negative values obtainable by symmetry.) (From Kendall, 1948, by courtesy of Charles Griffin & Co. Ltd)

	Values of n					Values of n		
S	4	5	8	9	S	6	7	10
0	0·625	0·592	0·548	0·540	1	0·500	0·500	0·500
2	0·375	0·408	0·452	0·460	3	0·360	0·386	0·431
4	0·167	0·242	0·360	0·381	5	0·235	0·281	0·364
6	0·042	0·117	0·274	0·306	7	0·136	0·191	0·300
8		0·042	0·199	0·238	9	0·068	0·119	0·242
10		$0{\cdot}0^{2}83$	0·138	0·179	11	0·028	0·068	0·190
12			0·089	0·130	13	$0{\cdot}0^{2}83$	0·035	0·146
14			0·054	0·090	15	$0{\cdot}0^{2}14$	0·015	0·108
16			0·031	0·060	17		$0{\cdot}0^{2}54$	0·078
18			0·016	0·038	19		$0{\cdot}0^{2}14$	0·054
20			$0{\cdot}0^{2}71$	0·022	21		$0{\cdot}0^{3}20$	0·036
22			$0{\cdot}0^{2}28$	0·012	23			0·023
24			$0{\cdot}0^{3}87$	$0{\cdot}0^{2}63$	25			0·014
26			$0{\cdot}0^{3}19$	$0{\cdot}0^{2}29$	27			$0{\cdot}0^{2}83$
28			$0{\cdot}0^{4}25$	$0{\cdot}0^{2}12$	29			$0{\cdot}0^{2}46$
30				$0{\cdot}0^{3}43$	31			$0{\cdot}0^{2}23$
32				$0{\cdot}0^{3}12$	33			$0{\cdot}0^{2}11$
34				$0{\cdot}0^{4}25$	35			$0{\cdot}0^{2}47$
36				$0{\cdot}0^{5}28$	37			$0{\cdot}0^{3}18$
					39			$0{\cdot}0^{4}58$
					41			$0{\cdot}0^{4}15$
					43			$0{\cdot}0^{5}28$
					45			$0{\cdot}0^{6}28$

Repeated zeros are indicated by powers, e.g. $0{\cdot}0^{3}47$ stands for $0{\cdot}00047$.

References

Aberdeen, J. E. C. (1954) Estimation of basal or cover areas in plant ecology. *Aust. J. Sci.* **17**, 35–6.

Aberdeen, J. E. C. (1955) Quantitative methods for estimating the distribution of soil fungi. *Pap. Dep. Bot. Univ. Qd*, **3**, 83–96.

Aberdeen, J. E. C. (1958) The effect of quadrat size, plant size and plant distribution on frequency estimates in plant ecology. *Aust. J. Bot.* **6**, 47–58.

Agnew, A. D. Q. (1961) The ecology of *Juncus effusus* L. in North Wales. *J. Ecol.* **49**, 83–102.

Anderson, A. J. B. (1971) Ordination models in ecology. *J. Ecol..* **59**, 713–26.

Anderson, D. J. (1963) The structure of some upland plant communities in Caernarvonshire. III. The continuum analysis. *J. Ecol.* **51**, 403–14.

Anderson, D. J. (1965) Studies on structure in plant communities. I. An analysis of limestone grassland in Monk's Dale, Derbyshire. *J. Ecol.* **53**, 97–107.

Archibald, E. E. A. (1948) Plant populations. I. A new application of Neyman's contagious distribution. *Ann. Bot.* **12**, 221–35.

Archibald, E. E. A. (1949a) The specific character of plant communities. I. Herbaceous communities. *J. Ecol.* **37**, 260–73.

Archibald, E. E. A. (1949b) The specific character of plant communities. II. A quantitative approach. *J. Ecol.* **37**, 274–88.

Archibald, E. E. A. (1950) Plant populations. II. The estimation of the number of individuals per unit area of species in heterogeneous plant populations. *Ann. Bot.* **14**, 7–21.

Archibald, E. E. A. (1952) A possible method for estimating the area covered by the basal parts of plants. *S. Afr. J. Sci.* **48**, 286–92.

Ashby, E. (1935) The quantitative analysis of vegetation. *Ann. Bot.* **49**, 779–802.

Ashby, E. (1936) Statistical ecology. *Bot. Rev.* **2**, 221–35.

Ashby, E. (1948) Statistical ecology. II. A reassessment. *Bot. Rev.* **14**, 222–34.

Ashby, W. C. (1972) Distance measurements in vegetation study. *Ecology*, **53**, 980–1.

Ashton, P. S. (1964) Ecological studies in the mixed dipterocarp forests of Brunei State. *Oxf. For. Mem.* **25**.

Austin, M. P. (1968a) Pattern in a *Zerna erecta* dominated community. *J. Ecol.* **56**, 197–218.

Austin, M. P. (1968b) An ordination study of a chalk grassland community. *J. Ecol.* **56**, 739–57.

Austin, M. P. (1971) Role of regression analysis in plant ecology. *Proc. ecol. Soc. Austr.* **6**, 63–75.

Austin, M. P. (1972) Models and analysis of descriptive vegetation data. In:

Mathematical Models in Ecology (Ed. by J. N. R. Jeffers), pp. 61–86. Blackwell Scientific Publications, Oxford.

Austin, M. P. (1976a) On non-linear species response models in ordination. *Vegetatio*, **33**, 33–41.

Austin, M. P. (1976b) Performance of four ordination techniques assuming three different non-linear species response models. *Vegetatio*, **33**, 43–9.

Austin, M. P. (1977) Use of ordination and other multivariate descriptive methods to study succession. *Vegetatio*, **35**, 165–75.

Austin, M. P. (1980a) Searching for a model for use in vegetation analysis. *Vegetatio*, **42**, 11–21.

Austin, M. P. (1980b) An exploratory analysis of grassland dynamics: an example of a lawn succession. *Vegetatio*, **43**, 87–94.

Austin, M. P. (1981) The role of certain diversity properties in vegetation classification. In: *Vegetation Classification in the Australian Region* (Ed. by A. N. Gillison & D. J. Anderson), pp. 125–40. C.S.I.R.O. and Australian National University Press, Canberra.

Austin, M. P. & Noy-Meir, I. (1971) The problem of non-linearity in ordination: experiments with two-gradient models. *J. Ecol.* **59**, 763–73.

Austin, M. P., Ashton, P. S. & Greig-Smith, P. (1972) The application of quantitative methods to vegetation survey. III. A re-examination of rain forest data from Brunei. *J. Ecol.* **60**, 305–24.

Austin, M. P. & Greig-Smith, P. (1968) The application of quantitative methods in vegetation survey. II. Some methodological problems of data from rain forest. *J. Ecol.* **56**, 827–44.

Austin, M. P. & Orloci, L. (1966) Geometric models in ecology. II. An evaluation of some ordination techniques. *J. Ecol.* **54**, 217–27.

Barkham, J. P. & Norris, J. M. (1970) Multivariate procedures in an investigation of vegetation and soil relations of two beech woodlands, Cotswold Hills, England. *Ecology*, **51**, 630–9.

Barnes, H. & Stanbury, F. A. (1951) A statistical study of plant distribution during the colonization and early development of vegetation on china clay residues. *J. Ecol.* **39**, 171–81.

Bartlett, M. S. (1936) Some examples of statistical methods of research in agriculture and applied biology. *Suppl. Jl R. statist. Soc.* **4**, 137–83.

Bartlett, M. S. (1963) The spectral analysis of point processes. *Jl R. statist. Soc. B*, **25**, 264–96.

Bartlett, M. S. (1964) The spectral analysis of two-dimensional point processes. *Biometrika*, **51**, 299–311.

Batcheler, C. L. (1971) Estimation of density from a sample of joint point- and nearest-neighbour distances. *Ecology*, **52**, 703–9.

Batcheler, C. L. & Bell, D. J. (1970) Experiments in estimating density from joint point- and nearest-neighbour distance samples. *Proc. N.Z. ecol. Soc.* **17**, 111–7.

Beals, E. W. (1960) Forest bird communities in the Apostle Islands of Wisconsin. *Wilson Bull.* **72**, 156–181.

Beals, E. W. (1965) Species patterns in a Lebanese Poterietum. *Vegetatio*, **13**, 68–87.

Beals, E. W. (1973) Ordination: mathematical elegance and ecological naïveté. *J. Ecol.* **61**, 23–35.

Beard, J. S. (1973) The physiognomic approach. In: *Ordination and Classification of Communities. Handbook of Vegetation Science*, **5,** (Ed. by R. H. Whittaker), pp. 355–86. Junk, The Hague.

Benninghoff, W. S. & Southworth, W. C. (1964) Ordering of tabular arrays of phytosociological data by digital computer. *Abstracts Xth Int. bot. Congr.* pp. 331–2.

Benzecri, J. P. (1973) *L'Analyse des Données.* Vol. 2. *L'Analyse des Correspondances.* Dunod, Paris.

Beshir, M. E. (1968) *A study of the accuracy of estimates of the specific composition of vegetation.* M.Sc. thesis, University of Wales.

Bieleski, R. L. (1959) Factors affecting growth and distribution of Kauri (*Agathis australis* Salisb.). I. Effect of light on the establishment of Kauri and of *Phyllocladus trichomanoides* D. Don. *Aust. J. Bot.* **7,** 252–67.

Birks, H. J. B. & Birks, H. H. (1980) *Quarternary Palaeoecology.* Arnold, London.

Bitterlich, W. (1948) Die Winkelzählprobe. *Allg. Forst- u. Holzw. Ztg,* **59,** 4–5.

Blackith, R. E. & Reyment, R. A. (1971) *Multivariate Morphometrics.* Academic Press, London.

Blackman, G. E. (1935) A study by statistical methods of the distribution of species in grassland associations. *Ann. Bot.* **49,** 749–77.

Blackman, G. E. (1942) Statistical and ecological studies in the distribution of species in plant communities. I. Dispersion as a factor in the study of changes in plant populations. *Ann. Bot.* **6,** 351–70.

Blackman, G. E. & Rutter, A. J. (1946) Physiological and ecological studies in the analysis of plant environment. I. The light factor and the distribution of the bluebell (*Scilla non-scripta*) in woodland communities. *Ann. Bot.* **10,** 361–90.

Bormann, F. H. (1953) The statistical efficiency of sample plot size and shape in forest ecology. *Ecology,* **34,** 474–87.

Bottomley, J. (1971) Some statistical problems arising from the use of the information statistic in numerical classification. *J. Ecol.* **59,** 339–42.

Bourdeau, P. F. (1953) A test of random versus systematic ecological sampling. *Ecology,* **34,** 499–512.

Bray, J. R. (1956) A study of mutual occurrence of plant species. *Ecology,* **37,** 21–8.

Bray, J. R. (1962) Use of non-area analytic data to determine species dispersion. *Ecology,* **43,** 328–33.

Bray, J. R. & Curtis, J. T. (1957) An ordination of the upland forest communities of southern Wisconsin. *Ecol. Monogr.* **27,** 325–49.

Brillouin, L. (1962) *Science and Information Theory,* 2nd edn. Academic Press, New York.

Brown, D. (1954) Methods of surveying and measuring vegetation. *Bull. Bur. Past., Hurley,* **42.**

Brown, R. T. & Curtis, J. T. (1952) The upland conifer-hardwood forests of northern Wisconsin. *Ecol. Monogr.* **22,** 217–34.

Bullock, J. A. (1971) The investigation of samples containing many species. I. Sample description. *Biol. J. Linn. Soc.* **3,** 1–21.

Bunce, R. G. H. (1968) An ecological study of Ysgolion Duon, a mountain cliff in Snowdonia. *J. Ecol.* **56,** 59–75.

Bunce, R. G. H. & Shaw, M. W. (1973) A standardized procedure for ecological survey. *J. environ. Mgmt*, **1**, 239–58.

Burr, E. J. (1970) Cluster sorting with mixed character types. II. Fusion strategies. *Aust. Comput. J.* **2**, 98–103.

Cain, S. A. (1934) Studies on virgin hardwood forest. II. A comparison of quadrat sizes in a quantitative phytosociological study of Nash's Woods, Posey County, Indiana. *Amer. Midl. Nat.* **15**, 529–66.

Cain, S. A. (1938) The species-area curve. *Amer. Midl. Nat.* **19**, 573–81.

Cain, S. A. (1943) Sample-plot technique applied to alpine vegetation in Wyoming. *Amer. J. Bot.* **30**, 240–47.

Carleton, T. J. (1979) Floristic variation and zonation in the boreal forest south of James Bay: a cluster seeking approach. *Vegetatio*, **39**, 147–60.

Carleton, T. J. (1980) Non-centered component analysis of vegetation data: a comparison of orthogonal and oblique rotation. *Vegetatio*, **42**, 59–66.

Carleton, T. J. & Maycock, P. F. (1978) Dynamics of the boreal forest south of James Bay. *Can. J. Bot.* **56**, 1157–73.

Carleton, T. J. & Maycock, P. F. (1980) Vegetation of the boreal forests south of James Bay: non-centered component analysis of the vascular flora. *Ecology*, **61**, 1199–212.

Cassie, R. M. & Michael, A. D. (1968) Fauna and sediments of an intertidal mud flat: a multivariate analysis. *J. exp. mar. Biol. Ecol.* **2**, 1–23.

Češka, A. & Roemer, H. (1971) A computer program for identifying species-relevé groups in vegetation studies. *Vegetatio*, **23**, 255–77.

Chadwick, M. J. (1960) *Nardus stricta*, a weed of hill grazings. In: *The Biology of Weeds* (Ed. by J. L. Harper), pp. 246–56. Blackwell Scientific Publications, Oxford.

Christian, C. S. & Perry, R. A. (1953) The systematic description of plant communities by the use of symbols. *J. Ecol.* **41**, 100–5.

Clapham, A. R. (1932) The form of the observational unit in quantitative ecology. *J. Ecol.* **20**, 192–97.

Clapham, A. R. (1936) Over-dispersion in grassland communities and the use of statistical methods in plant ecology. *J. Ecol.* **24**, 232–51.

Clark, P. J. (1956) Grouping in spatial distributions. *Science*, **123**, 373–4.

Clark, P. J. & Evans, F. C. (1954a) Distance to nearest neighbour as a measure of spatial relationships in populations. *Ecology*, **35**, 445–53.

Clark, P. J. & Evans, F. C. (1954b) On some aspects of spatial pattern in biological populations. *Science*, **121**, 397–8.

Clausen, J. J. (1957) A comparison of some methods of establishing plant community patterns. *Bot. Tidsskr.* **53**, 253–78.

Clements, F. E. (1916) Plant succession: an analysis of the development of vegetation. *Publ. Carneg. Instn.* **242**.

Cliff, A. D. & Ord, J. K. (1973) *Spatial Autocorrelation*. Pion, London.

Clymo, R. S. (1980) Preliminary survey of the peat-bog Hummell Knowe Moss using various numerical methods. *Vegetatio*, **42**, 129–48.

Cochran, W. G. (1954) Some methods for strengthening the common χ^2 tests. *Biometrics*, **4**, 417–51.

Cochran, W. G. (1963) *Sampling Techniques*, 2nd edn. Wiley, New York and London.

Cochran, W. G. & Cox, G. M. (1944) *Experimental Design*. Mimeographed. (Quoted from Snedecor, 1946).

Conway, V. M. (1938) Studies in the autecology of *Cladium mariscus* R. Br. IV. Growth rates of the leaves. *New Phytol.* **37,** 254–78.

Cooper, C. F. (1957) The variable plot method for estimating shrub density. *J. Range Mgmt,* **10,** 111–15.

Cooper, C. F. (1963) An evaluation of variable plot sampling in shrub and herbaceous vegetation. *Ecology,* **44,** 565–9.

Cormack, R. M. (1971) A review of classification. *Jl R. statist. Soc. A.* **134,** 321–67.

Cormack, R. M. & Ord, J. K. (Eds) (1979) *Spatial and Temporal Analysis in Ecology. Statistical Ecology,* **8.** International Co-operative Publishing House, Fairland, Md, U.S.A.

Cottam, G. (1947) A point method for making rapid surveys of woodlands. *Bull. ecol. Soc. Amer.* **28,** 60.

Cottam, G. & Curtis, J. T. (1949) A method for making rapid surveys of woodlands by means of pairs of randomly selected trees. *Ecology,* **30,** 101–4.

Cottam, G. & Curtis, J. T. (1955) Correction for various exclusion angles in the random pairs method. *Ecology,* **36,** 767.

Cottam, G. & Curtis, J. T. (1956) The use of distance measures in phytosociological sampling. *Ecology,* **37,** 451–60.

Cottam, G., Curtis, J. T. & Catana, A. J. (1957) Some sampling characteristics of a series of aggregated populations. *Ecology,* **38,** 610–22.

Cottam, G., Curtis, J. T. & Hale, B. W. (1953) Some sampling characteristics of a population of randomly dispersed individuals. *Ecology,* **34,** 741–57.

Cottam, G., Goff, F. G. & Whittaker, R. H. (1973) Wisconsin comparative ordination. In: *Ordination and Classification of Communities. Handbook of Vegetation Science,* **5,** (Ed. by R. H. Whittaker), pp. 195–221. Junk, The Hague.

Crawford, R. M. M. & Wishart, D. (1967) A rapid multivariate method for the detection and classification of groups of ecologically related species. *J. Ecol.* **55,** 505–24.

Crawford, R. M. M. & Wishart, D. (1968) A rapid classification and ordination method and its application to vegetation mapping. *J. Ecol.* **56,** 385–404.

Crawford, R. M. M., Wishart, D. & Campbell, R. M. (1970) A numerical analysis of high altitude scrub vegetation in relation to soil erosion in the Eastern Cordillera of Peru. *J. Ecol.* **58,** 173–91.

Crovello, T. J. (1970) Analysis of character variation in ecology and systematics. *A. Rev. Ecol. Syst.* **1,** 55–98.

Curtis, J. T. (1947) The palo verde forest type near Gonaives, Haiti, and its relation to the surrounding vegetation. *Caribb. Forester,* **8,** 1–26.

Curtis, J. T. (1955) A note on recent work dealing with the spatial distribution of plants. *J. Ecol.* **43,** 309.

Curtis, J. T. (1959) *The Vegetation of Wisconsin: An Ordination of Plant Communities.* University of Wisconsin Press, Madison, Wisconsin.

Curtis, J. T. & McIntosh, R. P. (1950) The interrelation of certain analytic and synthetic phytosociological characters. *Ecology,* **31,** 434–55.

Curtis, J. T. & McIntosh, R. P. (1951) An upland forest continuum in the prairie-forest border region of Wisconsin. *Ecology*, **32**, 476–96.

Czekanowski, J. (1913) *Zarys Metod Statystycznyck*. Warsaw.

Dagnelie, P. (1960) Contribution à l'étude des communautés végétales par l'analyse factorielle. *Bull. Serv. Carte phytogéogr. Sér. B*, **5**, 7–71 and 93–195.

Dagnelie, P. (1961) L'application de l'analyse multi-variable à l'étude des communautés végétales. *Bull. Inst. int. Statist.* **39**.

Dagnelie, P. (1962) Étude statistique d'une pelouse a *Brachypodium ramosum*. IV. La distribution des fréquences des espèces. *Bull. Serv. Carte phytogéogr. Sér. B*, **7**, 99–109.

Dagnelie, P. (1965a) L'étude des communautés végétales par l'analyse statistique des liaisons entre les espèces et les variables écologique: principes fondamentaux. *Biometrics*, **21**, 345–61.

Dagnelie, P. (1965b) L'étude des communautés végétales par l'analyse des liaisons entre les espèces et les variables écologiques: un example. *Biometrics*, **21**, 890–907.

Dagnelie, P. (1973) L'analyse factorielle. In: *Ordination and Classification of Communities. Handbook of Vegetation Science*, **5** (Ed. by R. H. Whittaker), pp. 223–48. Junk, The Hague.

Dale, M. B. (1971) Information analysis of quantitative data. In: *Statistical Ecology* (Ed. by G. P. Patil *et al.*), Vol. 3, pp. 133–48. Pennsylvania State University Press.

Dale, M. B. (1975) On objectives of methods of ordination. *Vegetatio*, **30**, 15–32.

Dale, M. B. & Anderson, D. J. (1972) Qualitative and quantitative information analysis. *J. Ecol.* **60**, 639–53.

Dale, M. B. & Anderson, D. J. (1973) Inosculate analysis of vegetation data. *Aust. J. Bot.* **21**, 253–76.

Dale, M. B. & Clifford, H. T. (1976) On the effectiveness of higher taxonomic ranks for vegetation analysis. *Aust. J. Ecol.* **1**, 37–62.

Dale, M. B., Lance, G. N. & Albrecht, L. (1971) Extensions of information analysis. *Aust. Comput. J.* **3**, 29–34.

Dale, M. B. & Quadraccia, L. (1970) Computer-assisted tabular sorting of phytosociological data. *Vegetatio*, **28**, 57–73.

Dale, M. B. & Webb, L. J. (1975) Numerical methods for the establishment of associations. *Vegetatio*, **30**, 77–87.

Dansereau, P. (Ed.) (1968) The continuum concept of vegetation: responses. *Bot. Rev.* **34**, 253–332.

David, F. N. & Moore, P. G. (1954) Notes on contagious distributions in plant populations. *Ann. Bot.* **18**, 47–53.

David, F. N. & Moore, P. G. (1957) A bivariate test for the clumping of supposedly random individuals. *Ann. Bot.* **21**, 315–20.

Davis, T. A. W. & Richards, P. W. (1933–4) The vegetation of Moraballi Creek, British Guiana: an ecological study of a limited area of tropical rain forest. *J. Ecol.* **21**, 350–84, **22**, 106–55.

Dawson, G. W. P. (1951) A method for investigating the relationship between the distribution of individuals of different species in a plant community. *Ecology*, **32**, 332–4.

De Jong, P., Aarssen, L. W. & Turkington, R. (1980) The analysis of contact sampling data. *Oecologia*, **45**, 322–4.

De Jong, P., Aarssen, L. W. & Turkington, R. (1983) The use of contact sampling in studies of association in vegetation. *J. Ecol.* **71** (in press).

De Vries, D. M. (1953) Objective combinations of species. *Acta bot. neerl.* **1**, 497–9.

De Vries, D. M. (1954) Constellation of frequent herbage plants, based on their correlation in occurrence. *Vegetatio*, **5–6**, 105–11.

De Vries, D. M. & De Boer, T. A. (1959) Methods used in botanical grassland research in the Netherlands and their application. *Herb. Abstr.* **29**, 1–7.

Del Moral, R. (1975) Vegetation clustering by means of ISODATA: revision by multiple discriminant analysis. *Vegetatio*, **29**, 179–90.

Del Moral, R. (1980) On selecting indirect ordination methods. *Vegetatio*, **42**, 75–84.

Del Moral, R. & Denton, M. F. (1977) Analysis and classification of vegetation based on family composition. *Vegetatio*, **34**, 155–65.

Denning, W. E. & Stephan, F. F. (1940) On a least squares adjustment of a samples frequency table when the expected marginal totals are known. *Ann. math. Statist.* **11**, 427–44.

Dice, L. R. (1945) Measures of the amount of ecologic association between species. *Ecology*, **26**, 297–302.

Dice, L. R. (1952) Measure of the spacing between individuals within a population. *Contr. Lab. Vertebr. Biol. Univ. Mich.* **55**, 1–23.

Dirven, J. G. P., Hoogers, B. J. & De Vries, D. M. (1969) Interrelation between frequency, dominance and dry-weight percentages of species in grassland vegetations. *Netherlands J. agric. Sci.* **17**, 161–6.

Dix, R. L. (1961) An application of the point-centred quarter method to the sampling of grassland vegetation. *J. Range Mgmt*, **14**, 63–9.

Dix, R. L. & Butler, J. E. (1960) A phytosociological study of a small prairie in Wisconsin. *Ecology*, **41**, 316–27.

Dixon, W. J. & Massey, F. J. (1957) *Introduction to Statistical Analysis*, 2nd edn. McGraw-Hill, New York.

Dony, J. G. (1977) Species-area relationships in an area of intermediate size. *J. Ecol.* **65**, 475–84.

Du Rietz, G. E. (1921) Zur methodologischen Grundlage der modernen Pflanzen-soziologie. *Akad. Afh., Uppsala.*

Du Rietz, G. E. (1931) Life-forms of terrestrial flowering plants. *Acta phytogeogr. suec.* **3**, 1–95.

Dye, P. J. & Walker, B. H. (1980) Vegetation-environment relations on sodic soils of Zimbabwe. *J. Ecol.* **68**, 589–606.

Eberhardt, L. L. (1967) Some developments in 'distance sampling'. *Biometrics*, **23**, 207–16.

Edwards, A. W. F. & Cavalli-Sforza, L. L. (1965) A method for cluster analysis. *Biometrics*, **21**, 362–75.

Ellison, L. (1942) A comparison of methods of quadratting short-grass vegetation. *J. agric. Res.* **64**, 595–614.

Emmett, H. E. G. & Ashby, E. (1934) Some observations on the relation between the hydrogen-ion concentration of the soil and plant distribution. *Ann. Bot.* **48**, 869–76.

Enright, N. & Ogden, J. (1979) Applications of transition matrix models in forest dynamics: *Araucaria* in Papua New Guinea and *Nothofagus* in New Zealand. *Aust. J. Ecol.* **4**, 3–23.

Erickson, R. O. & Stehn, J. R. (1945) A technique for analysis of population density data. *Amer. midl. Nat.* **33**, 781–7.

Errington, J. C. (1973) The effect of regular and random distributions on the analysis of pattern. *J. Ecol.* **61**, 99–105.

Evans, D. A. (1953) Experimental evidence concerning contagious distributions in ecology. *Biometrika*, **40**, 186–211.

Evans, F. C. (1952) The influence of size of quadrat on the distributional patterns of plant populations. *Contr. Lab. Vertebr. Biol. Univ. Mich*, **54**, 1–15.

Everitt, B. S. (1977) *The Analysis of Contingency Tables*. Chapman and Hall, London.

Ewer, S. J. (1932) Life forms of Illinois plants. *Trans. Ill. Acad. Sci.* **24**, 108–21.

Fager, E. W. (1972) Diversity: a sampling study. *Am. Nat.* **106**, 293–310.

Falinski, J. (1960) Zastosowanie taksonomii wroslawskiej do fitosocjologii. (Anwendung der sog. 'Breslauer Taxonomie' in der Pflanzensoziologie). *Acta Soc. Bot. Polon.* **29**, 333–61.

Fasham, M. J. R. (1977) A comparison of non-metric multidimensional scaling, principal components and reciprocal averaging for the ordination of simulated coenoclines and coenoplanes. *Ecology*, **58**, 551–61.

Feller, W. (1943) On a general class of 'contagious' distributions. *Ann. math. Statist.* **14**, 389–400.

Feoli, E. (1977) On the resolving power of principal component analysis in plant community ordination, *Vegetatio*, **33**, 119–25.

Feoli, E. & Orloci, L. (1979) Analysis of concentration and detection of underlying factors in structured tables. *Vegetatio*, **40**, 49–54.

Ferrari, H. P., Pijl, H. & Venekamp, J. T. N. (1957) Factor analysis in agricultural research. *Netherlands J. agric. Sci.* **5**, 211–21.

Field, J. G. (1969) The use of the information statistic in the numerical classification of heterogeneous systems. *J. Ecol.* **57**, 565–9.

Finney, D. J. (1948) Random and systematic sampling in timber surveys. *Forestry*, **22**, 64–99.

Finney, D. J. (1950) An example of periodic variation in forest sampling. *Forestry*, **23**, 96–111.

Finney, D. J., Latscha, R., Bennet, B. M. & Hsu, P. (1963) *Tables for Testing Significance in a 2 × 2 Contingency Table*. Cambridge University Press, London.

Fisher, R. A. (1941) *Statistical Methods for Research Workers*, 8th edn. Oliver and Boyd, Edinburgh.

Fisher, R. A., Corbet, A. S. & Williams, C. B. (1943) The relation between the number of species and the number of individuals in a random sample of an animal population. *J. anim. Ecol.* **12**, 42–58.

Fisher, R. A. & Yates, F. (1943) *Statistical Tables for Biological, Agricultural and Medical Research*, 2nd edn. Oliver and Boyd, Edinburgh.

Fisser, H. G. & Van Dyne, G. M. (1966) Influence of number and spacing of points on accuracy and precision of basal cover estimates. *J. Range Mgmt*, **19**, 205–11.

Fracker, S. B. & Brischle, H. A. (1944) Measuring the local distribution of *Ribes*. *Ecology*, **25**, 283–303.

Freeman, M. F. & Tukey, J. W. (1950) Transformations related to the angular and the square root. *Ann. math. Statist.* **21**, 607–11.

Garrison, G. A. (1949) Uses and modifications for the 'moosehorn' crown closure estimator. *J. For.* **47**, 733–5.

Gauch, H. G. (1973) The relationship between sample similarity and ecological distance. *Ecology*, **54**, 618–22.

Gauch, H. G. (1980) Rapid initial clustering of large data sets. *Vegetatio*, **42**, 103–11.

Gauch, H. G. & Chase, G. B. (1974) Fitting the Gaussian curve to ecological data. *Ecology*, **55**, 1377–81.

Gauch, H. G., Chase, G. B. & Whittaker, R. H. (1974) Ordination of vegetation samples by Gaussian species distribution. *Ecology*, **55**, 1382–90.

Gauch, H. G. & Scruggs, W. M. (1979) Variants of polar ordination. *Vegetatio*, **40**, 147–53.

Gauch, H. G. & Wentworth, T. R. (1976) Canonical correlation analysis as an ordination technique. *Vegetatio*, **33**, 17–22.

Gauch, H. G. & Whittaker, R. H. (1972a) Coenocline simulation. *Ecology*, **53**, 446–51.

Gauch, H. G. & Whittaker, R. H. (1972b) Comparison of ordination techniques. *Ecology*, **53**, 868–75.

Gauch, H. G. & Whittaker, R. H. (1976) Simulation of community patterns. *Vegetatio*, **33**, 13–16.

Gauch, H. G., Whittaker, R. H. & Wentworth, T. R. (1977) A comparative study of reciprocal averaging and other ordination techniques. *J. Ecol.* **65**, 157–74.

Ghent, A. W. (1963) Kendall's 'tau' coefficient as an index of similarity in comparisons of plant or animal communities. *Canad. Ent.* **95**, 568–75.

Ghent, A. W. (1972) A simplified computation procedure for Kendall's Tau suited to extensive species-density comparisons. *Am. Midl. Nat.* **87**, 459–71.

Gilbert, N. & Wells, T. C. E. (1966) Analysis of quadrat data. *J. Ecol.* **54**, 675–85.

Gimingham, C. H. (1961) Northern European heath communities: a 'network of variation'. *J. Ecol.* **49**, 655–94.

Gimingham, C. H., Gemmell, A. R. & Greig-Smith, P. (1948) The vegetation of a sand-dune system in the Outer Hebrides. *Trans. bot. Soc. Edinb.*, **25**, 82–96.

Gittins, R. (1965a) Multivariate approaches to a limestone grassland community. I. A stand ordination. *J. Ecol.* **53**, 385–401.

Gittins, R. (1965b) Multivariate approaches to a limestone grassland community. II. A direct species ordination. *J. Ecol.* **53**, 403–9.

Gittins, R. (1965c) Multivariate approaches to a limestone grassland community. III. A comparative study of ordination and association-analysis. *J. Ecol.* **53**, 411–25.

Gittins, R. (1968) Trend-surface analysis of ecological data. *J. Ecol.* **56**, 845–69.

Gittins, R. (1969) The application of ordination techniques. In: *Ecological Aspects of Mineral Nutrition in Plants* (Ed. by I. H. Rorison), pp. 37–66. Blackwell Scientific Publications, Oxford.

Gittins, R. (1979) Ecological applications of canonical analysis. In: *Multivariate Methods in Ecological Work* (Ed. by L. Orloci, C. R. Rao and W. M. Stiteler). *Statistical Ecology*, **7**, 309–535. International Co-operative Publishing House, Fairland, Md. U.S.A.

Gleason, H. A. (1926) The individualistic concept of the plant association. *Bull. Torrey bot. Cl.* **53**, 7–26.

Goff, F. G. (1968) Use of size stratification and differential weighting to measure forest trends. *Am. Midl. Nat.* **79**, 402–18.

Goff, F. G. & Cottam, G. (1967) Gradient analysis: the use of species and synthetic indices. *Ecology*, **48**, 793–806.

Goff, F. G. & Mitchell, R. (1975) A comparison of species ordination results from plot and stand data. *Vegetatio*, **31**, 15–22.

Goff, F. G. & Zedler, P. H. (1972) Derivation of species succession vectors. *Am. Midl. Nat.* **87**, 397–412.

Goff, F. G. & Zedler, P. H. (1968) Structural gradient analysis of upland forests in the western Great Lakes area. *Ecol. Monogr.* **38**, 65–86.

Goldsmith, F. B. (1973) The vegetation of exposed sea cliffs at South Stack, Anglesey. I. The multivariate approach. *J. Ecol.* **61**, 787–818.

Goldstein, R. A. & Grigal, D. F. (1972) Definition of vegetation structure by canonical analysis. *J. Ecol.* **60**, 277–84.

Good, I. J. (1953) The population frequencies of species and the estimation of population parameters. *Biometrika*, **40**, 237–64.

Goodall, D. W. (1952a) Quantitative aspects of plant distribution. *Biol. Rev.* **27**, 194–245.

Goodall, D. W. (1952b) Some considerations in the use of point quadrats for the analysis of vegetation. *Aust. J. sci. Res. Ser. B*, **5**, 1–41.

Goodall, D. W. (1953a) Objective methods for the classification of vegetation. I. The use of positive interspecific correlation. *Aust. J. Bot.* **1**, 39–63.

Goodall, D. W. (1953b) Objective methods for the classification of vegetation. II. Fidelity and indicator value. *Aust. J. Bot.* **1**, 434–56.

Goodall, D. W. (1953c) Point quadrat methods for the analysis of vegetation. The treatment of data for tussock grasses. *Aust. J. Bot.* **1**, 457–67.

Goodall, D. W. (1954a) Vegetational classification and vegetational continua. *Angew. Pflanzensoz. (Wien), Festschr. Aichinger*, **1**, 168–82.

Goodall, D. W. (1954b) Minimal area: a new approach. *Int. bot. Congr. 8, Rap.* Sect. 7 and 8, 19–21.

Goodall, D. W. (1954c) Objective methods for the classification of vegetation. III. An essay in the use of factor analysis. *Aust. J. Bot.* **2**, 304–24.

Goodall, D. W. (1961) Objective methods for the classification of vegetation. IV. Pattern and minimal area. *Aust. J. Bot.* **9**, 162–96.

Goodall, D. W. (1962) Bibliography of statistical plant sociology. *Excerpta bot.* Sect. B, **4**, 253–322.

Goodall, D. W. (1963) Pattern analysis and minimal area—some further comments. *J. Ecol.* **51**, 705–10.

Goodall, D. W. (1965) Plot-less tests of interspecific association. *J. Ecol.* **53**, 197–210.

Goodall, D. W. (1970) Statistical plant ecology. *A. Rev. Ecol. Syst.* **1**, 99–124.

Goodall, D. W. (1973) Sample similarity and species correlation. In: *Ordination and Classification of Communities, Handbook of Vegetation Science*, **5** (Ed. by R. H. Whittaker), pp. 105–56. Junk, The Hague.

Goodall, D. W. (1974) A new method for the analysis of spatial pattern by random pairing of quadrats. *Vegetatio*, **29**, 135–46.

Goodall, D. W. & West, N. E. (1979) A comparison of techniques for assessing dispersion pattern. *Vegetatio*, **40**, 15–27.

Gordon, A. D. (1981) *Classification: Methods for the Exploratory Analysis of Multivariate Data*. Chapman and Hall, London.

Gordon, A. D. & Birks, H. J. B. (1972) Numerical methods in quaternary palaeoecology. I. Zonation of pollen diagrams. *New Phytol.* **71**, 961–79.

Gordon, A. D. & Birks, H. J. B. (1974) Numerical methods in quaternary palaeoecology. II. Comparison of pollen diagrams. *New Phytol.* **73**, 221–49.

Gounot, M. (1962) Étude statistique d'une pelouse à *Brachypodium ramosum*. II. Étude de la distribution des espèces au moyen d'un test non-parametrique. *Bull. Serv. Carte phytogéogr. Sér. B*, **7**, 65–84.

Gounot, M. (1969) *Méthodes d'Étude Quantitative de la Végétation*. Masson, Paris.

Gower, J. C. (1966) Some distance properties of latent root and vector methods used in multivariate analysis. *Biometrika*, **53**, 325–38.

Gower, J. C. (1967a) Multivariate analysis and multidimensional geometry. *Statistician*, **17**, 13–28.

Gower, J. C. (1967b) A comparison of some methods of cluster analysis. *Biometrics*, **23**, 623–37.

Gower, J. C. (1971) A general coefficient of similarity and some of its properties. *Biometrics*, **27**, 857–74.

Gower, J. C. & Ross, G. J. S. (1969) Minimum spanning trees and single linkage cluster analysis. *Appl. Statist.* **18**, 54–64.

Greig-Smith, P. (1952a) The use of random and contiguous quadrats in the study of the structure of plant communities. *Ann. Bot.* **16**, 293–316.

Greig-Smith, (1952b) Ecological observations on degraded and secondary forest in Trinidad, British West Indies. II. Structure of the communities. *J. Ecol.* **40**, 316–30.

Greig-Smith (1961a) The use of pattern analysis is ecological investigations. *Recent Advances in Botany*, **2**, pp. 1354–8. University of Toronto Press, Toronto.

Greig-Smith (1961b) Data on pattern within plant communities. I. The analysis of pattern. J. Ecol. **49**, 695–702.

Greig-Smith (1965) Notes on the quantitative description of humid tropical forest. *Symposium on Ecological Research in Humid Tropics vegetation, Kuching, Sarawak*, pp. 227–34. UNESCO Science Co-operation Office for Southeast Asia.

Greig-Smith (1971a) Analysis of vegetation data: the user viewpoint. In: *Statistical Ecology* (Ed. by G. P. Patil *et al.*), Vol. 3, pp. 149–66. Pennsylvania State University Press.

Greig-Smith, P. (1971b). Application of numerical methods to tropical forests. In: *Statistical Ecology* (Ed. by G. P. Patil *et al.*), Vol. 3, pp. 195–206. Pennsylvania State University Press.

Greig-Smith, P. (1971c). Some problems of analytical data. *Statistician*, **21**, 215–19.

Greig-Smith, P. (1979) Pattern in vegetation: presidential address to the British Ecological Society. *J. Ecol.* **67**, 755–79.

Greig-Smith (1980) The development of numerical classification and ordination. *Vegetatio*, **42**, 1–9.

Greig-Smith, P., Austin, M. P. & Whitmore, T. C. (1967) The application of quantitative methods to vegetation survey. I. Association-analysis and principal component ordination of rain forest. *J. Ecol.* **55**, 483–503.

Greig-Smith, P. & Chadwick, M. J. (1965) Data on pattern within plant communities. III. *Acacia-Capparis* semi-desert scrub in the Sudan. *J. Ecol.* **53**, 465–74.

Greig-Smith, P., Kershaw, K. A. & Anderson, D. J. (1963) The analysis of pattern in vegetation: a comment on a paper by D. W. Goodall. *J. Ecol.* **51**, 223–9.

Grigal, D. F. & Goldstein, R. A. (1971) An integrated ordination-classification analysis of an intensively sampled oak-hickory forest. *J. Ecol.* **59**, 481–92.

Grime, J. P. & Lloyd, P. S. (1973) *An Ecological Atlas of Grassland Plants.* Arnold, London.

Groenewoud, H. van (1965) Ordination and classification of Swiss and Canadian coniferous forests by various biometric and other methods. *Ber. geobot. Inst. ETH, Stiftg. Rübel, Zürich,* **36**, 28–102.

Groenewoud, H. van (1976) Theoretical considerations on the covariation of plant species along ecological gradients with regard to multivariate analysis. *J. Ecol.* **64**, 837–47.

Grunow, J. O. (1967) Objective classification of plant communities: a synecological study in the sourish mixed bushveld of Transvaal. *J. Ecol.* **55**, 691–710.

Haberman, S. J. (1973) The analysis of residuals in cross-classified tables. *Biometrics,* **29**, 205–20.

Hairston, N. G. (1964) Species abundance and community organization. *J. Ecol. (Suppl.)*, **52**, 227–39.

Hall, A. V. (1970) A computer-based method for showing continua and communities in ecology. *J. Ecol.* **58**, 591–602.

Hall, J. B. & Swaine, M. D. (1976) Classification and ecology of closed-canopy forest in Ghana. *J. Ecol.* **64**, 913–51.

Hall, John B. & Okali, D. U. U. (1978) Observer-bias in a survey of complex tropical vegetation. *J. Ecol.* **66**, 241–9.

Hall, John B. & Okali, D. U. U. (1979) A structural and floristic analysis of woody fallow vegetation near Ibadan, Nigeria. *J. Ecol.* **67**, 321–46.

Hamilton, K. C. & Buchholtz, K. P. (1955) Effect of rhizomes of quackgrass (*Agropyron repens*) and shading on the seedling development of weedy species. *Ecology,* **36**, 304–8.

Hansen, K. & Jensen, J. (1974) Edaphic conditions and plant-soil relationships on roadsides in Denmark. *Dansk bot. Ark.* **28** (3), 1–143.

Harberd, D. J. (1962) Application of a multivariate technique to ecological survey. *J. Ecol.* **50**, 1–17.

Harman, H. H. (1976) *Modern Factor Analysis,* 3rd edn. Chicago University Press.

Hasel, A. A. (1938) Sampling error in timber surveys. *J. agric. Res.* **57**, 713–36.

Hazen, W. E. (1966) Analysis of spatial pattern in epiphytes. *Ecology,* **47**, 634–5.

Hecke, P. van, Impens, I., Goossens, R. & Hebrant, F. (1980) Multivariate analysis of multispectral remote sensing data on grasslands from different soil types. *Vegetatio,* **42**, 165–70.

Heslop-Harrison, J. (1964) Forty years of genecology. *Adv. ecol. Res.* **2**, 159–247.

Hill, M. O. (1973a) The intensity of spatial pattern in plant communities. *J. Ecol.* **61**, 225–35.

Hill, M. O. (1973b) Reciprocal averaging: an eigenvector method of ordination. *J. Ecol.* **61**, 237–249.

Hill, M. O. (1973c) Diversity and evenness: a unifying notation and its consequences. *Ecology,* **54**, 427–32.

Hill, M. O. (1974) Correspondence analysis: a neglected multivariate method. *Appl. Statist.* **23**, 340–54.

Hill, M. O. (1979a) *TWINSPAN—a FORTRAN program for arranging multivariate data in an ordered two-way table by classification of the individuals and attributes.* Ecology and Systematics, Cornell University, Ithaca, New York.

Hill, M. O. (1979b) *DECORANA—a FORTRAN program for detrended correspondence analysis and reciprocal averaging.* Ecology and Systematics, Cornell University, Ithaca, New York.

Hill, M. O., Bunce, R. G. H. & Shaw, M. W. (1975) Indicator species analysis, a divisive polythetic method of classification, and its application to a survey of native pinewoods in Scotland. *J. Ecol.* **63**, 597–613.

Hill, M. O. & Gauch, H. G. (1980) Detrended correspondence analysis: an improved ordination technique. *Vegetatio*, **42**, 47–58.

Hill, M. O. & Smith, A. J. E. (1976) Principal component analysis of taxonomic data with multi-state discrete characters. *Taxon*, **25**, 249–55.

Hole, F. D. & Hironaka, M. (1960) An experiment in ordination of some soil profiles. *Proc. Soil Sci. Soc. Am.* **24**, 309–12.

Holgate, P. (1965a) Some new tests of randomness. *J. Ecol.* **53**, 261–6.

Holgate, P. (1965b) Tests of randomness based on distance methods. *Biometrika*, **52**, 345–53.

Holland, P. G. (1969) The plant patterns of different seasons in two stands of Mallee vegetation. *J. Ecol.* **57**, 323–33.

Hope-Simpson, J. F. (1940) On the errors in the ordinary use of subjective frequency estimations in grassland. *J. Ecol.* **28**, 193–209.

Hopkins, B. (1954) A new method for determining the type of distribution of plant individuals. *Ann. Bot.* **18**, 213–27.

Hopkins, B. (1955) The species-area relations of plant communities. *J. Ecol.* **43**, 409–26.

Hopkins, B. (1957) Pattern in the plant community. *J. Ecol.* **45**, 451–63.

Hora, F. B. (1947) The pH range of some cliff plants on rocks of different geological origin in the Cader Idris area of North Wales. *J. Ecol.* **35**, 158–65.

Hughes, R. E. (1961) The application of certain aspects of multivariate analysis to plant ecology. *Recent Advances in Botany*, **2**, pp. 1350–4. University of Toronto Press, Toronto.

Hughes, R. E. & Lindley, D. V. (1955) Application of biometric methods to problems of classification in ecology. *Nature, Lond.* **175**, 806–7.

Hurlbert, S. H. (1971) The nonconcept of species diversity: a critique and alternative parameters. *Ecology*, **52**, 577–86.

Hutchings, S. S. & Holmgren, R. C. (1959) Interpretation of loop-frequency data as a measure of plant cover. *Ecology*, **40**, 668–77.

Ihm, P. & Groenewoud, H. van (1975) A multivariate ordering of vegetation data based on Gaussian type gradient response curves. *J. Ecol.* **63**, 767–77.

Ivimey-Cook, R. B. & Proctor, M. C. F. (1967) Factor analysis of data from an east Devon heath: a comparison of principal component and rotated solutions. *J. Ecol.* **55**, 405–13.

Ivimey-Cook, R. B., Proctor, M. C. F. & Wigston, D. L. (1969) On the problem of the 'R/Q' terminology in multivariate analyses of biological data. *J. Ecol.* **57**, 573–5.

Ivimey-Cook, R. B., Proctor, M. C. F. & Rowland, D. M. (1975) Analyses of the plant communities of a heathland site: Aylesbeare Common, Devon, England. *Vegetatio*, **31**, 33–45.

Jaccard, P. (1912) The distribution of the flora in the alpine zone. *New Phytol.* **11**, 37–50.

James, F. C. (1971) Ordination of habitat relationships among breeding birds. *Wilson Bull.* **83**, 215–36.

Jancey, R. C. (1979) Species ordering on a variance criterion. *Vegetatio*, **39**, 59–63.

Janssen, J. G. M. (1975) A simple clustering procedure for preliminary classification of very large sets of phytosociological relevés. *Vegetatio*, **30**, 67–71.

Jeffers, J. N. R. (1978) *An Introduction to Systems Analysis: with Ecological Applications.* Arnold, London.

Jeglum, J. K., Wehrhahn, C. F. & Swan, J. M. A. (1971) Comparisons of environmental ordinations with principal component vegetational ordinations for sets of data having different degrees of complexity. *Can. J. For. Res.* **1**, 99–112.

Jensen, D. R., Beus, G. B. & Storm, G. (1968) Simultaneous statistical tests on categorical data. *J. exp. Educ.* **36**(4), 46–56.

Jensen, S. (1978) Influences of transformation of cover values on classification and ordination of lake vegetation. *Vegetatio*, **37**, 19–31.

Johnson, R. W. & Goodall, D. W. (1979) A maximum likelihood approach to non-linear ordination. *Vegetatio*, **41**, 133–42.

Johnston, A. (1957) A comparison of the line interception, vertical point quadrat, and loop methods as used in measuring basal area of grassland vegetation. *Can. J. Pl. Sci.* **37**, 34–42.

Jones, E. W. (1955–6) Ecological studies on the rain forest of southern Nigeria. IV. The plateau forest of the Okomu Forest Reserve. *J. Ecol.* **43**, 564–94, **44**, 83–117.

Jowett, G. H. & Scurfield, G. (1949a) Statistical test for optimal conditions: note on a paper of Emmett and Ashby. *J. Ecol.* **37**, 65–7.

Jowett, G. H. & Scurfield, G. (1949b) A statistical investigation into the distribution of *Holcus mollis* L. and *Deschampsia flexuosa* (L.) Trin. *J. Ecol.* **37**, 68–81.

Kaiser, H. F. (1958) The Varimax criterion for analytic rotation in factor analysis. *Psychometrika*, **23**, 187–200.

Kemp, C. D. & Kemp, A. W. (1956) The analysis of point quadrat data. *Aust. J. Bot.* **4**, 167–74.

Kendall, M. G. (1948) *Rank Correlation Methods.* Griffin, London.

Kendall, M. G. (1957) *A Course in Multivariate Analysis.* Griffin, London.

Kendall, M. G. (1966) Discrimination and classification. *Proc. Symp. Multiv. Analysis, Drayton, Ohio*, (Ed. by P. R. Krishnaiah), pp. 165–85. Academic Press, New York.

Kercher, J. R. & Goldstein, R. A. (1977) Analysis of a East Tennessee oak hickory forest by canonical correlation of species and environmental parameters. *Vegetatio*, **35**, 153–63.

Kershaw, K. A. (1957a) The use of cover and frequency in the detection of pattern in plant communities. *Ecology*, **38**, 291–9.

Kershaw, K. A. (1957b) *A study of pattern in certain plant communities.* Ph.D. thesis, University of Wales.

Kershaw, K. A. (1958) An investigation of the structure of a grassland community. I. The pattern of *Agrostis tenuis. J. Ecol.* **46,** 571–92.

Kershaw, K. A. (1959) An investigation of the structure of a grassland community. II. The pattern of *Dactylis glomerata, Lolium perenne* and *Trifolium repens.* III. Discussion and conclusions. *J. Ecol.* **47,** 31–53.

Kershaw, K. A. (1960) The detection of pattern and association. *J. Ecol.* **48,** 233–42.

Kershaw, K. A. (1961) Association and co-variance analysis of plant communities. *J. Ecol.* **49,** 643–54.

Kershaw, K. A. (1968) Classification and ordination of Nigerian savanna vegetation. *J. Ecol.* **56,** 467–82.

Kershaw, K. A. (1970) An empirical approach to the estimation of pattern intensity from density and cover data. *Ecology,* **51,** 729–34.

Kilburn, P. D. (1966) Analysis of the species-area relation. *Ecology,* **47,** 831–43.

King, J. (1962) The *Festuca-Agrostis* grassland complex in south-east Scotland. *J. Ecol.* **50,** 321–56.

Knight, D. H. (1965) A gradient analysis of Wisconsin prairie vegetation on the basis of plant structure and function. *Ecology,* **46,** 744–7.

Knight, D. H. & Loucks, O. L. (1969) A quantitative analysis of Wisconsin forest vegetation on the basis of plant function and gross morphology. *Ecology,* **50,** 219–34.

Kooijman, S. A. L. M. (1976) Some remarks on the statistical analysis of grids especially with respect to ecology. *Anns Syst. Res.* **5,** 113–32.

Krishna Iyer, P. V. (1948) The theory of probability distribution of points on a line. *J. Indian Soc. agric. Statist.* **1,** 173–95.

Krishna Iyer, P. V. (1950) The theory of probability distributions of points on a lattice. *Ann. math. Statist.* **21,** 198–217.

Kruskal, J. B. (1964a) Multidimensional scaling by optimizing goodness of fit to a nonmetric hypothesis. *Psychometrika,* **29,** 1–27.

Kruskal, J. B. (1964b) Nonmetric multidimensional scaling: a numerical method. *Psychometrika,* **29,** 115–29.

Kruskal, J. B. & Carmone, F. (1971) *How to use M-D-SCAL (Version 5M) and other useful information.* Bell Telephone Laboratory, Murray Hill, New Jersey, U.S.A. and University of Waterloo, Waterloo, Ontario, Canada. (Mimeographed). (Quoted from Orloci, 1978).

Kruskal, J. B. & Carroll, J. D. (1969) Geometric models and badness of fit functions. In: *Multivariate Analysis II,* (Ed. by P. R. Krishnaiah), pp. 639–71. Academic Press, London.

Kullback, S. (1959) *Information Theory and Statistics.* Wiley, Chapman and Hall, New York.

Kylin, H. (1926) Über Begriffsbildung und Statistik in der Pflanzensoziologie. *Bot. Notiser* (1926), 81–180.

Lambert, J. M. (1972) Theoretical models for large-scale vegetation survey. In: *Mathematical Models in Ecology,* (Ed. by J. N. R. Jeffers), pp. 87–109. Blackwell Scientific Publications, Oxford.

Lambert, J. M. & Dale, M. B. (1964) The use of statistics in phytosociology. *Adv. ecol. Res.* **2,** 59–99.

Lambert, J. M., Meacock, S. E., Barrs, J. & Smartt, P. F. M. (1973) AXOR and MONIT: two new polythetic-divisive strategies for hierarchical classification. *Taxon*, **22**, 173-6.

Lambert, J. M. & Williams, W. T. (1962) Multivariate methods in plant ecology. IV. Nodal analysis. *J. Ecol.* **50**, 775–802.

Lambert, J. M. & Williams, W. T. (1966) Multivariate methods in plant ecology. VI. Comparison of information-analysis and association-analysis. *J. Ecol.* **54**, 635–64.

Lance, G. N., Milne, P. W. & Williams, W. T. (1968) Mixed-data classificatory programs. III. Diagnostic systems. *Aust. Comput. J.* **1**, 178–81.

Lance, G. N. & Williams, W. T. (1965) Computer programs for monothetic classification ("Association analysis"). *Comput. J.* **8**, 246–9.

Lance, G. N. & Williams, W. T. (1966a) Computer programs for hierarchical polythetic classification ("Similarity analyses")' *Comput. J.* **9**, 60–4.

Lance, G. N. & Williams, W. T. (1966b) A generalized sorting strategy for computer classifications. *Nature, Lond.* **212**, 218.

Lance, G. N. & Williams, W. T. (1967a) A general theory of classificatory sorting strategies. I. Hierarchical systems. *Comput. J.* **9**, 373–80.

Lance, G. N. & Williams, W. T. (1967b) Mixed-data classificatory programs. I. Agglomerative systems. *Aust. Comput. J.* **1**, 15–20.

Lance, G. N. & Williams, W. T. (1968a) Note on a new information-statistic classificatory program. *Comput. J.* **11**, 195.

Lance, G. N. & Williams, W. T. (1968b) Mixed-data classificatory programs. II. Divisive systems. *Aust. Comput. J.* **1**, 82–99.

Lange, R. T. (1968) Influence analysis in vegetation. *Aust. J. Bot.* **16**, 555–64.

Lange, R. T. (1971) Influence analysis and prescriptive management of rangeland vegetations. *Proc. ecol. Soc. Austr.* **6**, 153–8.

Lange, R. T., Stenhouse, N. S. & Offler, C. E. (1965) Experimental appraisal of certain procedures for the classification of data. *Aust. J. biol. Sci.* **18**, 1189–205.

Langlet, O. (1971) Two hundred years genecology. *Taxon*, **20**, 653–721.

Lawley, D. N. & Maxwell, A. E. (1963) *Factor Analysis as a Statistical Method.* Butterworth, London.

Lawson, G. W. (1978) The distribution of seaweed floras in the tropical and subtropical Atlantic Ocean: a quantitative approach. *Bot. J. Linn. Soc.* **76**, 177–93.

Lefkovitch, L. P. (1966) An index of spatial distribution. *Res. Popul. Ecol.* **8**, 89–92.

Legg, C. J. (1980) A Markovian approach to the study of heath vegetation dynamics. *Bull. Écol.* **11**, 393–404.

Levy, E. B. (1933) Technique employed in grassland research in New Zealand. I. Strain testing and strain building. *Bull. Bur. Pl. Genet. Aberystw.* **11**, 6–16.

Levy, E. B. & Madden, E. A. (1933) The point method of pasture analysis. *N. Z. Jl Agric.* **46**, 267–79.

Lieth, H. & Moore, G. W. (1971) Computerized clustering of species in phytosociological tables and its utilization for field work. In: *Statistical Ecology* (Ed. by G. P. Patil *et al.*), Vol. 1, pp. 403–22. Pennsylvania State University Press.

Lindsey, A. A., Barton, J. D. & Miles, S. R. (1958) Field efficiencies of forest sampling methods. *Ecology*, **39**, 428–44.

Lloyd, M. (1967) 'Mean crowding'. *J. anim. Ecol.* **36,** 1–30.

Lloyd, M. & Ghelardi, R. J. (1964) A table for calculating the equitability component of species diversity. *J. anim. Ecol.* **33,** 217–25.

Lloyd, M., Zar, J. H. & Karr, J. R. (1968) On the calculation of information— theoretical measures of diversity. *Am. Midl. Nat.* **79,** 257–72.

Lloyd, P. S. (1972) The grassland vegetation of the Sheffield region. II. Classification of grassland types. *J. Ecol.* **60,** 739–76.

Loetsch, F., Zöhrer, F. & Haller, K. E. (1973) *Forest Inventory* Vol. 2. BLV Verlagsgesellschaft, Munich.

Loucks, O. L. (1962) Ordinating forest communities by means of environmental scalars and phytosociological indices. *Ecol. Monogr.* **32,** 137–66.

Louppen, J. M. W. & van der Maarel, E. (1979) CLUSLA: a computer programme for the clustering of large phytosociological data sets. *Vegetatio,* **40,** 107–14.

Lowe, R. G. (1967) *Competition and Thinning Studies in Nigeria,* 16 pp. Federal Dept. of Forest Research, Ibadan, Nigeria.

Ludwig, J. A. & Goodall, D. W. (1978) A comparison of paired- with blocked-quadrat variance methods for the analysis of spatial pattern. *Vegetatio,* **38,** 49–59.

Lynch, D. W. & Schumacher, F. X. (1941) Concerning the dispersion of natural regeneration. *J. For.* **39,** 49–51.

Maarel, E. van der (1969) On the use of ordination models in phytosociology. *Vegetatio,* **19,** 21–46.

Maarel, E. van der (1972) Ordination of plant communities on the basis of their plant genus, family and order relationships. In: *Grundfragen und Methoden in der Pflanzensoziologie,* (Ed. By E. van der Maarel and R. Tüxen), pp. 183–92. Junk, The Hague.

Maarel, E. van der (1979) Transformation of cover-abundance values in phyto-sociology and its effects on community similarity. *Vegetatio,* **39,** 97–114.

Maarel, E. van der, Janssen, J. G. M. & Louppen, J. M. W. (1978) TABORD, a program for structuring phytosociological tables. *Vegetatio,* **38,** 143–56.

MacArthur, R. H. (1957) On the relative abundance of bird species. *Proc. natn. Acad. Sci. U.S.A.* **43,** 293–5.

MacArthur, R. H. (1960) On the relative abundance of species. *Am. Nat.* **94,** 25–36.

MacArthur, R. H. (1965) Patterns of species diversity. *Biol. Rev.* **40,** 510–33.

MacArthur, R. H. & MacArthur, J. W. (1961) On bird species diversity. *Ecology,* **42,** 594–8.

McGinnies, W. G. (1934) The relation between frequency index and abundance as applied to plant populations in a semi-arid region. *Ecology,* **15,** 263–82.

McIntosh, R. P. (1962) Pattern in a forest community. *Ecology,* **43,** 25–33.

McIntosh, R. P. (1967a) An index of diversity and the relation of certain concepts to diversity. *Ecology,* **48,** 392–404.

McIntosh, R. P. (1967b) The continuum concept of vegetation. *Bot. Rev.* **33,** 130–87.

McIntosh, R. P. (1973) Matrix and plexus techniques. In: *Ordination and Classification of Communities. Handbook of Vegetation Science,* **5** (Ed. by R. H. Whittaker), pp. 159–91. Junk, The Hague.

McIntyre, G. A. (1953) Estimation of plant density using line transects. *J. Ecol.* **41,** 319–30.

Macnaughton-Smith, P. (1963) The classification of individuals by the possession of attributes associated with a criterion. *Biometrics*, **19**, 364–6.

Macnaughton-Smith, P. (1965) Some statistical and other numerical techniques for classifying individuals. *Studies in the Causes of Delinquency and the Treatment of Offenders*, **6**, H.M.S.O., London.

Macnaughton-Smith, P., Williams, W. T., Dale, M. B. & Mockett, L. G. (1964) Dissimilarity analysis: a new technique of hierarchical sub-division. *Nature, Lond* **202**, 1034–5.

Madgwick, H. A. I. & Desrochers, P. A. (1972) Association-analysis and the classification of forest vegetation of the Jefferson National Forest. *J. Ecol.* **60**, 285–92.

Mainland, D., Herrera, L. & Sutcliffe, M. I. (1956) *Statistical Tables for Use with Binomial Samples—contingency tests, confidence limits and sample size estimates.* New York University College of Medicine.

Margalef, R. (1957) La teoria de la informacion en ecologia. *Mems R. Acad. Cienc. Artes Barcelona*, **32**, 373–449. (Translated (1958), Information theory in ecology. *Gen. Syst.* **3**, 36–71).

Matthews, J. A. (1978) An application of non-metric multidimensional scaling to the construction of an improved species plexus. *J. Ecol.* **66**, 157–73.

Matthews, J. A. (1979) A study of the variability of some successional and climax plant assemblage-types using multiple discriminant analysis. *J. Ecol.* **67**, 255–71.

Matuszkiewicz, W. (1948) Roslinnosc lasow okolic lwowa (The vegetation of the forests of the environs of Lvov). *Annls Univ. Mariae Curie-Sklodowska*, Sect. C, **3**, 119–93.

Maycock, P. F. (1967) Jozef Paczoski: founder of the science of phytosociology. *Ecology*, **48**, 1031–4.

Mead, R. (1971) A note on the use and misuse of regression models in ecology. *J. Ecol.* **59**, 215–19.

Mead, R. (1974) A test for spatial pattern at several scales using data from a grid of contiguous quadrats. *Biometrics*, **30**, 295–307.

Miller, J. B. (1967) A formula for average foliage density. *Aust. J. Bot.* **15**, 141–4.

Monk, C. D. (1965) Southern mixed hardwood forest of north central Florida. *Ecol. Monogr.* **35**, 335–56.

Moore, G. W., Benninghoff, W. F. & Dwyer, P. S. (1967) A computer procedure for the rearrangement of phytosociological tables. *Proc. Assn. Comput. Mach.* **20**, 297–300.

Moore, J. J., Fitzsimons, P., Lambe, E. & White, J. (1970) A comparison and evaluation of some phytosociological techniques. *Vegetatio*, **20**, 1–20.

Moore, P. G. (1953) A test for non-randomness in plant populations. *Ann. Bot.* **17**, 57–62.

Moore, P. G. (1954) Spacing in plant populations. *Ecology*, **35**, 222–7.

Morisita, M. (1954) Estimation of population density by spacing method. *Mem. Fac. Sci. Kyushu Univ.* Ser. E. **1**, 187–97.

Morisita, M. (1957) A new method for the estimation of density by the spacing method applicable to non-randomly distributed populations. *Seiro-Seitai*, **7**, 134–44.

Morisita, M. (1959) Measuring of the dispersion of individuals and analysis of the distributional patterns. *Mem. Fac. Sci. Kyushu Univ.* Ser. E, **2**, 215–35.

Morisita, M. (1962, I_δ index, a measure of dispersal of individuals. *Res. Popul. Ecol.* **4**, 1–7.

Morisita, M. (1971) Composition of the I_δ index. *Res. Popul. Ecol.* **13**, 1–27.

Moroney, M. J. (1951) *Facts from Figures.* Penguin Books, Harmondsworth.

Morrison, R. G. & Yarranton, G. A. (1970) An instrument for rapid and precise point sampling of vegetation. *Can. J. Bot.* **48**, 293–7.

Morton, A. J. (1974a) Ecological studies of a fixed dune grassland at Newborough Warren, Anglesey. I. The structure of the grassland. *J. Ecol.* **62**, 253–60.

Morton, A. J. (1974b) Ecological studies of a fixed dune grassland at Newborough Warren, Anglesey. II. Causal factors of the grassland structures. *J. Ecol.* **62**, 261–78.

Mosteller, F. & Youtz, C. (1961) Tables of the Freeman-Tukey transformations for the binomial and Poisson distributions. *Biometrika*, **48**, 433–40.

Motomura, I. (1932) (A statistical treatment of associations). (In Japanese). *Jap. J. Zool.* **44**, 379–83.

Motomura, I. (1952) Comparison of communities based on correlation coefficients. *Ecol. Rev.* **13**, 67–71. (Quoted from Dagnelie, 1960).

Motyka, J., Dobrzanski, B. & Zawadzski, S. (1950) Wstępne badania nad łagami południowowschodniej Lubelszczyzny. *Annls Univ. Mariae Curie-Sklodowska*, Sect. E, **5**, 367–447.

Mountford, M. D. (1961) On E. C. Pielou's index of non-randomness. *J. Ecol.* **49**, 271–6.

Mueller-Dombois, D. & Ellenberg, H. (1974) *Aims and Methods of Vegetation Ecology.* Wiley, New York.

Myers, E. & Chapman, V. J. (1953) Statistical analysis applied to a vegetation type in New Zealand. *Ecology*, **34**, 175–85.

Neal, M. W. & Kershaw, K. A. (1973) Studies on lichen-dominated systems. IV. The objective analysis of Cape Henrietta Maria raised-beach systems. *Can. J. Bot.* **51**, 1177–90.

Neyman, J. (1939) On a new class of contagious distributions applicable in entomology and bacteriology. *Ann. math. Statist.* **10**, 35–57.

Nichols, S. (1977) On the interpretation of principal components analysis in ecological contexts. *Vegetatio*, **34**, 191–7.

Nielen, G. C. J. F. & Dirven, J. G. P. (1950) De nauwkeurigheid van de plantensociologische 1/4 dm² frequentie-methode. *Versl. landbouwk. Onderz*, **56** (13), 1–27.

Norris, J. M. & Barkham, J. P. (1970) A comparison of some Cotswold beechwoods using multiple-discriminant analysis. *J. Ecol.* **58**, 603–19.

Noy-Meir, I. (1970) *Component analysis of semi-arid vegetation in southeastern Australia.* Ph.D. thesis, Australian National University.

Noy-Meir, I. (1971a) Multivariate analysis of desert vegetation. II. Qualitative/quantitative partition of heterogeneity. *Israel J. Bot.* **20**, 203–13.

Noy-Meir, I. (1971b) Multivariate analysis of the semi-arid vegetation in south-eastern Australia: nodal ordination by component analysis. *Proc. ecol. Soc. Austr.* **6**, 159–93.

Noy-Meir, I. (1973a) Data transformation in ecological ordination. I. Some advantages of non-centering. *J. Ecol.* **61**, 329–41.

Noy-Meir, I. (1973b) Divisive polythetic classification of vegetative data by optimized division on ordination components. *J. Ecol.* **61**, 753–60.

Noy-Meir, I. (1974a) Catenation: quantitative methods for the definition of coenoclines. *Vegetatio*, **29**, 89–99.

Noy-Meir, I. (1974b). Multivariate analysis of the semi-arid vegetation in south-eastern Australia. II. Vegetation catenae and environmental gradients. *Aust. J. Bot.* **22**, 115–40.

Noy-Meir, I. & Anderson, D. J. (1971) Multiple pattern analysis, or multiscale ordination: towards a vegetation hologram? In: *Statistical Ecology* (Ed. by G. P. Patil *et al.*), Vol. 3, pp. 207–31. Pennsylvania State University Press.

Noy-Meir, I. & Austin, M. P. (1970) Principal component ordination and simulated vegetational data. *Ecology*, **51**, 551–2.

Noy-Meir, I., Orshan, G. & Tadmor, N. H. (1973) Multivariate analysis of desert vegetation. III. The relation of vegetation units to habitat classes. *Israel J. Bot.* **22**, 239–57.

Noy-Meir, I., Tadmor, N. H. & Orshan, G. (1970) Multivariate analysis of desert vegetation. I. Association analysis at various quadrat sizes. *Israel J. Bot.* **19**, 561–91.

Noy-Meir, I., Walker, D. & Williams, W. T. (1975) Data transformation in ecological ordination. II. On the meaning of data standardization. *J. Ecol.* **63**, 779–800.

Noy-Meir, I. & Whittaker, R. H. (1977) Continuous multivariate methods in community analysis: some problems and developments. *Vegetatio*, **33**, 79–98.

Numata, M. (1949) The basis of sampling in the statistics of plant communities. Studies on the structure of plant communities. III. *Bot. Mag., Tokyo*, **62**, 35–8.

Numata, M. (1954) Some special aspects of the structural analysis of plant communities. *J. Coll. Arts Sci., Chiba Univ.* **1**, 194–202.

Ogawa, H., Yoda, K. & Kira, T. (1961) A preliminary survey on the vegetation of Thailand. In: *Nature and Life in Southeast Asia* (Ed. by T. Kira and T. Umesao), Vol. 1, pp. 22–157. Kyoto.

Orloci, L. (1966) Geometric models in ecology. I. The theory and application of some ordination methods. *J. Ecol.* **54**, 193–215.

Orloci, L. (1967a) An agglomerative method for classification of plant communities. *J. Ecol.* **55**, 193–205.

Orloci, L. (1967b) Data centering: a review and evaluation with reference to component analysis. *Syst. Zool.* **16**, 208–12.

Orloci, L. (1968) Information analysis in phytosociology: partition, classification and prediction. *J. theor. Biol.* **20**, 271–84.

Orloci, L. (1971) An information theory model for pattern analysis. *J. Ecol.* **59**, 343–9.

Orloci, L. (1972a) On objective functions of phytosociological resemblance. *Am. Midl. Nat.* **88**, 28–55.

Orloci, L. (1972b) On information analysis in phytosociology. In: *Grundfragen und Methoden in den Pflanzensoziologie* (Ed. by E. van der Maarel and R. Tüxen), pp. 75–88. Junk, The Hague.

Orloci, L. (1973) Ordination by resemblance matrices. In: *Ordination and Classification of Communities. Handbook of Vegetation Science*, **5**, (Ed. by R. M. Whittaker), pp. 249–86. Junk. The Hague.

Orloci, L. (1974) Revision for the Bray and Curtis ordination. *Can. J. Bot.* **52**, 1773–6.

Orloci, L. (1976) Ranking species by an information criterion. *J. Ecol.* **64**, 417–19.

Orloci, L. (1978) *Multivariate Analysis in Vegetation Research*, 2nd edn. Junk, The Hague.

Osvald, H. (1947) Växternas vapen i kampen om utrymet. *Växtodling*, **2**, 288–303.

Parker, K. W. (1950) Report on 3-step method for measuring condition and trend of forest ranges. U.S. Forest Service, Washington D.C., 68pp. (processed). (Quoted from Hutchings & Holmgren, 1959).

Parker, K. W. (1951) A method for measuring trend in range condition on national forest ranges. U.S. Forest Service, Washington D.C., 26pp. (processed). (Quoted from Hutchings & Holmgren, 1959).

Patil, G. P., Pielou, E. C. & Waters, W. E. (Eds) (1971) *Statistical Ecology.* Vol. 1: *Spatial Patterns and Statistical Distributions.* Vol. 2: *Sampling and Modeling Biological Populations and Population Dynamics.* Vol. 3: *Many Species Populations, Ecosystems and Systems Analysis.* Pennsylvania State University Press, University Park, Pa.

Pearsall, W. H. (1924) The statistical analysis of vegetation: a criticism of the concepts and methods of the Uppsala school. *J. Ecol.* **12,** 135–9.

Pearson, E. S. & Hartley, H. O. (1954) *Biometrika Tables for Statisticians.* Vol. 1, 4th edn. Cambridge University Press, Cambridge.

Pearson, K. (1934) *Tables of the Incomplete Beta-Function.* Biometrika, London.

Pechanec, J. F. & Stewart, G. (1940) Sagebrush-grass range sampling studies: size and structure of sampling unit. *J. Amer. Soc. Agron.* **32,** 669–82.

Peet, R. K. (1974) The measurement of species-diversity. *A. R. Ecol. Syst.* **5,** 285–307.

Peet, R. K. (1975) Relative diversity indices. *Ecology,* **56,** 496–8.

Pemadasa, M. A., Greig-Smith, P. & Lovell, P. H. (1974) A quantitative description of the distribution of annuals in the dune system at Aberffraw, Anglesey. *J. Ecol.* **62,** 379–402.

Perring, F. (1959) Topographical gradients of chalk grassland. *J. Ecol.* **47,** 447–81.

Persson, O. (1971) The robustness of estimating density by distance measurements. In: *Statistical Ecology* (Ed. by G. P. Patil *et al.*), Vol. **2,** pp. 175–90. Pennsylvania State University Press.

Peterson, C. H. (1976) Measurement of community pattern by indices of local segregation and species diversity. *J. Ecol.* **64,** 157–69.

Philip, J. R. (1965a) The distribution of foliage density with foliage angle estimated from inclined point quadrat observations. *Aust. J. Bot.* **13,** 357–66.

Philip, J. R. (1965b) The distribution of foliage density on single plants. *Aust. J. Bot.* **13,** 411–18.

Philip, J. R. (1966) The use of point quadrats, with special reference to stem-like organs. *Aust. J. Bot.* **14,** 105–25.

Phillips, M. E. (1953) Studies in the quantitative morphology and ecology of *Eriophorum angustifolium* Roth. I. The rhizome system. *J. Ecol.* **41,** 295–318.

Phillips, M. E. (1954a) Studies in the quantitative morphology and ecology of *Eriophorum angustifolium* Roth. II. Competition and dispersion. *J. Ecol.* **42,** 187–210.

Phillips, M. E. (1954b) Studies in the quantitative morphology and ecology of *Eriophorum angustifolium* Roth. III. The leafy shoot. *New Phytol.* **53,** 312–43.

Pidgeon, I. M. & Ashby, E. (1940) Studies in applied ecoly. I. A statistical analysis of regeneration following protection from grazing. *Proc. Linn. Soc. N.S.W.* **65,** 123–43.

Pidgeon, I. M. & Ashby, E. (1942) A new quantitative method of analysis of plant communities. *Aust. J. Sci.* **5,** 19–21.

Pielou, E. C. (1957) The effect of quadrat size on the estimation of the parameter of Neyman's and Thomas's distributions. *J. Ecol.* **45**, 31–47.

Pielou, E. C. (1959) The use of point-to-plant distances in the study of the pattern of plant populations. *J. Ecol.* **47**, 607–13.

Pielou, E. C. (1960) A single mechanism to account for regular, random and aggregated populations. *J. Ecol.* **48**, 575–84.

Pielou, E. C. (1961) Segregation and symmetry in two-species populations as studied by nearest neighbour relations. *J. Ecol.* **49**, 255–69.

Pielou, E. C. (1962a) The use of plant-to-neighbour distances for the detection of competition. *J. Ecol.* **50**, 357–68.

Pielou, E. C. (1962b) Runs of one species with respect to another in transects through plant populations. *Biometrics*, **18**, 579–93.

Pielou, E. C. (1964) The spatial pattern of two-phase patchworks of vegetation. *Biometrics*, **20**, 156–67.

Pielou, E. C. (1965) The concept of randomness in the patterns of mosaics. *Biometrics*, **21**, 908–20.

Pielou, E. C. (1966) The measurement of diversity in different types of biological collections. *J. theor. Biol.* **13**, 131–41.

Pielou, E. C. (1967) A test for random mingling of the phases of a mosaic. *Biometrics*, **23**, 657–70.

Pielou, E. C. (1969) Association tests versus homogeneity tests: their use in subdividing quadrats into groups. *Vegetatio*, **18**, 4–18.

Pielou, E. C. (1977) *Mathematical Ecology*. Wiley, New York.

Pielou, E. C. (1979) *Biogeography*. Wiley, New York.

Podani, J. (1979a) Association-analysis based on the use of mutual information. *Acta bot. hung.* **25**, 125–30.

Podani, J. (1979b) Generalized strategy for homogeneity-optimizing hierarchical classificatory methods. In: *Multivariate Methods in Ecological Work* (Ed. by L. Orloci, C. R. Rao and W. M. Stiteler). *Statistical Ecology*, **7**, 203–9. International Co-operative Publishing House, Fairland, Md., U.S.A.

Poissonet, P. S., Daget, P. M., Poissonet, J. A. & Long, G. A. (1972) Rapid point survey by bayonet blade. *J. Range Mgmt*, **25**, 313.

Pólya, G. (1930) Sur quelques points de la théorie des probabilités. *Ann. Inst. Poincaré*, **1**, 117–61.

Poore, M. E. D. (1955a) The use of phytosociological methods in ecological investigations. II. Practical issues involved in an attempt to apply the Braun-Blanquet system. *J. Ecol.* **43**, 245–69.

Poore, M. E. D. (1955b) The use of phytosociological methods in ecological investigations. III. Practical applications. *J. Ecol.* **43**, 606–51.

Prentice, I. C. (1977) Non-metric ordination methods in ecology. *J. Ecol.* **65**, 85–94.

Prentice, I. C. (1980) Vegetation analysis and order invariant gradient models. *Vegetatio*, **42**, 27–34.

Preston, F. W. (1948) The commonness, and rarity, of species. *Ecology*, **29**, 254–83.

Preston, F. W. (1962) The canonical distribution of commonness and rarity. *Ecology*, **43**, 185–215 and 410–31.

Pritchard, N. M. & Anderson, A. J. B. (1971) Observations on the use of cluster analysis in botany with an ecological example. *J. Ecol.* **59**, 727–47.

Quenouille, M. H. (1949) Problems in plane sampling. *Ann. math. Statist.* **20,** 355–75.

Rao, C. R. (1952) *Advanced Statistical Methods in Biometric Research.* Wiley, New York.

Raunkiaer, C. (1934) *The Life Forms of Plants and Statistical Plant Geography.* Oxford University Press, Oxford.

Reynolds, K. C. & Edwards, K. (1977) A short-focus telescope for ground cover estimation. *Ecology,* **58,** 939–41.

Rice, E. L. & Penfound, W. T. (1955) An evaluation of the variable-radius and paired-tree methods in the black-jack post oak forest. *Ecology,* **36,** 315–20.

Richards, P. W. (1952) *The Tropical Rain Forest.* Cambridge University Press, Cambridge.

Richards, P. W., Tansley, A. G. & Watt, A. S. (1940) The recording of structure, life form and flora of tropical forest communities as a basis for their classification. *J. Ecol.* **28,** 224–39.

Ripley, B. D. (1978) Spectral analysis and the analysis of pattern in plant communities. *J. Ecol.* **66,** 965–81.

Risser, P. G. & Zedler, P. H. (1968) An evaluation of the grassland quarter method. *Ecology,* **49,** 1006–9.

Robertson, E. F. (1977) *The use of line transects in vegetation survey.* M.Sc. dissertation, University of Wales.

Robertson, P. A. (1978) Comparisons of techniques for ordinating and classifying old-growth floodplain forests in southern Illinois. *Vegetatio,* **37,** 43–51.

Robinson, P. (1954) The distribution of plant populations. *Ann. Bot.* **18,** 35–45.

Robinson, P. (1955) The estimation of ground cover by the point quadrat method. *Ann. Bot.* **19,** 59–66.

Rohlf, F. J. (1974) Methods of comparing classifications. *A. Rev. Ecol. Syst.* **5,** 101–13.

Rutter, A. J. (1955) The composition of wet-heath vegetation in relation to the water-table. *J. Ecol.* **43,** 507–43.

Ryland, J. S. (1972) The analysis of pattern in communities of bryozoa. I. Discrete sampling methods. *J. exp. mar. Biol. Ecol.* **8,** 277–97.

Salisbury, E. J. (1925) The incidence of species in relation to soil reaction. *J. Ecol.* **13,** 149–60.

Sampford, M. R. (1962). *An Introduction to Sampling Theory with Applications to Agriculture.* Oliver and Boyd, Edinburgh.

Sarmiento, G. (1972) Ecological and floristic convergences between seasonal plant formations of tropical and subtropical South America. *J. Ecol.* **60,** 367–410.

Sarmiento, G. & Monasterio, M. (1969) Studies on the savanna vegetation of the Venezuelan llanos. I. The use of association-analysis. *J. Ecol.* **57,** 579–98.

Schnell, G. D., Risser, P. G. & Helsel, J. F. (1977) Factor analysis of tree distribution patterns in Oklahoma. *Ecology,* **58,** 1345–55.

Scott, G. A. M. (1966) The quantitative description of New Zealand bryophyte communities. *Proc. N.Z. ecol. Soc.* **13,** 8–11.

Seal, H. L. (1964) *Multivariate Statistical Analysis for Biologists.* Methuen, London.

Shanks, R. E. (1954) Plotless sampling trials in Appalachian forest types. *Ecology,* **35,** 237–44.

Shannon, C. E. & Weaver, W. (1949) *The Mathematical Theory of Communication.* University of Illinois Press, Urbana.

Sheldon, A. L. (1969) Equitability indices: dependence on the species count. *Ecology,* **50,** 466–7.

Shepard, R. N. & Carroll, J. D. (1966) Parametric representation of non-linear data structures. In: *Multivariate Analysis* (Ed. by P. R. Krishnaiah), pp. 561–92. Academic Press, London.

Sibson, R. (1971) Some observations on a paper by Lance and Williams. *Comput. J.* **14,** 156–7.

Sibson, R. (1972) Order invariant methods for data analysis. *Jl R. statist. Soc. B,* **34,** 311–49.

Simberloff, D. (1979) Nearest neighbor assessments of spatical configurations of circles rather than points. *Ecology,* **60,** 679–85.

Simpson, E. H. (1949) Measurement of diversity. *Nature, Lond.* **163,** 688.

Singh, B. N. & Das. K. (1938) Distribution of weed species on arable land. *J. Ecol.* **26,** 455–66.

Skellam, J. G. (1952) Studies in statistical ecology. I. Spatial pattern. *Biometrika,* **39,** 346–62.

Smartt, P. F. M. & Grainger, J. E. A. (1974) Sampling for vegetation survey: some aspects of the behaviour of unrestricted, restricted and stratified techniques. *J. Biogeog.* **1,** 193–206.

Smartt, P. F. M., Meacock, S. E. & Lambert, J. M. (1974) Investigations into the properties of quantitative vegetation data. I. Pilot study. *J. Ecol.* **62,** 735–59.

Smartt, P. F. M., Meacock, S. E. & Lambert, J. M. (1976) Investigations into the properties of quantitative vegetational data. II. Further data type comparisons. *J. Ecol.* **64,** 41–78.

Smith, A. D. (1944) A study of the reliability of range vegetation estimates. *Ecology,* **25,** 441–8.

Sneath, P. H. A. & Sokal, R. R. (1962) Numerical Taxonomy. *Nature, Lond.* **193,** 855–60.

Sneath, P. H. A. & Sokal, R. R. (1973) *Numerical Taxonomy.* Freeman, San Francisco.

Snedecor, G. W. (1946) *Statistical Methods Applied to Experiments in Agriculture and Biology,* 4th edn. Iowa State College Press, Ames, Iowa.

Sobolev, L. N. & Utekhin, V. D. (1973) Russian (Ramensky) approaches to community systematization. In: *Ordination and Classification of Communities, Handbook of Vegetation Science,* **5,** (Ed. by R. H. Whittaker) pp. 75–103. Junk, The Hague.

Sokal, R. R. & Michener, C. D. (1958) A statistical method for evaluating systematic relationships, *Kans. Univ. Sci. Bull.* **38,** 1409–37.

Sokal, R. R. & Oden, N. L. (1978) Spatial autocorrelation in biology. I. Methodology. *Biol. J. Linn. Soc.* **10,** 199–228.

Sokal, R. R. & Rohlf, F. J. (1981) *Biometrics,* 2nd edn. Freeman, San Francisco.

Sørensen, T. (1948) A method of establishing groups of equal amplitude in plant sociology based on similarity of species content. *Biol. Skr.* **5** (4), 1–34.

Spatz, G. & Siegmund, J. (1973) Eine method zur tabellarischen ordination, klassifikation und ökologischen auswertung von pflanzensoziologischen beslandsaufnahmen durch den computer. *Vegetatio,* **28,** 1–17.

Steiger, T. L. (1930) Structure of prairie vegetation. *Ecology*, **11**, 170–217.

Stevens, W. L. (1937) Significance of grouping. *Ann. Eug., Lond.* **8**, 57–69.

Stowe, L. G. & Wade, M. J. (1979) The detection of small-scale patterns in vegetation. *J. Ecol.* **67**, 1047–64.

Strong, C. W. (1966) An improved method of obtaining density from line-transect data. *Ecology*, **47**, 311–13.

'Student' (1919) An explanation of deviations from Poisson's law in practice. *Biometrika*, **12**, 211–15.

Svedberg, T. (1922) Ett bidrag till de statistika metodernas användning inom växtbiologien. *Svensk bot. Tidskr.* **16**, 1–8.

Swaine, M. D. & Greig-Smith, P. (1980) An application of principal components analysis to vegetation change in permanent plots. *J. Ecol.* **68**, 33–41.

Swaine, M. D. & Hall, J. B. (1976) An application of ordination to the identification of forest types. *Vegetatio*, **32**, 83–6.

Swan, J. M. A. (1970) An examination of some ordination problems by use of simulated vegetation data. *Ecology*, **51**, 89–102.

Swan, J. M. A., Dix, R. L. & Wehrhahn, C. F. (1969) An ordination technique based on the best possible stand-defined axes and its application to vegetational analysis. *Ecology*, **50**, 206–12.

Tallis, J. H. (1969) The blanket bog vegetation of the Berwyn Mountains, North Wales. *J. Ecol.* **57**, 765–87.

Tansley, A. G. (1920) The classification of vegetation and the concept of development. *J. Ecol.* **8**, 118–49.

Tharu, J. & Williams, W. T. (1966) Concentration of entries in binary arrays. *Nature, Lond.* **210**, 549.

Thomas, M. (1949) A generalization of Poisson's binomial limit for use in ecology. *Biometrika*, **36**, 18–25.

Thompson, H. R. (1955) Spatial point processes, with application to ecology. *Biometrika*, **42**, 102–15.

Thompson, H. R. (1956) Distribution of distance to nth neighbour in a population of randomly distributed individuals. *Ecology*, **37**, 391–4.

Thompson, H. R. (1958) The statistical study of plant distribution patterns using a grid of quadrats. *Aust. J. Bot.* **6**, 322–43.

Thomson, G. W. (1952) Measures of plant aggregation based on contagious distributions. *Contr. Lab. Vertebr. Biol. Univ. Mich*, **53**, 1–16.

Tidmarsh, C. E. M. & Havenga, C. M. (1955) The wheel-point method of survey and measurement of semi-open grasslands and karoo vegetation in South Africa. *Mem. bot. Surv. S. Afr.* **29**, pp. iv + 49.

Tinney, F. W., Aamodt, O. S. & Ahlgren, H. L. (1937) Preliminary report of a study on methods used in botanical analyses of pasture swards. *J. Amer. Soc. Agron.* **29**, 835–40.

Turkington, R., Cavers, P. B. & Aarssen, L. W. (1977) Neighbour relationships in grass-legume communities. I. Interspecific contacts in four grassland communities near London, Ontario. *Can. J. Bot.* **55**, 2701–11.

Turkington, R. & Harper, J. L. (1979) The growth, distribution and neighbour relationships of *Trifolium repens* in a permanent pasture. II. Inter- and intra-specific contact. *J. Ecol.* **67**, 219–30.

Usher, M. B. (1969) The relation between mean square and block size in the analysis of similar patterns. *J. Ecol.* **57,** 505–14.

Usher, M. B. (1975) Analysis of pattern in real and artificial plant populations. *J. Ecol.* **63,** 569–85.

Usher, M. B. (1981) Modelling ecological succession, with particular reference to Markovian models. *Vegetatio,* **46,** 11–18.

Vandermeer, J. H. & MacArthur, R. H. (1966) A reformulation of alternative (b) of the broken stick model of species abundance. *Ecology,* **47,** 139–40.

Vestal, A. G. (1949) Minimum areas for different vegetations. *Illinois biol. Monogr.* **20** (3).

Volk, O. H. (1931) Beiträge zur Ökologie der Sandvegetation der oberrheinischen Tiefebene. *Z. Bot.* **24,** 81–185.

Walker, B. H. & Wehrhahn, C. F. (1971) Relationships between derived vegetation gradients and measured environmental variables in Saskatchewan wetlands. *Ecology,* **52,** 85–95.

Walker, J., Noy-Meir, I., Anderson, D. J. & Moore, R. M. (1972) Multiple pattern analysis of a woodland in south central Queensland. *Aust. J. Bot.* **20,** 105–18.

Warren, W. G. & Batcheler, C. L. (1979) The density of spatial patterns: robust estimation through distance methods. In: *Spatial and Temporal Analysis in Ecology* (Ed. By R. M. Cormack and J. K. Ord). pp. 247–69. International Cooperative Publishing, Fairland, Md., U.S.A.

Warren Wilson, J. (1959a) Analysis of the spatial distribution of foliage by two-dimensional point quadrats. *New Phytol.* **58,** 92–101.

Warren Wilson, J. (1959b) Analysis of the distribution of foliage area in grassland. In: *The Measurement of Grassland Productivity* (Ed. by J. D. Ivins), pp. 51–61. Butterworth, London.

Warren Wilson, J. (1960) Inclined point quadrats. *New Phytol.* **59,** 1–8.

Warren Wilson, J. (1963a) Estimation of foliage denseness and foliage angle by inclined point quadrats. *Aust. J. Bot.* **11,** 95–105.

Warren Wilson, J. (1963b) Errors resulting from thickness of point quadrats. *Aust. J. Bot.* **11,** 178–88.

Warren Wilson, J. (1965) Point quadrat analysis of foliage distribution for plants growing singly or in rows. *Aust. J. Bot.* **13,** 405–9.

Watt, A. S. (1947) Pattern and process in the plant community. *J. Ecol.* **35,** 1–22.

Webb, D. A. (1954) Is the classification of plant communities either possible or desirable? *Bot. Tidsskr.* **51,** 362–70.

Webb, L. J., Tracey, J. G., Williams, W. T. & Lance, G. N. (1967a) Studies in the numerical analysis of complex rain-forest communities. I. A comparison of methods applicable to site/species data. *J. Ecol.* **55,** 171–91.

Webb, L. J., Tracey, J. G., Williams, W. T. & Lance, G. N. (1967b) Studies in the numerical analysis of complex rain-forest communities. II. The problem of species-sampling. *J. Ecol.* **55,** 525–38.

Webb, L. J., Tracey, J. G., Williams, W. T. & Lance, G. N. (1970) Studies in the numerical analysis of complex rain-forest communities. V. A comparison of the properties of floristic and physiognomic-structural data. *J. Ecol.* **58,** 203–32.

Welbourn, R. M. & Lange, R. T. (1967) Subdividing vegetation on interspecific association. *Vegetatio,* **15,** 129–36.

Welch, J. R. (1960) Observations on deciduous woodland in the Eastern Province of Tanganyika. *J. Ecol.* **48**, 557–73.

Went, F. W. (1942) The dependence of certain annual plants on shrubs in Southern California deserts. *Bull. Torrey. bot. Cl.* **69**, 100–14.

West, O. (1937) An investigation of the methods of botanical analysis of pasture. *S. Afr. J. Sci.* **33**, 501–59.

Westhoff, V. & van der Maarel, E. (1973) The Braun-Blanquet approach. In: *Ordination and Classification of Communities. Handbook of Vegetation Science*, **5** (Ed. by R. H. Whittaker), pp. 617–726. Junk, The Hague.

Westman, W. E. (1971) Mathematical models of contagion and their relation to density and basal area sampling techniques. In: *Statistical Ecology* (Ed. by G. P. Patil *et al.*) Vol. 1, pp. 515–36. Pennsylvania State University Press.

Westman, W. E. (1980) Gaussian analysis: identifying environmental factors influencing bell-shaped species distributions. *Ecology*, **61**, 733–9.

Whitford, P. B. (1949) Distribution of woodland plants in relation to succession and clonal growth. *Ecology*, **30**, 199–208.

Whittaker, R. H. (1952) A study of summer foliage insect communities in the Great Smoky Mountains. *Ecol. Monogr.* **22**, 1–44.

Whittaker, R. H. (1953) A consideration of climax theory: the climax as a population and pattern. *Ecol. Monogr.* **23**, 41–78.

Whittaker, R. H. (1954) Plant populations and the basis of plant indication. *Angew. Pflanzensoz. (Wien), Festschr. Aichinger*, **1**, 183–206.

Whittaker, R. H. (1956) Vegetation of the Great Smoky Mountains. *Ecol. Monogr.* **26**, 1–80.

Whittaker, R. H. (1960) Vegetation of the Siskiyou Mountains, Oregon and California. *Ecol. Monogr.* **30**, 279–338.

Whittaker, R. H. (1965) Dominance and diversity in land plant communities. *Science, N.Y.* **147**, 250–60.

Whittaker, R. H. (1967) Gradient analysis of vegetation. *Biol. Rev.* **42**, 207–64.

Whittaker, R. H. (1969) Evolution of diversity in plant communities. *Brookhaven Symp. Biol.* **22**, 178–96.

Whittaker, R. H. (1972) Evolution and measurement of species diversity. *Taxon*, **21**, 213–51.

Whittaker, R. H. (Ed.) (1973a) *Ordination and Classification of Communities. Handbook of Vegetation Science* (Ed. by R. Tüxen), **5**. Junk, The Hague.

Whittaker, R. H. (1973b) Direct gradient analysis: techniques. In: *Ordination and Classification of Communities. Handbook of Vegetation Science*, **5**, 7–31. Junk, The Hague.

Whittaker, R. H. & Fairbanks, C. W. (1958) A study of plankton copepod communities in the Columbia Basin, southeastern Washington. *Ecology*, **39**, 46–65.

Whittaker, R. H. & Gauch, H. G. (1973) Evaluation of ordination techniques. In: *Ordination and Classification of Communities. Handbook of Vegetation Science*, **5**, 287–321. Junk, The Hague.

Whittaker, R. H., Gilbert, L. E. & Connell, J. H. (1979) Analysis of two-phase pattern in a mesquite grassland, Texas. *J. Ecol.* **67**, 935–52.

Williams, C. B. (1947) The logarithmic series and the comparison of island floras. *Proc. Linn. Soc. Lond.* **158**, 104–8.

Williams, C. B. (1949) Jaccard's generic coefficient and coefficient of floral community, in relation to the logarithmic series and the index of diversity. *Ann. Bot.* **13**, 53–8.

Williams, C. B. (1950) The application of the logarithmic series to the frequency of occurrence of plant species in quadrats. *J. Ecol.* **38**, 107–38.

Williams, C. B. (1964) *Patterns in the Balance of Nature.* Academic Press, London.

Williams, W. T. (1972) Partition of information. *Aust. J. Bot.* **20**, 235–40.

Williams, W. T. (1973) Partition of information: the CENTPERC problem. *Aust. J. Bot.* **21**, 277–81.

Williams, W. T. (Ed.) (1976) *Pattern Analysis in Agricultural Science.* C.S.I.R.O., Melbourne and Elsevier, Amsterdam.

Williams, W. T., Clifford, H. T. & Lance, G. N. (1971) Group-size dependence: a rationale for choice between numerical classifications. *Comput. J.* **14**, 157–62.

Williams, W. T. & Dale, M. B. (1962) Partition correlation matrices for heterogeneous quantitative data. *Nature, Lond.* **196**, 602.

Williams, W. T. & Dale, M. B. (1965) Fundamental problems in numerical taxonomy. *Adv. bot., Res.* **2**, 35–68.

Williams, W. T., Dale, M. B. & Lance, G. N. (1971) Two outstanding ordination problems. *Aust. J. Bot.* **19**, 251–8.

Williams, W. T., Dale, M. B. & Macnaughton-Smith, P. (1964) An objective method of weighting in similarity analysis. *Nature, Lond.* **201**, 426.

Williams, W. T. & Lambert, J. M. (1959) Multivariate methods in plant ecology. I. Association-analysis in plant communities. *J. Ecol.* **47**, 83–101.

Williams, W. T. & Lambert, J. M. (1960) Multivariate methods in plant ecology. II. The use of an electronic digital computer for association-analysis. *J. Ecol.* **48**, 689–710

Williams, W. T. & Lambert, J. M. (1961a) Multivariate methods in plant ecology. III. Inverse association-analysis. *J. Ecol.* **49**, 717–29.

Williams, W. T. & Lambert, J. M. (1961b) Nodal analysis of associated populations, *Nature*, Lond. **191**, 202.

Williams, W. T., Lambert, J. M. & Lance, G. N. (1966) Multivariate methods in plant ecology. V. Similarity analyses and information-analysis. *J. Ecol.* **54**, 427—45.

Williams, W. T. & Lance, G. N. (1958) Automatic subdivision of associated populations. *Nature, Lond.* **182**, 1755.

Williams, W. T. & Lance, G. N. (1968) Choice of strategy in the analysis of complex data. *Statistician*, **18**, 31–43.

Williams, W. T., Lance, G. N., Dale, M. B. & Clifford, H. T. (1971) Controversy concerning the criteria for taxonometric strategies. *Comput. J.* **14**, 162–5.

Williams, W. T., Lance, G. N., Webb, L. J. & Tracey, J. G. (1973) Studies in the numerical analysis of complex rain-forest communities. VI. Models for the classification of quantitative data. *J. Ecol.* **61**, 47–70.

Williams, W. T., Lance, G. N., Webb, L. J., Tracey, J. G. & Connell, J. H. (1969) Studies in the numerical analysis of complex rain-forest communities. IV. A method for the elucidation of small-scale forest pattern. *J. Ecol.* **57**, 635–54.

Williamson, M. H. (1972) The relation of principal component analysis to the analysis of variance. *Int. J. Math. Educ. Sci. Technol.* **3**, 35–42.

Williamson, M. H. (1978) The ordination of incidence data. *J. Ecol.* **66**, 911–20.

Wilson, M. V. (1981) A statistical test of the accuracy and consistency of ordination. *Ecology*, **62**, 8–12.

Winkworth, R. E. (1955) The use of point quadrats for the analysis of heathland. *Aust. J. bot.* **3**, 68–81.

Winkworth, R. E. & Goodall, D. W. (1962) A crosswise sighting tube for point quadrat analysis. *Ecology*, **43**, 342–3.

Yarranton, G. A. (1966) A plotless method of sampling vegetation. *J. Ecol.* **54**, 229–37.

Yarranton, G. A. (1967a) Principal components analysis of data from saxicolous bryophyte vegetation at Steps Bridge, Devon. I. A quantitative assessment of variation in the vegetation. *Can. J. Bot.* **45**, 93–115.

Yarranton, G. A. (1967b) Principal components analysis of data from saxicolous bryophyte vegetation at Steps Bridge, Devon. II. An experiment with heterogeneity. *Can. J. Bot.* **45**, 229–47.

Yarranton, G. A. (1967c) Principal components analysis of data from saxicolous bryophyte vegetation at Steps Bridge, Devon. III. Correlation of variation in vegetation with environmental variables. *Can. J. Bot.* **45**, 249–58.

Yarranton G. A. (1967d) Organismal and individualistic concepts and the choice of methods of vegetation analysis. *Vegetatio*, **15**, 113–16.

Yarranton, G. A. (1969a) Pattern analysis by regression. *Ecology*, **50**, 390–5.

Yarranton, G. A. (1969b) Plant ecology: a unifying model. *J. Ecol.* **57**, 245–50.

Yarranton, G. A. (1970) Towards a mathematical model of limestone pavement vegetation. III. Estimation of the determinants of species frequencies. *Can. J. Bot.* **48**, 1387–1404.

Yarranton, G. A. (1971) Mathematical representations and models in plant ecology: response to a note by R. Mead. *J. Ecol.* **59**, 221–4.

Yarranton, G. A., Beasleigh, W. J., Morrison, R. G. & Shafi, M. I. (1972) On the classification of phytosociological data into nonexclusive groups with a conjecture about determining the optimum number of groups in a classification. *Vegetatio*, **24**, 1–12.

Zahl, S. (1974) Application of the S-method to the analysis of spatial pattern. *Biometrics*, **30**, 513–24.

Zahl, S. (1977) A comparison of three methods for the analysis of spatial pattern. *Biometrics*, **33**, 681–92.

Author Index

Subject Index

354